A TEACHER'S GUIDE TO
ELEMENTARY
ALGEBRA

A TEACHER'S GUIDE TO
ELEMENTARY
ALGEBRA

Harold R. Jacobs

W. H. FREEMAN AND COMPANY
New York

The cover illustration and the frontispiece are reproductions of a periodic
drawing by Maurits Escher.
Reproduced with the permission of the Escher Foundation,
Haags Gemeentemuseum, The Hague.

Printed in the United States of America

Fourteenth printing

Contents

Introductory Comments

In his book *The Search for Pattern* (Penguin Books, Baltimore, 1970), W. W. Sawyer tells the following story:

> Some years ago I was told of a primary school teacher in England who felt it would be good if her children learnt some French. An inspector happened to visit her room as she was giving her first lesson. She was on her bicycle and was cycling around the room; the children were standing up like traffic policemen and as she approached would call out, "À droite!" or "À gauche!" and she would turn right or left accordingly. It would, I imagine, be rather exhausting to design and conduct an entire French course on these lines, but as an initial lesson this was wonderful. One can imagine the children going home and telling their families "If we said *à droite* she went right and if we said *à gauche* she went left." The lesson would create the impression that learning French and probably all the other lessons with this teacher were going to be interesting, and this feeling would persist even when some fairly routine piece of work was being done. The first lesson in any subject tends to fix the emotional colour of that subject, and it should be designed at least as much for its dramatic effect as for its intellectual content.

Although I have never ridden a bicycle around my classroom (it is too crowded with desks in any case), I feel that the point of Sawyer's story is well made. What a wonderful thing it would be if we could make each and every one of our lessons entertaining as well as informative!

Having made this a major goal as a beginning teacher, I soon realized that to achieve it in teaching a subject for the first time required a great deal of time and energy: time to think about effective ways of introducing and developing each lesson and energy to find the resources and prepare the materials with which to accomplish it. My purpose in writing this guide is to share some of these ways and materials with you. I have described many of the lessons in great detail in an attempt to convey a feeling for the plans that have worked well for me in the classroom, even to the extent of including the dialogue with my students (which is set in italic type). As a beginning teacher, I wrote detailed "scenarios" such as those presented in this guide for many of my lessons. Sometimes what I had written down actually took place in the classroom; at other times, however, it did not. Students, for example, frequently asked much better questions about the topic being discussed than those for which I had planned in advance. And, once in a while, what looked good to me on paper simply didn't work at all. I am telling you this because, if you have never taught algebra before, you might make the mistake of trying to follow the plans in this guide rigidly. In practice, I am continually changing them, either trying to make them better or replacing them with others. I am sure that as you try them out, you too will make improvements.

Another goal in teaching that I consider extremely important is to help students learn how to teach themselves. In some mathematics classes, the students are never expected to look at anything in the textbook except the exercises. This is unfortunate. I think that reading the lesson should be an important part of every assignment and have tried to write each one in such a way that it can and will be read with understanding.

The exercises for each lesson are arranged in four sets. Those in Set I review ideas from preceding lessons, affording continual practice with material considered earlier in the course. Sets II and III enable the student to apply the concepts of the new lesson. Because both sets have the same content, you will probably prefer to assign one or the other. (Answers to all of the Set II exercises are given at the back of the textbook. Answers to Sets I, III, and IV are given in this guide.) Most of the Set IV exercises are intended to challenge the better students, although some are simple enough that everyone can be successful with them. I make the Set IV exercises optional and count them as extra credit.

The overhead transparencies, duplication masters, and student worksheets illustrated in the lesson plans are available in book form. The pages can be removed from the book and put through a Thermofax machine to make projection transparencies or duplication masters. You may use as many or as few of these masters as you wish or you may present the ideas contained in them on the board or in other ways. Tests for the seventeen chapters, together with midyear and final examinations, are also available. Each test comes in four forms, which can either be made into duplication masters or be used merely as a source of ideas.

Harold R. Jacobs

Ways to Use ELEMENTARY ALGEBRA

Elementary Algebra can be used by students of differing backgrounds and needs and in courses of different lengths. For example, the schedules given below indicate how the book can be used in a regular two-semester course, in an abbreviated two-semester course, and in a three-semester course. They are designed to fit within the number of class days available in the usual secondary-school year. Notice that in the abbreviated course and in the three-semester course the Midterm Review is taken out of sequence to fit the calendar. This will require the addition or deletion of problems in both the Midterm and the Final Reviews. Midterm Review exercises may be borrowed from the Final Review material in the text or from the Test Masters.

A Regular One-Year Algebra Course

	First semester			Second semester	
	Chapter	*Class days*		*Chapter*	*Class days*
	Introduction	1	9	Polynomials	10
1	Fundamental Operations	12	10	Factoring	11
2	Functions and Graphs	9	11	Fractions	11
3	The Integers	10	12	Square Roots	10
4	The Rational Numbers	8	13	Quadratic Equations	12
5	Equations in One Variable	11	14	The Real Numbers	8
6	Equations in Two Variables	9	15	Fractional Equations	8
7	Simultaneous Equations	10	16	Inequalities	7
8	Exponents	10	17	Number Sequences	7
	Midterm Review	2		Final Review	2
	TOTAL	82		TOTAL	86

An Abbreviated One-Year Algebra Course

First semester		Class days		Second semester	Class days
	Chapter			Chapter	
	Introduction	1		Midterm Review	2
1	Fundamental Operations	13	8	Exponents	11
2	Functions and Graphs	11	9	Polynomials	12
3	The Integers	11	10	Factoring	12
4	The Rational Numbers	9	11	Fractions	12
5	Equations in One Variable	13	12	Square Roots	12
6	Equations in Two Variables	11	13	Quadratic Equations	12
7	Simultaneous Equations	14	15	Fractional Equations	10
	TOTAL	83		Final Review	2
				TOTAL	85

NOTE: This program omits: Chapter 10, Lesson 8; Chapter 11, Lesson 8; Chapter 13, Lessons 2, 3, 8, and 9; all of Chapter 14; Chapter 15, Lesson 5; and all of Chapters 16 and 17.

A Full Algebra Course in Three Semesters

First semester		Class days		Second semester	Class days
	Chapter			Chapter	
	Introduction	1	7	Simultaneous Equations	17
1	Fundamental Operations	14	8	Exponents	13
2	Functions and Graphs	14	9	Polynomials	16
3	The Integers	12	10	Factoring	17
4	The Rational Numbers	10	11	Fractions	17
5	Equations in One Variable	16		Midterm Review	2
6	Equations in Two Variables	15		TOTAL	82
	TOTAL	82			

Third semester		Class days
	Chapter	
12	Square Roots	16
13	Quadratic Equations	19
14	The Real Numbers	12
15	Fractional Equations	13
16	Inequalities	10
17	Number Sequences	11
	Final Review	2
	TOTAL	83

In a three-semester course, a teacher would probably assign almost all of the Set I, II, and III exercises for additional review and practice. In an abbreviated course, a teacher would probably assign only Sets I and II or I and III. In either case, the Set IV exercises could be used to challenge students.

For college students or students who have recently had a prealgebra course, Chapters 1, 3, and 4 could be covered lightly or omitted. Because college students are generally better prepared for this course than are high-school students and because college instructors rely on the ability of students to study material outside the classroom, this course can be covered more rapidly in college. A schedule for covering the material in a college semester follows.

ELEMENTARY ALGEBRA
for College Students

	Chapter	Class days
	Introduction, Chapter 1	1
2	Functions and Graphs	3
5	Equations in One Variable	4
6	Equations in Two Variables	3
7	Simultaneous Equations	4
8	Exponents	3
	Midterm Review	1
9	Polynomials	3
10	Factoring	3
11	Fractions	4
12	Square Roots	4
13	Quadratic Equations	4
15	Fractional Equations	2
	Final Review	2
	TOTAL	41

With adjustments for individual schools, this outline should fit a one-semester course allowing five days for examinations. Depending on the courses that follow this one, an instructor may want to cover inequalities (especially absolute value), review the integers and rational numbers, and do some work on the real numbers. On the other hand, some instructors may wish to de-emphasize the treatment of exponents and scientific notation and do less on dividing polynomials and the solution of higher-degree equations if these topics will be covered in other courses.

A TEACHER'S GUIDE TO
ELEMENTARY
ALGEBRA

LESSON PLANS

A NUMBER TRICK

George Polya has formulated three principles of learning:*

1. The best way to learn anything is to discover it by yourself.

2. For efficient learning, the learner should be interested in the material to be learned and find pleasure in the activity of learning.

3. An exploratory phase should precede the phase of verbalization and concept formation and, eventually, the material learned should be merged in, and contribute to, the integral mental attitude of the learner.

The first lesson begins with a number trick whose analysis illustrates all three of these principles. (They have been stated here for your consideration only and are not intended to be part of the lesson presented to the class.)

Simple number tricks, such as the one discussed on the first three pages of the text, are often encountered by students in elementary school. To begin the lesson, you might use the following trick, which is somewhat more complex.

Have a student write a three-digit number on the board, but do not look at the number. Then ask him to rearrange the digits of the number in any order he chooses to form another three-digit number. He should then subtract the smaller of the two numbers he has written on the board from the larger. Next, he is to multiply the result by any number except zero. As all of this is being done, the rest of the class should watch carefully to make sure that the student does not make any mistakes.

When this has been done, ask the student to circle any digit in the final answer except zero and to tell you the remaining digits in the answer. By knowing what these digits are, you can figure out what the missing digit is.* To do so, repeatedly add the digits until you obtain a single-digit number. If that number is nine, the missing digit is nine. However, if the single-digit number that you obtain is any number except nine, you can find the missing digit by subtracting that number from nine.

---

*Mathematical Discovery, volume 2, by George Polya (Wiley, 1965), pp. 102-104.

---

*For your information, but not for your students, this trick works by casting out nines. Many books describe this process and its basis is discussed on page 22 of the *Teacher's Guide* for *Mathematics: A Human Endeavor.* Casting out nines shows that the difference that the student finds must be a multiple of nine. Then the product found is also a multiple of nine. If a number is a multiple of nine, the sum of its digits is also a multiple of nine.

TRANSPARENCY 0-1   DUPLICATION MASTER 1

The following example illustrates both calculations:

$$732 - 273 = 459$$

$$459 \times 2 = 918$$

91⑧   ⑨18

Add digits:	$9 + 1 = 10$	$1 + 8 = 9$ (The single-digit
Add again:	$1 + 0 = 1$	number obtained is nine, and so the
Circled digit $= 9 - 1 = 8$		missing digit is nine.)

Repeat the trick either by having several other students do it simultaneously at the board or by having the entire class do it on paper. After a few examples, someone should be able to guess how the circled digit is obtained. Let that student explain the method to the class. In the unlikely event that no one is able to come up with an explanation (I have never had a class yet that, given sufficient time, could not), you might explain the procedure and suggest that your students try the trick on their families or friends. In any event, point out that an understanding of why this trick works requires some knowledge of algebra.

Follow up this trick with the simpler trick discussed in the first three pages of the text. Transparency 0-1 can be used to illustrate how the trick works with several specific numbers and why it works with boxes and circles.

Distribute either the textbooks or copies of the exercises for the introductory lesson. Duplication Master 1 can be used to make copies of the exercises if you are unable to issue books at this time.

Chapter 1

FUNDAMENTAL OPERATIONS

In his book *Vision in Elementary Mathematics,* W. W. Sawyer tells of a visitor to a school in England many years ago who asked the students the following question.

> A shepherd had 80 sheep. 16 died. How many were left?

A quarter of the class added 16 to 80, a quarter subtracted 16 from 80, a quarter multiplied 80 by 16, and a quarter divided 80 by 16. Sawyer remarks: "The mechanical part of the calculation was perfect; the children obtained correct answers so far as carrying out the operation was concerned. But it is clear they had not the slightest understanding of what they were doing."

This story illustrates a problem that I think is often ignored in an elementary algebra course: although students may be able to perform the basic operations correctly, many of them do not have a clear idea of what the operations mean. I have found, for example, that, when beginning geometry students are asked how to represent the length of a line segment consisting of two parts having lengths labeled a and b, many will respond with the answer: "ab." Even when prompted with the remark: "ab means a times b; is that what you mean?" some will continue to think that their response is correct.

For this reason, each of the first four lessons in this chapter reviews the meaning of one of the basic operations. After a lesson on exponentiation, an entire lesson deals with the behavior of zero and one, because a lack of attention to these seemingly very simple numbers can also lead to difficulties later. The remaining lessons present the rules for performing a series of operations.

Lesson 1
Addition

Before discussing the exercises in the introductory lesson, you might have your students try to explain the following number trick (on Transparency 1-1). They will be interested to know that it is from a French book of puzzles (*Problèmes plaisans et délectables* by Claude Gaspar Bachet) published in 1612. Ask your students to show, by drawing boxes and circles, how the trick works.

> Think of a number.
>
> Multiply by six.
>
> Add twelve.
>
> Divide by three.
>
> Subtract twice the number first thought of.

Transparencies 1-2, 1-3, 1-4, and 1-5 can be used to discuss the exercises in the introductory lesson. Students might explain exercises 1 and 3 using the appropriate transparency at the overhead projector, and several of those who solved exercise 4 might write their solutions on the board.

To introduce the lesson on addition, you might have the class consider the question shown on Transparency 1-6. The answer to the question is not as obvious as it first seems because the diagram contains squares of more than one size. Although the squares can be counted individually, it is probably simpler to count the number of squares of each size and add:

$$16 + 9 + 4 + 1 = 30$$

Some appropriate questions: *What is the answer to an addition problem called? Does the answer depend on the order in which the numbers are added?*

Notice that, although we can find the sum of as many numbers as we wish, we do addition by pairs. This is because the addition table that we learned in elementary school was arranged in pairs. Show Transparency 1-7. Although the answer to a problem such as 3 + 5 seems obvious to you now, it is not at all obvious to a child. How did you learn it? Some other words used to mean "add" in the past are "assemble" and "join." These words are suggested by the way in which a child learns to add. Draw a loop around the row 1 + 5, 2 + 5, 3 + 5, . . . What do these addition problems have in common? All of them consist of 5 added to some number. We can represent them by the expression x + 5, in which x represents the number. Because x can be replaced by various numbers, it is called a variable. Draw a loop around the column 3 + 1, 3 + 2, 3 + 3, . . . What do these addition problems have in common? How could you represent them by a single expression? Can you think of a way in which all of the problems in the table can be represented by a single expression?

These questions should help you understand the following definition of algebra, which appears in The American Heritage Dictionary: *"a generalization of arithmetic in which symbols, usually letters of the alphabet, represent numbers . . . and are related by operations that hold for all numbers in the set."*

If the students have been issued books, they can be assigned Lesson 1 (pages 6–8) to read and the exercises that you have selected (see the remarks about the exercises in the Introductory Comments of this guide). If books have not been issued, you can distribute copies of the Set I, II, and IV exercises of Lesson 1 made from Duplication Masters 2 and 3.

TRANSPARENCY 1-1

A number trick from

Problèmes plaisans et délectables
by Claude Bachet, 1612.

Think of a
number.

Multiply by six.

Add twelve.

Divide by three.

Subtract twice
the number
first thought of.

TRANSPARENCY 1-2

Think of a
number.

Double it.

Add six.

Divide by two.

Subtract the
number first
thought of.

TRANSPARENCY 1-3

What is happening in each step?

Step 1. □

Step 2. □□□□

Step 3. □□□□ ∘∘∘∘∘∘∘∘

Step 4. □ ∘∘

Step 5. □ ∘∘∘∘∘

Step 6. ∘∘∘∘∘

TRANSPARENCY 1-4

Think of a
number.

Divide by two.
Subtract the
number first
thought of.

TRANSPARENCY 1-5

Think of a
number.

Triple it.

Add twelve.

TRANSPARENCY 1-6

How many squares?

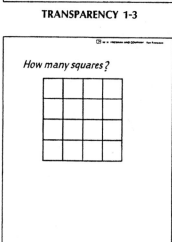

TRANSPARENCY 1-7

DUPLICATION MASTER 2

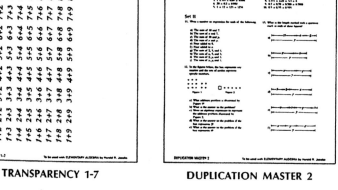

DUPLICATION MASTER 3

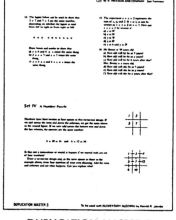

TRANSPARENCY 1-8	WORKSHEET 1	TRANSPARENCY 1-9

Lesson 2
Subtraction

Another number trick*

Before class, put a card with the number 73 on it above the front board with the number facing the wall.

Show the grid at the top of Transparency 1-8 (mask the rest of the transparency) and ask a student to choose one of the numbers. Circle it and cover the other numbers in the same row and the same column with pennies. Ask another student to pick one of the remaining numbers. Circle it and again cover the other numbers in the same row and column. After this has been done two more times, all but five of the numbers will have been covered. If they are added, their sum will be the number on the card, which you can now reveal to the class.

Have each student repeat the trick. (Either have the students copy the pattern at the lower left of Transparency 1-8 or hand out copies of Worksheet 1.) *Circle any number and cross out the other numbers in the same row and column. Circle another number and do the same thing. Circle the number that remains. The sum of the three circled numbers will be 36.*

---

*This trick has as its basis an article by Martin Gardner entitled "Magic with a Matrix," in his book *The Scientific American Book of Mathematical Puzzles and Diversions* (Simon and Schuster, 1959), pp. 15–22.

In both versions of the trick, the patterns of numbers form addition tables. Add the overlay to Transparency 1-8 to reveal this. *The elimination process results in each of the final numbers being in a different row and column. Because each is the sum of two different numbers along the border, their total is simply the sum of the six numbers along the border.* Filling in the diagram at the lower right of Transparency 1-8 (also reproduced on Worksheet 1), as

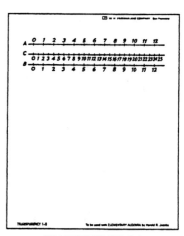

shown above, should help to make this clear.

A nomograph is a diagram, usually consisting of three parallel scales, used to make mathematical calculations. A simple example of a nomograph, shown on Transparency 1-9 and Worksheet 2, is especially appropriate for introducing the lesson on subtraction.

First, have the students observe how the three scales are numbered.* *Can you figure out how the*

---

*This nomograph computes the average of the two numbers being added and displays it on a 2× scale. Algebraically, it gives $2\left(\dfrac{a+b}{2}\right) = a + b$.

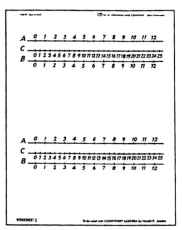

WORKSHEET 2

*A man bought 6 apples
and 5 pears. He ate
all but 4 apples and
2 pears.*

*Did he have more apples
or pears left?*

TRANSPARENCY 1-10

*nomograph can be used to add 5 and 5? 2 and 4? 10
and 7? Note that if* a, c, *and* b *represent numbers that
are in line with each other on the three respective
scales,* c = a + b.

*The nomograph can also be used to subtract num-
bers. Do you see how it can be used to subtract 11
from 19? 6 from 6? Suppose that 3 on the A-scale
were lined up with 8 on the C-scale and 5 on the
B-scale. What problems would this illustrate?*
(3 + 5 = 8, 8 − 3 = 5, *and* 8 − 5 = 3.) *Every
alignment of three numbers on the scales, in fact,
illustrates three problems: one in addition and two in
subtraction. Can you represent each problem in symbols?* (a + b = c, c − a = b, *and* c − b = a.) *Use
these examples to review the terms* sum *and* difference.

It would be well to remind your students that
they should read the new lesson before doing any of
the exercises.

Lesson 3
Multiplication

*The following problem (reproduced on Transpar-
ency 1-10) is from* More Posers, *by Philip Kaplan
(Harper & Row, 1964). The author of the book gives
his readers a time limit of two minutes in which to
solve it. See if you can do it in less time than that.*

A man bought 6 apples and 5 pears. He
ate all but 4 apples and 2 pears. Did he have
more apples or pears left?

*The moral of this exercise is: "Don't be too quick to
do the first thing that comes into your head" or "If
you don't read a problem carefully, you will probably
miss it!"*

According to David Eugene Smith (*History of
Mathematics,* volume 2, Dover, 1958, page 101),
"multiplication was developed as an abridgement of
addition. It was simply a folding together of many
equal addends." The word multiply, in fact, comes
from the Latin words *multus* (many) and *plicare* (to
fold): thus multiply means many folds. This derivation suggests the following. *Suppose that we have
a strip of eight 15-cent stamps. How much are they*

*worth altogether? A child first learns to get the an-
swer by adding 8 fifteens:*

$$15 + 15 + 15 + 15 + 15 + 15 + 15 + 15$$

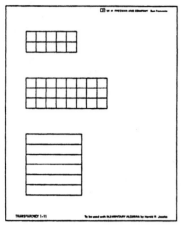

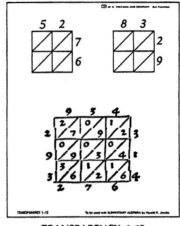

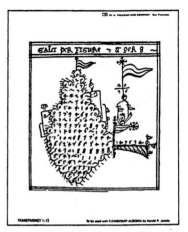

| TRANSPARENCY 1-11 | TRANSPARENCY 1-12 | TRANSPARENCY 1-13 |

Later, he learns the shortcut of multiplying

$$8 \times 15$$

We can extend our interpretation of multiplication by considering the area of a rectangle. Use Transparency 1-11 to illustrate that $5 + 5 = 2 \times 5 = 10$, $2 + 2 + 2 + 2 + 2 = 5 \times 2 = 10$, *and so forth. This interpretation suggests that the product of two numbers does not depend on their order. Use the third figure to introduce the notation* $6x$ *(comparing it with* $6 \times x$ *and* $x6$*). Also point out that* $6x = x + x + x + x + x + x.$

Lesson 4
Division

In one of my geometry classes we were figuring out the volumes of several boxes of different shapes. As I walked around the room observing the students' work, I noticed that, in doing a problem that required multiplying 108 by 4, one student had written the following on her paper:

$$\begin{array}{r} 3 \\ 108 \\ \times\ \ 4 \\ \hline 522 \end{array}$$

When I asked her to check her work, she said that she was certain that she had not made a mistake. Can

you figure out what she was doing? After carrying the 3, she multiplied it by 4, getting 12 and carrying the resulting 1. *The student was a little rusty on how to multiply. How could someone get the correct answer to the problem without multiplying?* Write 108 four times and add, or write 4 one hundred and eight times and add. *Compare the procedures:*

$$\begin{array}{r} 3 \\ 108 \\ 108 \\ 108 \\ +\ 108 \\ \hline 432 \end{array} \qquad \begin{array}{r} 3 \\ 108 \\ \times\ \ 4 \\ \hline 432 \end{array}$$

Transparency 1-12 can be used to discuss the Set IV exercise of Lesson 3. The method described is thought to have originated in India. It became very popular in both Asia and Europe and eventually became the basis for Napier's rods. (See Martin Gardner's "Mathematical Games" column in the March 1973 issue of *Scientific American*.)

The diagram at the bottom of the transparency appeared in the first printed arithmetic book, in 1478. After discussing the two multiplication problems in Set IV, you might ask what problem (and answer) are illustrated in the diagram. ($934 \times 314 = 293,276$.)

To introduce the lesson on division, you might show Transparency 1-13, which contains a reproduction of a division problem from a sixteenth-

century manuscript. *The division was done by a method called the "galley" method, so named because the outline of the work was thought to resemble a boat. Students of the time were expected to draw pictures of the sort shown here around their work. If anyone should ask, the problem seems to be to divide 965347655446, the number in the thirteenth row, by 654321888888, the number in the fourteenth row. The answer, to six figures, is written on the "tail" of the boat: 1.47534.*

When we divide one number by another, what does the answer mean? An example such as 105 divided by 15 can be used to illustrate the interpretation of division as repeated subtraction and as finding the number that, when multiplied by the divisor, will produce the dividend. Review the term *quotient* and the use of fractional notation to indicate division.

Lesson 5
Raising to a Power

There is an old comedy routine about two simple-minded fellows who are trying to divide 28 by 7. Martin Gardner included it, together with some of its history, in his "Mathematical Games" column in the April 1974 issue of *Scientific American. Flip Wilson did the routine once on his television show and was sued by the daughter of the man who developed it for using it without permission. The case was apparently settled out of court. The routine goes like*

this. (Use Transparency 1-14 to illustrate the three methods.)

One fellow got out a sheet of paper and reasoned as follows. Seven won't go into 2 but it goes into 8 once. Write down 1, 1 times 7 is 7, subtract; 7 into 21 is 3, write down 3, the answer is 13.

$$
\begin{array}{r}
13 \\
7{\overline{\smash{\big)}\,28}} \\
-7 \\
\hline
21 \\
-21 \\
\hline
0
\end{array}
$$

Although the other fellow couldn't figure out anything wrong with this method, he thought they ought to check the answer to be certain that it was correct. How can you check a division problem? Multiplying 13 by 7, as shown here, they got 28.

$$
\begin{array}{rr}
& 13 \\
13 & 13 \\
\times\ 7 & 13 \\
\hline
21 & 13 \\
+\ 7 & 13 \\
\hline
28 & 13 \\
& +\ 13 \\
\hline
& 28
\end{array}
$$

Can you think of a way to check that this multiplication is correct? Add 7 thirteens. Start at the bottom of the right-hand column and go up, counting 3, 6, 9, 12, 15, 18, 21; then continue down the left-hand column counting 22, 23, 24, 25, 26, 27, 28.

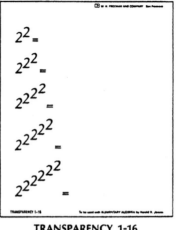

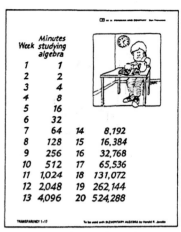

TRANSPARENCY 1-15 **TRANSPARENCY 1-16** **TRANSPARENCY 1-17**

Transparency 1-16 shows:

$$2^2 =$$
$$2^{2^2} =$$
$$2^{2^{2^2}} =$$
$$2^{2^{2^{2^2}}} =$$
$$2^{2^{2^{2^{2^2}}}} =$$

Transparency 1-17 shows:

Week	Minutes studying algebra	Week	Minutes studying algebra
1	1		
2	2		
3	4		
4	8		
5	16		
6	32		
7	64	14	8,192
8	128	15	16,384
9	256	16	32,768
10	512	17	65,536
11	1,024	18	131,072
12	2,048	19	262,144
13	4,096	20	524,288

Ask your students to try to figure out what went wrong in each method. (The original division is essentially correct but we have subtracted 4 sevens, not 13; the answer is $1 + 3$. In the two methods of checking it, we should be thinking of 70 rather than 7.)

Norman Rockwell's painting *A Family Tree*, for the *Saturday Evening Post*, is reproduced on Transparency 1-15. *The picture shows a small boy at the top, his parents, his grandparents, and some of his other ancestors for several generations back. Notice that the part of the tree shown is not complete. The branches that have been omitted, in fact, far outnumber those that have been included.* This illustration is a natural way of leading into the raising-to-a-power lesson, which can now be presented to the class. Several examples comparable to those at the end of the lesson should be included.

Lesson 6
Zero and One

Show Transparency 1-16, keeping each line covered until you get to it. *The symbol 2^2 means $2 \cdot 2$, or 4. What do you think the symbol 2^{2^2} means?* Starting at the bottom and working up, we get $4^2 = 4 \cdot 4 = 16$. Starting at the top and working down, we get $2^4 = 2 \cdot 2 \cdot 2 \cdot 2 = 16$.

What does the symbol $2^{2^{2^2}}$ mean? Starting at the bottom and working up, we get $4^{2^2} = 16^2 = 256$. Starting at the top and working down, we get $2^{2^4} = 2^{16} = 65,536$. *Because mathematicians don't like getting uncertain answers such as those that we have gotten, they have agreed that the value of a number that includes a ladder of exponents like this is to be found by starting at the top and working down.*

To figure out the digits of $2^{2^{2^{2^2}}}$ requires long computation. In his "Mathematical Games" column in the May 1973 issue of *Scientific American*, Martin Gardner reports that *the number starts 20035. . . and contains 19,729 digits. No one will ever work out all the digits of $2^{2^{2^{2^{2^2}}}}$ because it would take a computer as many years as the universe is old to find it and as much space as the universe occupies to print it out in its entirety!*

Transparency 1-17 can be used to discuss the Set IV exercise on Obtuse Ollie's plan for studying algebra. Students having calculators should figure out how many minutes there are in a week (10,080) and how many minutes there are in a 365-day year (525,600). You might also ask which numbers in the "minutes studying algebra" column can be expressed as powers of 2. (Don't tell them that $1 = 2^0$ at this point, but ask what would fit the rest of the pattern.)

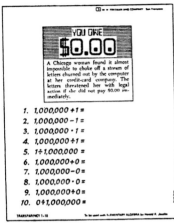

TRANSPARENCY 1-18   TRANSPARENCY 1-19

The news clipping reproduced at the top of Transparency 1-18 can be used to introduce the lesson on zero and one. *It took human beings a long time to accept zero as a number (when we count on our fingers, we begin with one), and operations with it still confuse many people.* Before discussing the properties of zero and one, have your students take the following quiz (reproduced on the bottom of Transparency 1-18). Don't collect their answers, but don't reveal any of the answers until the students have finished the quiz.

1. $1,000,000 + 1 = ?$

2. $1,000,000 - 1 = ?$

3. $1,000,000 \cdot 1 = ?$

4. $1,000,000 \div 1 = ?$

5. $1 \div 1,000,000 = ?$

6. $1,000,000 + 0 = ?$

7. $1,000,000 - 0 = ?$

8. $1,000,000 \cdot 0 = ?$

9. $1,000,000 \div 0 = ?$

10. $0 \div 1,000,000 = ?$

A discussion of the correct answers, together with appropriate generalizations (for example, in question 3, $1,000,000 \cdot 1 = 1,000,000$; $x \cdot 1 = x$ and $1x = x$ for every number x), will help the students understand the lesson when they read it.

Lesson 7
Several Operations

If you have access to an old rotary calculator (perhaps you can borrow one from your business education department), you might demonstrate the following division problems (reproduced on Transparency 1-19) on it to the class:

Problem 1 $97\overline{)2522}$

Problem 2 $2\overline{)806}$

On a Monroe calculator, the first problem took 8 seconds and the second took 41 seconds. The second problem seems much simpler—why does it take the calculator so much longer? (Because the calculator divides by repeated subtraction. In the first problem, it subtracts 97 from 2522, 97 from the result, 97 from the next result, and so on until it comes to 0. You can see the intermediate results appear if you watch the dials closely. In the second problem, it subtracts 2 over and over again in the same fashion.) How many subtractions are there in each case? (26 and 403.) How long do you suppose it would take the calculator to do the following problems? Students who have been assigned exercise 18 in Set III will have already thought about dividing by zero on such a calculator.

Problem 3 $1\overline{)806}$

Problem 4 $0\overline{)806}$

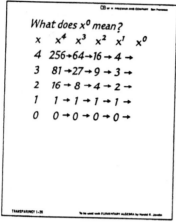

TRANSPARENCY 1-20

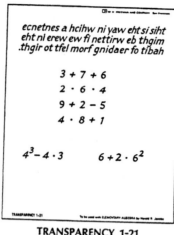

TRANSPARENCY 1-21

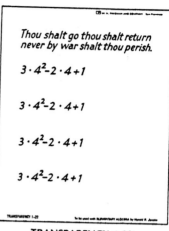

TRANSPARENCY 1-22

Electronic calculators, whose prevalence is rendering the mechanical calculator obsolete, will give the answers to the first three problems seemingly instantaneously. Unlike the rotary calculator, an electronic one will not even attempt the fourth problem but will indicate that it cannot be done.

Transparency 1-20 can be used to discuss the Set IV exercise of Lesson 6. Note that, according to the pattern in the bottom row, 0^0 should equal 0, whereas, the pattern in the last column indicates that 0^0 should equal 1. Because of inconsistencies such as this, mathematicians do not assign a meaning to 0^0.

The sentence at the top of Transparency 1-21 can be used to introduce the new lesson. After giving your students a chance to decipher it, comment on the fact that the practice of reading English from left to right is so ingrained that most people find it quite confusing to try to read it in the opposite direction. *What about mathematics? Consider the following problems* (reproduced on the center of Transparency 1-21). *Does the answer to each problem depend on the direction in which it is read?*

$3 + 7 + 6$ No; the answer is 16 in each direction.

$2 \cdot 6 \cdot 4$ No; the answer is 48 in each direction.

$9 + 2 - 5$ Yes; the answer from left to right is 6; the answer from right to left is 12.

$4 \cdot 8 + 1$ Yes; the answer from left to right is 33; the answer from right to left is 36.

Why doesn't the reversed order make any difference in the first two problems? (They are limited to additions and multiplications and we know that the order in which a set of numbers is added or multiplied has no effect on the answer.)

If more than one operation occurs, then the answer may depend on the order in which the operations are performed. Because we read English from left to right, it might seem reasonable to think that we should also do a series of mathematical operations from left to right and, to a certain extent, this is true. There is a catch, however. Mathematicians have found it convenient to alter this procedure in the following way. Explain that powers are to be computed first, multiplications and divisions next, and additions and subtractions last.* A couple of examples with which to illustrate these rules are at the bottom of Transparency 1-21.

---

*The rules for order of operations are essential for correct computation. They must be learned by practice and this course will be crucial to the students' correct application of these rules in all subsequent courses.

Find the value of
1. $4 \cdot 5 + 8$
2. $4 + 5 \cdot 8$
3. $12 - 2^3$
4. $12 \cdot 2^3$
5. $30 \cdot 3 - 7 \cdot 9$
6. $30 - 3 \cdot 7 - 9$
7. $3x + 2$ if x is 10
8. $3 + x^2$ if x is 10
9. $x^3 - 2$ if x is 4
10. $2x^3$ if x is 4

TRANSPARENCY 1-23

$$1^1 + 2^2 + 3^3 + 4^4$$

$$1^1 + 2^2 + 3^3 + 4^4$$

$$1^1 + 2^2 + 3^3 + 4^4$$

TRANSPARENCY 1-24

1 2 3 4 5 6 7 8 9 10   1 2 3 4 5 6 7 8 9 10

TRANSPARENCY 1-25

Lesson 8
Parentheses

Show Transparency 1-22. *According to legend, an ancient Greek went to the oracle at Delphi and asked whether he should go to war. He was told*

> *Thou shalt go thou shalt return*
> *never by war shalt thou perish.*

Thinking that, if this were punctuated, commas would go after the words "go" and "return," he went off to battle and was promptly slain. If he had punctuated the message differently, the man would probably have never gone to war. Do you see why? The second comma belonged after the word "never."

Just as punctuation in an English sentence can change its meaning, punctuation can be used in a mathematical expression to change its meaning. The punctuation symbols most frequently used for this purpose are parentheses; they change the usual order of operations.

Consider the expression $3 \cdot 4^2 - 2 \cdot 4 + 1$ (reproduced on Transparency 1-22). *Use the rules established in yesterday's lesson to find its value.* Add parentheses to the expression, thus, $(3 \cdot 4)^2 - 2 \cdot 4 + 1$; explain their meaning; and have the students evaluate. Also $3(4^2 - 2 \cdot 4) + 1$ and $3 \cdot 4^2 - 2(4 + 1)$.

A brief quiz (Transparency 1-23) on order of operations might be suitable for the end of the period.

Lesson 9
The Distributive Rule

Show the expression at the top of Transparency 1-24. *Here is a simple pattern consisting of the numbers 1, 2, 3, and 4 raised to the first, second, third, and fourth powers respectively. What is its value?* (288) Show the second expression. *Although we have just shown that* $1^1 + 2^2 + 3^3 + 4^4 = 288$, *it is possible to change the expression to one whose value is 292 by putting a pair of parentheses in it. Can you figure out where?* (Around the first 1 and the first 2.) Show the third expression. *Can you figure out where to put a pair of parentheses in* $1^1 + 2^2 + 3^3 + 4^4$ *in order to make the value of the resulting expression as large as possible? If so, can you figure out its value?* (1,679,616, from putting parentheses around the first 1 and the first 4.)

Because the Set IV exercise of Lesson 8 is similar to the foregoing material, you may want to discuss it next. Transparency 1-25 can be used for this purpose.

In 1940, a severe earthquake struck the Imperial Valley region of California. It took place along a fault that runs through the orange grove pictured in

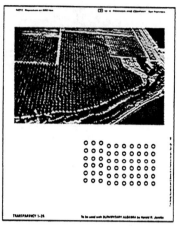

TRANSPARENCY 1-26

this photograph (on Transparency 1-26; the overlay should be in place when the picture is shown). *The land shifted along the fault, carrying the trees on each side of it in opposite directions and breaking the regular pattern of rows in which the trees had been planted.*

The circles in the diagram below the photograph represent trees in another orange grove that has been altered by an earthquake. How many trees are shown? Rather than counting all of them, we can find the number by means of a shortcut. There are $6 \cdot 3$ trees in the region on the left and $6 \cdot 7$ in the region on the right: $6 \cdot 3 + 6 \cdot 7 = 18 + 42 = 60.$

Because the number of trees in every column is the same, 6, and there are $3 + 7$ columns, we could also have gotten the answer by multiplying 6 by $3 + 7$: $6(3 + 7) = 6 \cdot 10 = 60.$ Slide the overlay up so that the trees form a 6 by 10 array. *We see from this example that $6(3 + 7)$ and $6 \cdot 3 + 6 \cdot 7$ represent the same number.*

The distributive rules of multiplication over addition and subtraction (subtraction because we have not yet studied negative numbers) can now be considered, with additional examples.

Review of Chapter 1

A number trick from the eighteenth century

Martin Gardner described the following number trick in his "Mathematical Games" column in the May 1967 issue of *Scientific American*. It is from an Italian book published in 1747 (*I giochi numerici: fatti arcani* By G. A. Alberti). The trick should be reviewed carefully before you attempt to present it to your class.

Have a student write a two-digit number twice on the front board as you and the rest of the class watch. Then have the student write another two-digit number below the number on the left. Without explaining what you are doing, subtract this number from 99 and write the result below the two-digit number on the right. An example of this is shown here.

$$43 \qquad 43$$
$$85 \qquad 14$$

The rest of the class should now copy these numbers on their papers. Tell the students that you are going to ask them to make some calculations for which you already know the answer and that you are writing it down. To find the answer, multiply the original number by 100 and then subtract it from the result. In the example above,

Answers to Set I

1. a) $4 \cdot 7$
   b) 7^4
   c) $x+x$
   d) $y \cdot y \cdot y \cdot y \cdot y$
2. a) w^2
   b) $3x$
   c) $17-y$
   d) z^5
3. a) [diagram]
   b) 1
   c) 3 and 4
   d) add 5
4. a) Fig. 3
   b) Fig. 1
   c) Fig. 2
5. a) $4^2=16$
   $4^3=64$
   $4^4=256$
   $4^5=1024$
   $4^6=4096$
   b) last is 6
6. a) 64
   b) 8^2
7. a) $22,28$
   b) $2x+6; 3x$
   c) $4y; y^2$
8. a) 14
   b) 28
   c) 65
   d) 125
9. a) $x+151$
   b) $160-(x+y)$ or $160-x-y$
10. a) If $\frac{2}{0}=n$, then $0 \cdot n=2$, but $0 \cdot n=0$.
    b) No
11. a) $600x$
    b) $\frac{10,000}{x}$
12. a) $5x+1$
    b) $(x+3)^2$
13. a) $7a+14$
    b) $b-b^2$
    c) $5c+45$
14. a) $3x; 8x$
    b) $3x+8x$
    c) 11
    d) $11x$

TRANSPARENCY 1-27

Match the expressions with the pictures.

A. $4+3$ F. x^2 K. $3x$
B. $a \cdot b$ G. $2 \cdot 4+3$ L. x^2+2x+3
C. $x+y$ H. $a \cdot b+c$ M. $a+b$
D. $4 \cdot 3$ I. x^3 N. xy
E. $a(b+c)$ J. $2(4+3)$ O. x^2+x+1

1. ___   10. ___
2. ___   11. ___
3. ___   12. ___
4. ___   13. ___
5. ___   14. (a Cube) ___
6. ___   15. ___
7. ___
8. ___
9. ___

TRANSPARENCY 1-28

$4300 - 43 = 4257$. Write this number on a large card or on an overhead transparency with the projector light turned off.

Now ask the students to multiply each pair of numbers and add the results, as shown in the example below.

$$\begin{array}{r} 43 \\ \times\, 85 \\ \hline 215 \\ 344 \\ \hline 3655 \end{array} \qquad \begin{array}{r} 43 \\ \times\, 14 \\ \hline 172 \\ 43 \\ \hline 602 \end{array} \qquad \begin{array}{r} 3655 \\ +\ 602 \\ \hline 4257 \end{array}$$

Reveal the answer and tell the students that it was possible to guess it as soon as the original two-digit number had been chosen. At this point, you may wish to repeat the trick with a different student working at the board. If so, write the final result in the corner of the board so that the class can see it before carrying out the rest of the procedure.

If someone can guess how the result is determined, let him explain before discussing the trick with the class. It is based on the distributive rule. Writing the original version on a line, we have

$$43 \cdot 85 + 43 \cdot 14 =$$
$$43(85 + 14) =$$
$$43 \cdot 99 =$$
$$43(100 - 1) =$$

$$43 \cdot 100 - 43 \cdot 1 =$$
$$4300 - 43 =$$
$$4257$$

After discussing the exercises assigned for Lesson 9, you might have the class try either Set I or Set II of the review lesson as a practice test. Call attention to the outline of the chapter on pages 59–60. If your students do not seem to need much review, you might have them do Set I in class, checking answers as each exercise is completed, and Set II at home for additional practice (since answers for Set II are given at the back of the book.) If you want to spend an extra day on review, you might assign Set II as classwork to be completed at home if necessary. The next day you could answer any questions on Set II. Set I could then be done in class as a practice test, the students being given the answers (Transparency 1-27) and correcting their papers in the last part of the period.

If you spend the extra day in review, you might want to open the class with the matching test reproduced on Transparency 1-28. It was devised by W. W. Sawyer, who says that "until a pupil can score 100 percent on such a test, it is unwise to try to teach him formal algebra, for he has no understanding of what you are talking about." *

---

* W. W. Sawyer, *Vision in Elementary Mathematics* (Penguin, 1964) pp. 193–194.

Chapter 2
FUNCTIONS AND GRAPHS

Although some analytic geometry is now included in many geometry courses, I feel that it is more appropriate to include many of the simpler concepts in the first-year algebra course so that they can be used to reinforce the corresponding algebraic content. The coordinate graph, for example, can be used to suggest some of the properties of negative numbers. A knowledge of linear functions helps to enhance the understanding of linear equations, both single and simultaneous, and the graphs of quadratic functions clarify the nature of the solutions of quadratic equations.

Most of my students do very well with the material in this chapter, and this is good for building morale, something that is extremely important in the early stages of the course.

Lesson 1
An Introduction to Functions

A good way to introduce Lesson 1 is with the following game. Tell your class that each student is to choose a number and tell what it is when called upon. For each number, you will respond with another.

As each student in the first row calls out a number, respond with the number that is 6 larger. Before every one in the row has given their number, the "add 6" rule will have become obvious to the class. Now ask everyone to take out a sheet of paper and construct a table of numbers as the first-row students repeat them.

Example:	5	2	10	25	200	0
	11	8	16	31	206	6

Label the two sets of numbers x *and* y, *and write the rule "the* y*-number is found by adding 6 to the* x*-number" and the formula "*y = x + 6.*"*

Continue the game with a new rule for each row of students. The other rules that I use are:

2. $y = 2x + 1$

3. $y = x^3$ (It is probably best to limit the numbers chosen here and in parts 4 and 6 to those no larger than 12.)

4. $y = x^2 - 3$

5. $y = 5$ (This always gets quite a reaction.)

6. $y = x(x + 1)$, or $y = x^2 + x$ (The two seemingly different rules here usually puzzle everyone until someone explains their equivalence by the distributive rule.)

The meaning attributed by mathematicians to the word "function" has undergone many refinements through the years.* Descartes used the term to mean any positive integral power of x. Leibniz used it in a very general way to refer to quantities related to curves, such as slope, radius of curvature, and so forth. Johann Bernoulli thought of a function as an algebraic expression. Euler thought of it as an equation. In the nineteenth century, Dirichlet formulated the following definition, from which our modern definitions have evolved:

If two variables x and y are so related that whenever a value is assigned to x there is automatically assigned, by some rule or correspondence, a value to y, then we say y is a function of x.

In those exercises in this lesson that require finding a formula from a table, the student is directed to guess *a* formula rather than *the* formula because, for any finite set of ordered pairs, there are infinitely many formulas. The answer expected in each case is the "simplest formula." For a table such as

$$\begin{array}{cccc} x & 1 & 2 & 3 \\ y & 2 & 4 & 6 \end{array}$$

the simplest formula is $y = 2x$. The formula could also be $y = x^3 - 6x^2 + 13x - 6$, which is easily

*The information in this paragraph is not intended to be part of the class lesson but rather to be of interest to you. It is abstracted from *An Introduction to the History of Mathematics*, third edition, by Howard Eves (Holt, Rinehart and Winston, 1969).

seen by observing that $x^3 - 6x^2 + 13x - 6$ is equivalent to $(x - 1)(x - 2)(x - 3) + 2x$. It could also be $y = (x - 1)(x - 2)(x - 3)(x - 4) + 2x$, or $y = (x - 1)(x - 2)(x - 3)(x + 10) + 2x$, and so forth. So a function defined by a table may have many formulas.

Lesson 2
The Coordinate Graph

If your students enjoyed the function game used to introduce Lesson 1 as much as mine have, they will find the following examples interesting. Before beginning, tell them that the rules for the functions in this game are still very simple but are not as easy to figure out.

For the first three parts, it is best to restrict the numbers chosen to those between 1 and 10 to keep things from getting too complicated. As the numbers are called, enter them in a table. Let the students choose as many numbers as are necessary to guess the rules. Tables for functions 2 and 3 are given for your reference.

1. $y = 17 - x$

2. $y = \dfrac{120}{x}$

x	1	2	3	4	5	6	7	8	9	10
y	120	60	40	30	24	20	$\approx$17.1	15	$\approx$13.3	12

3. $y = 2^x$

x	1	2	3	4	5	6	7	8	9	10
y	2	4	8	16	32	64	128	256	512	1024

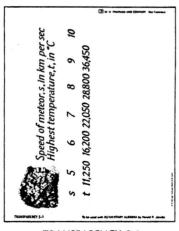

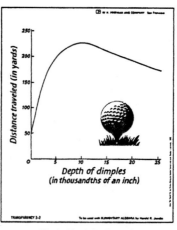

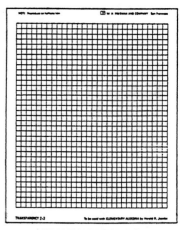

TRANSPARENCY 2-1   TRANSPARENCY 2-2   TRANSPARENCY 2-3

For part 4, tell the students that they may choose any number between 1 and 100.

    4. y is the digital root of x. This example emphasizes the fact that a function need not be expressible in terms of a formula.

Transparency 2-1 is for use in discussing the Set IV exercise of Lesson 1.

To introduce the lesson on the coordinate graph, you might show Transparency 2-2. *Anyone who has ever looked at a golf ball knows that it is not smooth, but covered with a lot of small dimples. The dimples enable the ball to travel farther than it would if it were perfectly smooth. How much farther depends on their number and shape. This drawing* (on Transparency 2-2) *shows the relationship between the depth of the dimples and the distance that the ball will travel. From a book on golf titled* The Search for the Perfect Swing, *it shows that the distance traveled by the ball increases as the depth of the dimples is increased to about ten thousandths of an inch. Increasing the depth further causes the distance traveled by the ball to decrease.*

    Explain that the picture is an example of a coordinate graph, the subject of the new lesson. Demonstrate how to draw and label the axes and how to plot points, given their coordinates. (Because negative numbers have not been introduced yet, graphs at this point consist of just the first quadrant.)

Transparency 2-3 can be used for this purpose. (I do not recommend reproducing it on standard "black on clear" film because it is difficult to see graphs drawn on the resulting grid.)

    Show how to graph a couple of simple functions, such as $y = x + 3$ and $y = 3x$, by making a table, plotting points, and connecting them.

Lesson 3
More on Functions

The following exercise (on Transparency 2-4) is a brief review of plotting points. *Copy the axes and follow the instructions given.*

    1. Connect (0, 8) to (1, 6).
    2. Connect (1, 5) to (0, 5) to (0, 3).
    3. Connect (0, 10) to (1, 11) to (1, 9).
    4. Connect (1, 0) to (1, 2).
    5. Connect (0, 1) to (1, 1).

Now finish the figure. Some students may need a hint: *Think of the* y-*axis as a mirror.*

The figures for the answers to the Set III exercises are given on Transparency 2-5. Select a graph done by one of the students for Set IV and hold it up for the rest of the class to see.

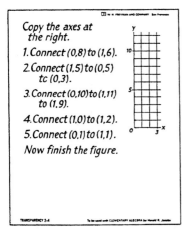

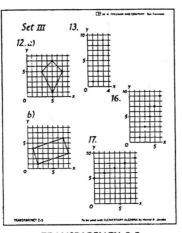

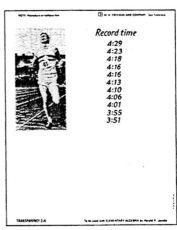

TRANSPARENCY 2-4   TRANSPARENCY 2-5   TRANSPARENCY 2-6

If you look in the latest edition of the Guinness Book of World Records, *you will probably find that most of the track and field records were established within the past five years. An interesting example of how a record has changed with the passage of time is the mile. An article* (titled "Future Performance in Footracing," June 1976) *in* Scientific American *contains a list of the records in the mile from 1864 (Charles Lawes, 4:56) to 1975 (John Walker, 3:49). This table* (Transparency 2-6) *lists the record time every ten years. (It is also on page 78 of the textbook.) Roger Bannister, the first person to run the mile in less than four minutes (1954, 3:59), has predicted that the record will be down to 3:30 within 30 or 40 years and he may be right!*

Use the tables and graphs (Transparency 2-6 with Overlay A, followed by Overlay B) to review the definition of function as done in the text. You may also wish to demonstrate the steps in graphing a function with an example such as $y = x^2 - x$.

Lesson 4
Direct Variation

For this lesson you need a Super Ball, an ordinary rubber ball of the same size, and a scale on the wall marked in intervals of 3 inches from the floor to a height of five feet.

In 1964, a research chemist in Whittier, California, was experimenting with synthetic rubber when

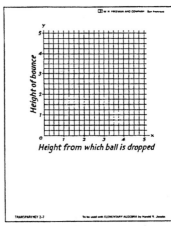

TRANSPARENCY 2-7

he made a ball that had an extraordinarily good bounce. He took it to the Wham-O Company and, after a few improvements, the ball was introduced in toy stores as the Super Ball.

Show your students the Super Ball and ordinary rubber ball and drop them from the same height so that the class can see the difference in bounciness. *The height to which a ball bounces is determined by the material it is made of, the height from which it is thrown, how hard it is thrown, and the type of surface it hits. Let's consider the effect of just one of these factors: the height from which the ball is thrown.* Have the class copy and label the pair of axes shown on Transparency 2-7. Drop the Super Ball from heights of 5 feet, 4 feet, 3 feet, 2 feet, and

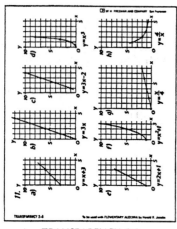

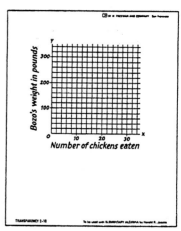

TRANSPARENCY 2-8   TRANSPARENCY 2-9   TRANSPARENCY 2-10

1 foot, asking the students to estimate the height of the first bounce in each case. *Plot a point corresponding to each of the five heights. The origin can be considered a sixth point of the graph. Can you explain why? What do you notice about the five points we have plotted and the origin? (They are approximately collinear.)* Have the students draw a line "through" the six points and explain that they have drawn the graph of a function. Drop the ball from a couple of other heights, such as $3\frac{1}{2}$ feet and $4\frac{1}{4}$ feet, to see how close the points are to being on the line. Observe that we may consider the line, rather than the points, to be the graph of the function because the ball can be dropped from any height and to every height from which it is dropped there corresponds a height to which it bounces.

Explain that the function is an example of a *direct variation. We say that the variables vary directly because, if one is doubled, so is the other, and so forth. Write an equation for the function;* for the floor in my classroom, it is approximately $y = 0.9x$. Comment on the form of the general equation for a direct variation and on the *constant of variation.*

The figures for the answers to exercise 11 in Set III are given on Transparency 2-8. Transparency 2-9 can be used to discuss Set IV.

Lesson 5
Linear Functions

There is a full-page photograph on page 116 of the Mathematics volume of the *Life Science Library* which would be suitable as a transparency with which to introduce this lesson. It shows a man sitting behind a table laden with food, the amount that he consumes in just one day.

Another man who really likes to eat is Edward "Bozo" Miller of Oakland, California, who, according to the Guinness Book of World Records, *consumes 25,000 calories a day! He has a 57-inch waist and his weight varies, but it is never less than 280 pounds.*

In 1963, Mr. Miller ate twenty-seven 2-pound chickens at one meal at Trader Vic's in San Francisco! While eating this marathon meal, he gained an incredible amount of weight. Let's graph his weight as a function of the number of chickens eaten.

Have your students draw and label the axes shown on Transparency 2-10. *Plot points corresponding to Mr. Miller's weight at the beginning of the meal, 280 pounds, and his weight after eating 10, 20, and 27 chickens. Label the last point "Bozo quit here." Draw a line through the points. Did Mr. Miller's weight vary directly with the number of chickens he had eaten? Even though this is not a*

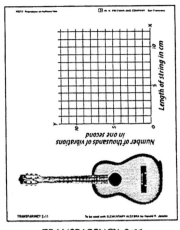

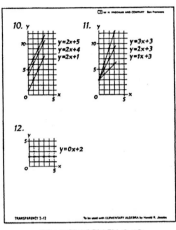

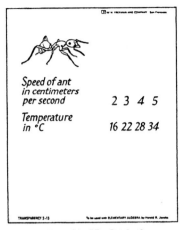

| TRANSPARENCY 2-11 | TRANSPARENCY 2-12 | TRANSPARENCY 2-13 |

direct variation, the function is linear. *Can you write a formula for it?* ($y = 280 + 2x$, or $y = 2x + 280$). *Every function having an equation of the form* $y = ax + b$ *has a graph that is a straight line. This is the reason that such functions are called* linear. Emphasize the fact that all direct variations are a special type of linear function.

Lesson 6
Inverse Variation

Show Transparency 2-11. *The sound produced by one of the strings of a guitar depends on the rate at which it vibrates. Someone playing a guitar can control this rate by varying the length of the part of the string that is free to vibrate. The frets along the fingerboard guide the player in doing this.*

A typical formula for the behavior of a vibrating string is $y = \dfrac{12}{x}$, *in which* x *represents the length of the string in centimeters and* y *represents the number of thousands of times that it vibrates in one second.*

Have students draw and label axes as shown on the transparency. Make a table, plot some points, and connect them with a smooth curve. *What happens to the length of the string as we look at the graph from left to right? What happens to the number of*

thousands of vibrations that it makes per second? What happens to y *if* x *is doubled? What happens if* x *is tripled?* Explain that the variables are said to vary *inversely,* that this function is an example of an *inverse variation,* and that its graph is typical of the graphs of inverse variations.

The figures for the answers to the Set III exercises are given on Transparency 2-12. Transparency 2-13 is for use in discussing Set IV.

Review of Chapter 2

First day

The following demonstration is a spectacular illustration of inverse variation in chemistry. Unfortunately, it requires two liquid solutions that take some time to prepare. If a student assistant working in the science department can make them for you, I recommend doing the demonstration. If not, you can use the alternative exercise that follows this one.

The demonstration is of a reaction called "the iodine clock." The following directions for making the two solutions are also given on Duplication

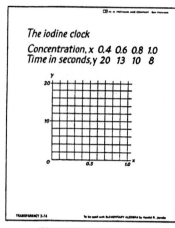

The iodine clock
Concentration, x 0.4 0.6 0.8 1.0
Time in seconds, y 20 13 10 8

TRANSPARENCY 2-14

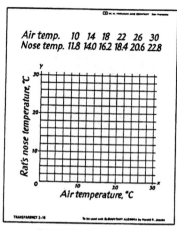

Air temp. 10 14 18 22 26 30
Nose temp. 11.8 14.0 16.2 18.4 20.6 22.8

TRANSPARENCY 2-15

Air temp. 10 14 18 22 26 30
Nose temp. 11.8 14.0 16.2 18.4 20.6 22.8

Rat's nose temperature, °C

Air temperature, °C

TRANSPARENCY 2-16

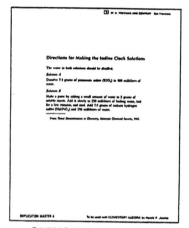

Directions for Making the Iodine Clock Solutions

DUPLICATION MASTER 4

Master 4 (so that the page can be torn out and given to the person making the solutions).

Solution A. Dissolve 7.5 grams of potassium iodate (KIO_3) in 500 milliliters of water (the water in both solutions should be distilled).

Solution B. Make a paste by adding a small amount of water to 2 grams of soluble starch. Add it slowly to 250 milliliters of boiling water, boil for a few minutes, and cool. Add 7.5 grams of sodium hydrogen sulfite ($NaHSO_3$) and 250 milliliters of water.

The demonstration itself is very simple. Mix equal quantities of the two solutions, pouring back and forth several times to mix thoroughly. Tell the students to watch the mixture closely. In a short time, it will suddenly change from being clear to a blue-black color.

The time that it takes for the reaction to take place depends on the concentration of the solutions and the temperature. Here are experimental data showing the time as a function of the concentration.

Concentration, x	0.4	0.6	0.8	1.0
Time in seconds, y	20	13	10	8

What happens to the time as the concentration is increased? Have the students draw and label the axes shown on Transparency 2-14. *Plot the four points corresponding to the numbers in the table and connect them with a smooth curve.*

What happens to the time when the concentration is doubled? (Compare $x = 0.4$ and $x = 0.8$, $x = 0.5$ and $x = 1.0$.) *What kind of function do you think this might be? Can you figure out a formula for it?* $\left(y = \dfrac{8}{x}. \right)$

The following is an alternative to the iodine clock demonstration. Show Transparency 2-15. *The*

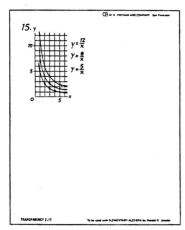

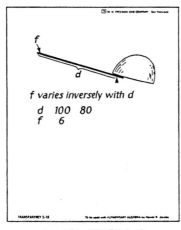

f varies inversely with d

d	100	80
f	6	

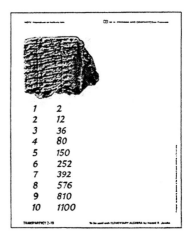

1	2
2	12
3	36
4	80
5	150
6	252
7	392
8	576
9	810
10	1100

TRANSPARENCY 2-17   TRANSPARENCY 2-18   TRANSPARENCY 2-19

noses of many animals are colder than the surrounding air. Perhaps you've noticed this if you've ever touched the nose of a dog.

Biologists have found that, for many animals, including the kangaroo rat shown in this photograph, nose temperature is a function of air temperature.* Add overlay. *Here is a table for this function. What do you notice?* (Among other things, the rat's nose is colder than the surrounding air for temperatures above 14°C, and the higher the air temperature, the greater the difference.) Show Transparency 2-16 and have the students copy the table and axes. *Plot the six points corresponding to the table. What do you notice? How could we have told that the function is linear from the table?* (For equal increases in x, there are equal increases in y.) *Is the function a direct variation? How could we have told from the table that it is not?* (As x is tripled from 10 to 30, for example, y is not tripled.)

Transparencies 2-17 and 2-18 can be used to discuss exercise 15, in Set III, and Set IV. *Archimedes, who is credited with discovering the inverse relationship between force and distance, is supposed*

---

*How Animals Work by Knut Schmidt-Nielsen (Cambridge University Press, 1972).

to have said, " Give me a place to stand and I will move the earth."

Second day

In the latter part of the nineteenth century, archeologists digging in Egypt discovered thousands of clay tablets left there by the Babylonians in about 1700 B.C. Show Transparency 2-19. *Some of the tablets, including the one in this photograph, contain mathematical tables. One of these tables is shown below the photograph. It might be considered a table for a function because it is a pairing of two sets of numbers so that to each number in the first set there corresponds exactly one number in the second set. The numbers in the first column need no explanation. What is the significance of the numbers in the second column? Can you discover a pattern? Here is something interesting: $12 = 3 \cdot 4$, $36 = 4 \cdot 9$, and $80 = 5 \cdot 16$. Letting x represent any number in the first column and y represent the corresponding number in the second column, $y = (x + 1)x^2 = x^3 + x^2$. The second column is a list of sums of cubes and squares: $2 = 1^3 + 1^2$, $12 = 2^3 + 2^2$, $36 = 3^3 + 3^2$, and so forth. The table was evidently of use to the Babylonians in solving problems of some sort.*

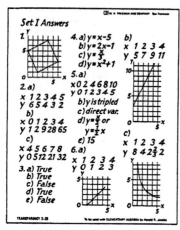

TRANSPARENCY 2-20

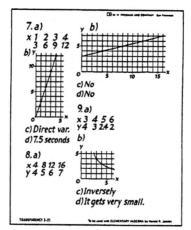

TRANSPARENCY 2-21

Transparencies 2-20 and 2-21 can be used for checking the answers to the Set I review exercises.

Chapter 3

THE INTEGERS

In an essay titled "The Imaginary That Isn't," Isaac Asimov tells of one of his experiences in college in which a sociology professor referred to mathematicians as mystics "because they believe in numbers that have no reality." Asimov asked him what numbers those were and the professor answered, "The square root of minus one." When Asimov responded by saying that the square root of minus one is just as real as any other number, the professor asked him to prove it by handing him the square root of minus one pieces of chalk. Asimov said he would if the professor would hand him one-half piece of chalk. At this point, the professor broke a fresh piece of chalk in half and handed Asimov one of the halves. Asimov tells the rest of the story like this:*

"Ah, but wait," I said: "you haven't fulfilled your end. This is one piece of chalk you've handed me, not a one-half

piece." I held it up for the others to see "Wouldn't you all say this was one piece of chalk? It certainly isn't two or three."

Now the professor was smiling. "Hold it. One piece of chalk is a piece of regulation length. You have one that's half the regulation length."

I said, "Now you're springing an arbitrary definition on me. But even if I accept it, are you willing to maintain that this is a one-half piece of chalk and not a 0.48 piece or a 0.52 piece? And can you really consider yourself qualified to discuss the square root of minus one, when you're a little hazy on the meaning of one-half?"

But by now the professor had lost his equanimity altogether and his final argument was unanswerable. He said, "Get the hell out of here!"

A moral to be drawn from this story is that the meaning and "reality" of a number depend on the context in which it is used. Perhaps Asimov's

---

*Excerpted from *Asimov on Numbers* by Isaac Asimov. Copyright © 1977 by Isaac Asimov. Reprinted by permission of Doubleday & Company, Inc.

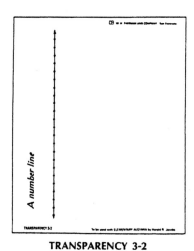

TRANSPARENCY 3-1   TRANSPARENCY 3-2

point in his argument with the professor might have been more clearly made had he asked the man to break the piece of chalk into $2\frac{1}{2}$ pieces. The number $2\frac{1}{2}$ clearly has no meaning in such a context. In the same way, there are situations in which it is meaningless to speak of negative numbers, whereas there are others in which they are every bit as "real" as positive numbers. As for the "square root of minus one," in a context such as the theory of electricity, it is just as meaningful as any counting number.

It is hard to believe the difficulty that mathematicians had in the past in accepting negative numbers. Descartes avoided using them, calling them "false." Cardan considered negative solutions to equations to be fictitious, whereas Euler had the idea that negative numbers were greater than infinity. Algebra books were being published as late as 1796, in fact, protesting against their use!

Fortunately, most students taking elementary algebra come to the course having heard of negative numbers and having some curiosity about them. The goal in Chapter 3 is to acquaint the students with the rules for computing with positive and negative numbers by extending the models of the basic operations introduced in Chapter 1. Because this is most easily done with the integers, work with rational numbers is postponed until Chapter 4.

It would be well to emphasize from the outset that a thorough mastery of the arithmetic of positive and negative numbers is essential for anyone who hopes to succeed in learning algebra.

Lesson 1
The Integers

There is a card on the wall of my classroom stating that *"healthful room temperature is between 68 and 72 degrees Fahrenheit." Unfortunately, I have no way of maintaining the temperature within those limits. I became especially aware of this while teaching in a bungalow for several years. One morning the heater would not work and the temperature remained a chilly 56 degrees. Another time, during a heat wave, the temperature rose to a sweltering 104 degrees in the afternoon.*

On the Celsius scale, these extremes are 13 degrees and 40 degrees, comfortable room temperature being about 20 degrees. What do you suppose are the highest and lowest temperatures to have been reached on the earth? Show Transparency 3-1. *According to the* Guinness Book of World Records, *the highest natural temperature on record, 136 degrees Fahrenheit (58 degrees Celsius), was recorded in the Libyan desert in 1922. The lowest, –127 degrees Fahrenheit (–88 degrees Celsius), was recorded in the Antarctic in 1960. The highest man-made temperature in the laboratory, 60,000,000 degrees Celsius, was achieved for a fraction of a second in 1978. (This is hotter*

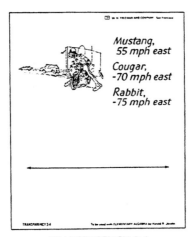

Mustang,
55 mph east

Cougar,
-70 mph east

Rabbit,
-75 mph east

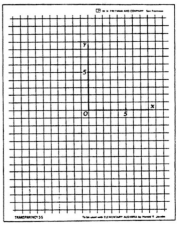

| TRANSPARENCY 3-3 | TRANSPARENCY 3-4 | TRANSPARENCY 3-5 |

than the center of the sun!) The lowest, −273 degrees Celsius, was achieved in 1969 in France.

Although the record for the highest temperature may be broken, the one for the lowest temperature will never be changed significantly. The reason for this is that there is no "hottest" temperature but there is a "coldest" one. Temperature is a consequence of energy. Add energy and the temperature goes up. Remove energy and it goes down. Remove it all and the lowest temperature possible has been achieved.*

To compare temperatures on the Celsius scale, we need both positive and negative numbers. Distribute either copies of Transparency 3-2 or rulers so that the students can draw an accurate number line. Use it to discuss the terms *counting numbers* and *integers* and to introduce the symbols for *is larger than* and *is smaller than*.

Lesson 2
More on the Coordinate Graph

Almost every day hundreds of positive and negative numbers can be found in the newspaper. Can you think of where? (On the pages of the financial section.) Show Transparency 3-3. Pictured on this

---

*Even the idea of absolute temperature can be extended, however. See "Negative Absolute Temperatures" by Warren G. Proctor, in the August 1978 issue of *Scientific American*, in which these negative temperatures are related to Euler's notion of negatives being higher than infinity.

transparency is part of the stock market report for October 29, 1929, the worst day of the Wall Street crash. The positive and negative numbers are in the columns headed "Net Change." On this particular day, most of the numbers were negative.

The net changes for seven of the stocks are shown in this list (on Transparency 3-3.) The smallest change in the list is also the largest number. Can you explain why?

Allied Chemical and Dye	−35
Continental Insurance	−11
Electric Auto-Lite	−45
Fox Films	−20
Midland Steel	−60
Nash Motors	−15
National Cash Register	+1

Transparency 3-4 is for use in discussing the Set IV exercise of Lesson 1.

In the new lesson, treatment of the coordinate graph is extended to all four quadrants. To introduce the lesson, you might have your students draw a pair of axes on graph paper as shown on Transparency 3-5 while you hand out compasses. Draw a circle whose center is at the origin and whose radius is 5 units. Mark and label the points (5, 0), (0, 5), (3, 4), and (4, 3). Then extend the axes and mark and label the other points on the circle whose coor-

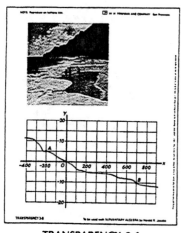

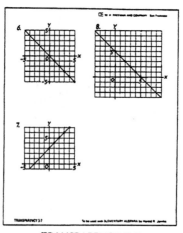

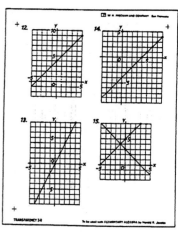

TRANSPARENCY 3-6   **TRANSPARENCY 3-7**   **TRANSPARENCY 3-8**

dinates are integers. Introduce the term *quadrant* and explain how the quadrants are numbered.

Lesson 3
Addition

Show Transparency 3-6. *If you have been to a variety of beaches, you know that some beaches slope gently into the ocean, whereas others become deep very quickly. The graph on this transparency shows the profile of one particular beach. What represents the shoreline of the beach in the graph?* (The origin.) *The numbers along the x-axis represent distances in feet from the shoreline and those along the y-axis represent distances in feet above sea level. Notice that the scales on the two axes are not the same. What are the coordinates of the point labeled A?* (-200, 5.) *What do they mean?* (Point A is 200 feet before the shoreline and 5 feet above it.) *What are the coordinates of the point labeled B?* (700, -10.) *What do they mean?* (Point B is 700 feet beyond the shoreline and 10 feet below it.) *In which quadrant is the part of the beach where you would sunbathe represented?* (The second.) *In which quadrant is the part of the beach where you would go swimming represented?* (The fourth.)

Transparencies 3-7 and 3-8 can be used to discuss some of the exercises of Sets II and

III of Lesson 2. Show a student's graph for Set IV.

The graphs in exercises 6 through 8 and 12 through 14 can be used to guess some of the results of operating with positive and negative numbers and should be discussed in this light. The graph of $y = (-1)x$ in exercise 6 suggests that $(-1)(-5) = 5$, $(-1)(-4) = 4$, and so forth. The graph of $y = x + 3$ in exercise 7 suggests that $-4 + 3 = -1, -3 + 3 = 0$, and so forth. Similarly, exercise 8 suggests that $5 - -4 = 9, 5 - -3 = 8, \ldots, 5 - 6 = -1, 5 - 7 = -2$, and so forth.

There is a book by Hannes Alfven, one of the winners of the Nobel Prize (1970, physics), titled Worlds-Antiworlds (W. H. Freeman and Company, 1966). "Antiworlds" refers to worlds made of "antimatter." Can someone explain what "antimatter" is? (After physicists discovered that atoms consist of negatively charged electrons surrounding a nucleus containing positively charged protons, they wondered if the opposite could be possible. Could atoms consisting of positively *charged particles surrounding a nucleus containing* negatively *charged particles exist?*) *In 1932, the positively charged electron, the "positron," was discovered. If an electron and a positron collide, they "annihilate" each other. We will use this as our model for understanding the addition of positive and negative numbers.*

Examples comparable to those in the textbook are given on Transparency 3-9 to show the students

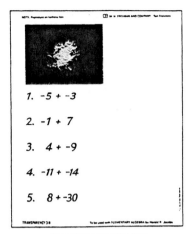

1. -5 + -3

2. -1 + 7

3. 4 + -9

4. -11 + -14

5. 8 + -30

TRANSPARENCY 3-9

If the ground air temperature is 31 degrees and it is -7 degrees atop a nearby mountain, how many degrees difference is there between the two?

TRANSPARENCY 3-10

TRANSPARENCY 3-11

how to draw pictures of (or simply imagine, for examples 4 and 5) particles and antiparticles to represent adding positive and negative numbers. A parenthetical note: It seems to me that the "matter-antimatter" model for addition is much more natural and straightforward for integers than thinking of displacements along a number line. Although 3 and –5 can be added, for example, by starting at 3 on a number line and moving 5 units to the left, I don't think most people who add positive and negative numbers in fact do it that way. Instead, the reasoning is

$$3 + -5 = 3 + (-3 + -2) = -2$$

which is exactly what the open and solid circles illustrate.

Lesson 4
Subtraction

Several years ago a test was given to 640,000 people across the nation by an organization called the National Assessment of Educational Progress. The test included questions on reading, writing, literature, mathematics, music, science, and social studies and was given to people in four age groups: 9, 13, 17, and 26–35. In a moment, I am going to show *you one of the mathematics questions.* Pass out slips of paper and then show Transparency 3-10. Have each student write his or her answer on the slip of paper. Collect the slips, make a tally of the answers on the lower part of the transparency, and then reveal the correct answer. *The problem can be solved either by thinking of a distance on a number line or by subtracting. In today's lesson, we will learn how to subtract positive and negative numbers.*

Be sure in discussing exercises 5 and 10 of Lesson 3 that you emphasize the fact that *x* and –*x* do not necessarily represent positive and negative numbers, respectively, and that, if *x* represents a negative number, –*x* represents a positive number.

Transparency 3-11, with its overlays, is for use in discussing the Set IV exercise of Lesson 3.

In developing the new lesson, the particle-antiparticle model is used to illustrate subtraction before the principle that subtracting a number gives the same result as adding its opposite is introduced. The cartoon on page 125 of the text might be used to introduce the method.* Transpar-

---

*A large copy of this cartoon is on page 90 of the Peanuts book titled *Fly, You Stupid Kite, Fly!* (Peanuts Parade Paperback No. 6).

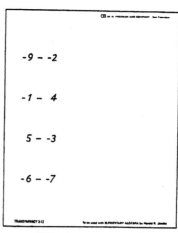

ency 3-12 gives examples with which to explain the method derived from this principle. After working each one by drawing circles, do the corresponding addition problems in the space at the right.

Lesson 5
Multiplication

A slide rule for subtracting numbers

Although the slide rule has become commercially obsolete, there is still a lot to be learned from making one. This exercise reinforces the lesson on subtraction and your students should enjoy doing it. Mine were so intrigued by the result that they asked if they could use their slide rules on the Chapter 3 test.

Give each student two strips of card stock measuring 1 by 6 inches. (Cut them in advance from unlined 4-by-6-inch file cards.) Have the students mark and number scales on each strip at $\frac{1}{4}$-inch intervals as shown below.

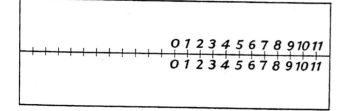

Placing each strip on a graph paper grid is more convenient than using a ruler. Demonstrate all of this at the overhead projector, using strips cut from heavy plastic.

Show how to subtract 2 from 7 on the slide rule by putting the 7 on the top scale above the 2 on the bottom scale; the answer is the number on the top scale that is above the 0 on the bottom scale. Have the students show how to subtract 3 from 10 by using the same method. *Now write "Answer is above this number" under the 0 on the bottom scale.*

Consider the problem of subtracting 8 from 5. Does

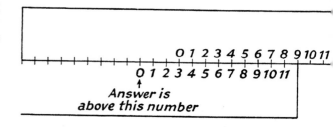

the rule give the correct answer? (*Yes, if we add negative numbers to the scales.*) *After extending both scales, use the slide rule to do these problems:*

$$9 - -2$$

$$-4 - 6$$

$$-1 - -8$$

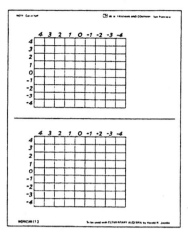

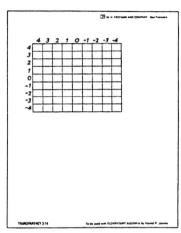

TRANSPARENCY 3-13	WORKSHEET 3	TRANSPARENCY 3-14

Finally, have the students put the 2 on the top scale above the 0 on the bottom scale and ask them "what subtraction problem" is illustrated. They should discover from this example

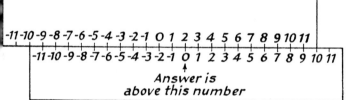

that each positioning of the scales solves many subtraction problems simultaneously.

Transparency 3-13 can be used to discuss the Set IV exercise of Lesson 4.

I like what W. W. Sawyer has to say concerning teaching the multiplication of negative numbers:

It is the experience of many teachers that addition of positive and negative numbers can be taught quite easily . . . and subtraction with a little more difficulty. The serious difficulties were felt to arise when multiplication was reached, in particular with the maxim "minus times minus is plus", so that, for example, –3 times –4 is +12. Here

again the wisest treatment is to build up the pupil's faith in this result by a series of widely different experiences, in each of which it is found that minus times minus being plus gives the most reasonable or the most natural arrangement. The aim all the time is not to tell the children the result, for the natural obstinacy and suspiciousness of human nature is such that the more we are told something, the less willing we are to believe it. Rather, we try to get the children to tell us. . . . Children have a natural liking for pattern and order, and they will always try to give an answer that preserves and continues a pattern. In this respect, they have the same instinct as research mathematicians.*

One approach suggested by Sawyer affords a nice way to enable the students to guess the rules for multiplication of positive and negative numbers. Although it admittedly does not prove anything, it does suggest the correct rules very strongly.

Hand out copies of the multiplication table reproduced on Worksheet 3 (also on Transparency 3-14), and ask the students to try to fill in the

---

*W. W. Sawyer, *Vision in Elementary Mathematics* (Penguin, 1964), p. 297.

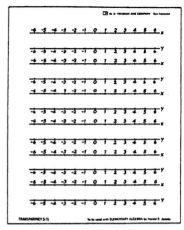

empty spaces. When the work is complete, explain that this is just one of many patterns that suggest that the product of a positive and a negative number is negative and that the product of two negative numbers is positive.

Lesson 6
Division

Give each student a copy of Transparency 3-15 and a ruler. The idea is to use each pair of number lines to illustrate one of the following operations:

1. Adding 2.
2. Subtracting 2.
3. Multiplying by 2.
4. Multiplying by –2.
5. Dividing by 2.
6. Dividing by –2.

This is done by connecting points on the first line to the appropriate points on the second line. After getting your students started by illustrating the first pattern on the screen, let them attempt the others on their own. (The finished sheet is shown at the left.) Although we have not yet considered the rules for dividing positive and negative numbers, the symmetries of these patterns strongly

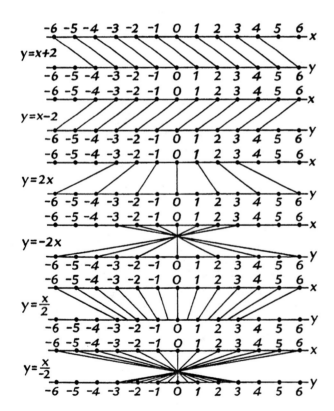

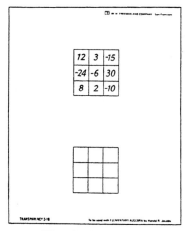

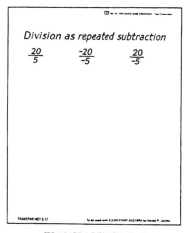

1. $8 - 6(2)$ 6. $-9(-5)^2$

2. $8 - 6(-2)$ 7. $(-1)^3 + (2)^3$

3. $(3-7) + 4$ 8. $(-1+2)^3$

4. $3 - (7+4)$ 9. $(-4)^2 + 5(-4) - 10$

5. $9 - (-5)^2$ 10. $(6+-7)(6--7)$

TRANSPARENCY 3-16	**TRANSPARENCY 3-17**	**TRANSPARENCY 3-18**

suggest the correct answers and give a nice preview of the new lesson.

Transparency 3-16 is for use in discussing the Set IV exercise of Lesson 5.

Because the lesson on division is especially simple and straightforward, you might assign it to be read without any further introduction.

Lesson 7
Several Operations

We began the course with several number tricks, one of which we will look at again today. Show Transparency 0-1. Have your students copy the directions in a column on the left side of their papers. Then have them perform each operation indicated, starting with the numbers 1 and 10. We proved, using boxes and circles, that the trick will give five beginning with every counting number. Because the rules for operating with negative numbers are consistent with those for positive numbers, the final answer should be five even if you begin with a negative integer. Give the students time to do the trick with –1 before checking their calculations. Then have them do it with –7, –10, and –1,000.

To be able to perform a series of operations with positive and negative numbers accurately is a very important skill in algebra. The new lesson will enable you to practice doing this.

Transparency 3-17 is for use in discussing the set IV exercise of Lesson 6. It is interesting to observe that, although the method does not work for the last problem in the usual sense, it *does* work if one subtracts –5 going upward rather than downward. We can subtract –5 "–4 times" to arrive at zero.

Additional exercises on performing several operations with positive and negative numbers are given on Transparency 3-18.

TRANSPARENCY 3-19

TRANSPARENCY 3-20

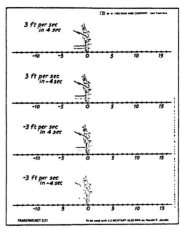

TRANSPARENCY 3-21

Review of Chapter 3

First day: Can time go backward?

Show Transparency 3-19, on which is reproduced the first page of an article by Martin Gardner in the January 1967 issue of *Scientific American,* and explain that, although there are many things that seem to make time moving backward an impossibility, scientists are not certain.

The *Backward March!* comic strip (Transparency 3-20) was drawn as an illustration for the article. Read it to the class and ask if the marcher obeyed the commander's direction to march forward. (The answer is yes. If time is moving backward as we read the strip, he is marching forward—the commander has given the command at the end. If we read the strip backwards, then the man seems to be marching backward.)

Show the top number line on Transparency 3-21. (The four soldier figures on the last page of the book of transparency masters should be cut out and stored in an envelope taped to the frame of the transparency. Place the first cut-out soldier so that it coincides with the drawing of the soldier on the number line.)

Suppose that the soldier marches to the right at the rate of 3 feet per second. Where would he be after 4 seconds? Move the cut-out soldier to 12 on the number line. *Notice that* $3 \cdot 4 = 12$*. In general,*

$r \cdot t = d$, *in which* r *represents the rate,* t *the time, and* d *the distance.* Show the second number line. *Where would the soldier be after –4 seconds?* Place the second cut-out soldier at –12 on the second line. *Notice that* $3(-4) = -4 + -4 + -4 = -12$*. Now suppose that the soldier marches to the right at the rate of –3 feet per second.* Illustrate where he would be after 4 seconds and –4 seconds on the remaining lines, writing the corresponding product equation in each case. Observe that the last case is in agreement with the rule that the product of two negative numbers is positive.

Second day

The following activity was inspired by an attempt by one of my students to make some tables showing the types of numbers produced when positive and negative numbers are added, subtracted, multiplied, and divided. I decided to include zero for good measure.

Have your students draw four tic-tac-toe patterns, one for each operation, and label them as shown. The symbols along the top represent the three possibilities for the first number: positive,

Answers to Set I

1. a) False   c) x-3<2   b) >
   b) True   c) >
   c) True   d) =
   d) False   e) <
   e) True

5. a)
```
 2  0 -6 -12
 8  6  0  -6
16 12  0 -12
-2 -6 -18 -30
-1 -3 -9 -15
 4  2 -4 -10
 2  2  2   2
```
8. a)
```
x 2 3 4 5
y 0 1 2 3
```
b) y=x-2
c) y (graph)

2. a) -12
   b) -18   b) (box figures)
   c) -45   c) y (graph)
   d) -5
   e) -18
   f) -12
   g) 45
   h) 5

6. a) 11
   b) -1
   c) -8   d) x -2 -1 0 1
   d) -12   y -4 -3 -2 -1
   e) 6   9. a) 2
   f) -2   b) 10
3. a) 3+7   7. a) <   c) -5
   b) -8+-4   d) 22
   c) x+-y
4. a) x>0
   b) x³=9x

TRANSPARENCY 3-22

---

10. a) -8   e) -23
    b) 5   14. a) 2(-4+-6) = 2(-4)+2(-6)
    c) x   2(-10)   = -8 + -12
    d) -1   -20   = -20
11. a) 60   b) -5(8+-1) = -5(8)+-5(-1)
    b) -19   -5(7)   = -40 + 5
    c) -71   -35   = -35
    d) -2,002   c) -7(-3+9) = -7(-3)+-7(9)
    e) 17   -7(6)   = 21 +-63
12. a) He got   -42   = -42
    5 more wrong
    than he got right.
    b) 11
    c) 8
13. a) 17
    b) 17
    c) 6
    d) -6

TRANSPARENCY 3-23

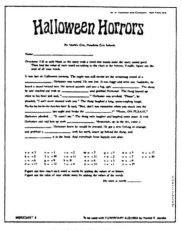

WORKSHEET 4

---

zero, or negative. The symbols at the left represent the possibilities for the second number. The idea is to show by using these symbols in filling in the grid the type of answer produced where possible. For example, the sum of two positive numbers is always positive, the sum of a positive number and zero is always positive, and the sum of a positive number and a negative number may be either positive, zero, or negative; to show this, we can fill in the first column of the grid as follows:

```
    +  0  -
+ +
0 +
- ?
```

Finished grids for all four operations are shown below. The NP indicates that division by zero is not possible.

Addition
```
   +  0  -
+  +  +  ?
0  +  0  -
-  ?  -  -
```

Subtraction
```
   +  0  -
+  ?  -  -
0  +  0  -
-  +  +  ?
```

Multiplication
```
   +  0  -
+  +  0  -
0  0  0  0
-  -  0  +
```

Division
```
   +  0   -
+  +  0   -
0  NP NP NP
-  -  0   +
```

The answers to the Set I exercises of the review lesson are provided on Transparencies 3-22 and 3-23.

Halloween horrors

If Halloween occurs sometime soon after Lesson 3 of this chapter has been completed, your students might enjoy the following activity.

Hand out copies of Worksheet 4 on which there is a halloween story by Shirley Cox of the Pasadena City Schools. The directions at the top and bottom of the sheet explain what to do. I suggest making this exercise strictly optional and, because it is rather time-consuming, not collecting it until the day of the test. You might offer a small prize or some extra credit to the person who turns in the story worth the most points.

Chapter 4

THE RATIONAL NUMBERS

The primary purpose of this chapter is to teach the student how to compute with positive and negative rational numbers in decimal form. To do this, we first take the particle-antiparticle model used to add positive and negative integers and extend it to lengths and antilengths. This extension gives meaning to the rules for addition in terms of absolute values, which are now introduced. (It is interesting to note that the idea of absolute value was not included in algebra texts until the present century.*) The rules for subtraction, multiplication and division easily follow.

Rational numbers whose decimal forms do not terminate are not considered until the lesson on approximations and a discussion of the theory of such numbers is postponed until Chapter 14.

The chapter closes with another lesson on graphing. In doing the only lesson dealing with negative numbers in graphs so far (Chapter 3, Lesson 2), the student had not yet learned how to compute with positive and negative numbers. For this lesson, such computations are necessary in

order to be able to produce the tables by means of which a variety of linear and nonlinear functions are graphed.

Lesson 1
The Rational Numbers

NBC began television broadcasting in 1930, followed by CBS in 1931. Show Transparency 4-1. *The photograph at the top of this transparency shows NBC's original television station. A model of Felix the Cat stands on a turntable in front of the camera. The picture produced, shown in the photograph below, consisted of 120 lines, a very crude image in comparison with the modern television picture, which consists of 525 lines.*

It takes the electron beam inside a television tube 0.033 second to produce the picture. The number, 0.033, is neither a counting number nor an integer. It is an example of a rational number *because it can be written as the quotient of two integers:* $\frac{33}{1000}$. *Written as 0.033, it is said to be in* decimal form.

How many pictures are produced on a television

---

*For example, it is not discussed in a book published in 1905, titled *Advanced Algebra* by Herbert E. Hawkes of Yale University.

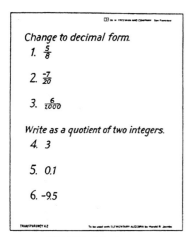

Change to decimal form.

1. $\frac{5}{8}$

2. $\frac{-7}{20}$

3. $\frac{6}{1000}$

Write as a quotient of two integers.

4. 3

5. 0.1

6. -9.5

TRANSPARENCY 4-2

TRANSPARENCY 4-3

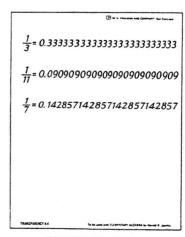

$\frac{1}{3} = 0.333333333333333333333333$

$\frac{1}{11} = 0.09090909090909090909090909$

$\frac{1}{7} = 0.142857142857142857142857$

TRANSPARENCY 4-4

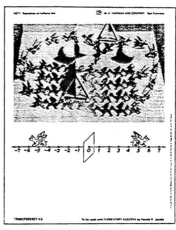

TRANSPARENCY 4-5

screen in one second? $\left(\dfrac{1}{0.033} \text{ or } \dfrac{1000}{33}; \text{ approxi-} \right.$ mately 30. $\Big)$

More exercises with rational numbers are included on Transparency 4-2.

Lesson 2
Absolute Value and Addition

According to an article in Newsweek *several years ago* (January 28, 1974, page 69), *the number of colds that a person catches in a year depends on his or her age.* Show Transparency 4-3. *Infants less than one year old catch the most: about 6. People more than sixty catch the least: less than 2.*

Notice that none of the numbers of colds per year listed are integers. For people between fifteen and nineteen years of age, for example, the number is not 2 or 3 but 2.4. It doesn't seem reasonable to speak of someone catching 2.4 colds. What kind of a number is 2.4 and what does it mean? (It is a rational number, obtained as a quotient of two integers: the total number of colds caught in a year by a group of fifteen-to-nineteen-year-olds and the number of people in the group.)

Transparency 4-4 is for use in discussing the Set IV exercise of Lesson 1.

The purpose of introducing the concept of absolute value in Lesson 2 is to enable the student to understand the rules for the addition of positive and negative rational numbers. You might begin the lesson by showing the class Escher's lithograph, *Magic Mirror*, which is reproduced at the top of Transparency 4-5. *It shows a row of animals emerging from a mirror and walking to the right as their reflections walk to the left. Each animal on one side of the mirror seems to have an "opposite" on the other side. The drawing below the lithograph shows a number line with a mirror at zero separating the positive numbers at the right from their opposites, the negative numbers at the left. The distance of each number from the mirror is equal to the distance of its*

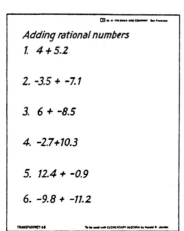

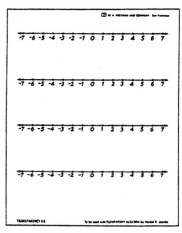

TRANSPARENCY 4-6   TRANSPARENCY 4-7   TRANSPARENCY 4-8

opposite from the mirror. This distance is called the absolute value *of the number.*

Develop the rest of the lesson in accord with the text. Examples of adding positive and negative rational numbers are given on Transparency 4-6. Strips for illustrating the first four examples are on the last page of the transparency book. I suggest covering the four negative strips with a piece of colored adhesive film before cutting them out.

Lesson 3
More on Operations with Rational Numbers

Several years ago a fast-food chain advertised a "0.1133975 kilogram" hamburger. Show Transparency 4-7. Given that 1 pound = 0.45359 kilogram, can you figure out what the ad means? (The hamburger weighs a quarter of a pound.) Although your students may be using calculators for problems such as this one, it might be well to review how to obtain the answer without a calculator. We can either divide one number into the other or observe that 0.1133975 is about one-fourth of 0.45359 and divide 0.45359 by 4.

To translate the ad into more familiar terms, we had to find the quotient of two rational numbers. In

yesterday's lesson, we learned how to add such numbers. Today we will review subtraction, multiplication, and division.

Transparency 4-8 can be used to discuss the Set IV exercise of Lesson 2 by illustrating several possible two-part trips on the lines.

Examples for use in presenting Lesson 3 are given on Transparency 4-9.

Lesson 4
Approximations

Another calculator riddle

To find the answer to the following riddle (Transparency 4-10), do what was done to find the answer to the Set IV exercise of Lesson 3: solve both problems with a calculator and turn it upside-down to read the answer.

What did John Travolta say when asked if he would be willing to star in Saturday Night Fever?

1. $(-4.15)(-19.2) - 2.58$

2. $\dfrac{53.37 - (98.1)^2}{-0.03}$

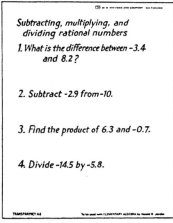

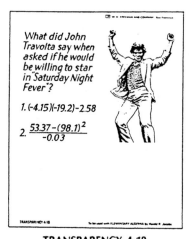

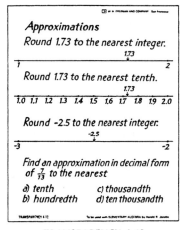

TRANSPARENCY 4-9	TRANSPARENCY 4-10	TRANSPARENCY 4-11

TRANSPARENCY 4-12

The solutions are 77.1 and 319008: I'LL BOOGIE.

To introduce the lesson on approximations, show Transparency 4-11. *The height of Mount Everest, the highest place on the earth, was first measured accurately by the Survey Department of the Government of India in 1852.* (Add overlay.) *They made six measurements, each from a different place. When they found the average of the six numbers, it turned out to be exactly 29,000 feet. Unhappy with this result, the surveyors secretly added 2 feet to it and for many years the height of Mount Everest was reported as 29,002 feet. Can you figure out why the surveyors did this?* They were afraid that people might think that the result was more approximate than it really was. *The number 29,000 might mean that they had only found the height of Mount Everest to the nearest thousand feet, whereas their work was much more precise than that.* This can be illustrated on the number line at the bottom of the overlay. Indicate the interval of numbers on the line that are closer to 29,000 than to 28,000 or 30,000 and compare it with the interval containing the six measurements.

The examples on Transparency 4-12 can be used to develop the rest of the lesson on approximations.

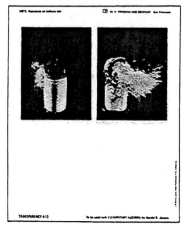

TRANSPARENCY 4-13

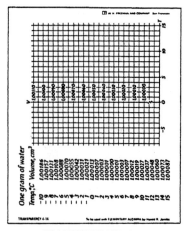

TRANSPARENCY 4-14

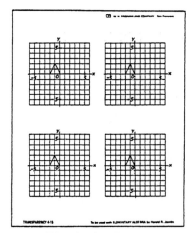

TRANSPARENCY 4-15

Lesson 5
More on Graphing Functions

A water bomb that is dangerous

Show Transparency 4-13. *When water changes in temperature, it also changes in volume. The first photograph shows a cast-iron container with walls a quarter of an inch thick filled with water and placed in a beaker of dry ice and alcohol. The second photograph shows what happened to the container when the water froze. It exploded, hurling pieces of glass and steel as far as 20 feet away.*

The volume of a given weight of water is a function of its temperature. To graph this function requires working with rational numbers, the subject of our lesson for today. Hand out duplicated copies of Transparency 4-14. Explain the numbering of the scales and the meaning of the zig-zag on the y-axis. Show how to plot some of the points of the table that fit within the scales chosen. Ask the students to plot the rest of the points and connect them with a smooth curve. The result, which strongly resembles a parabola, is shown in the next column.

After discussing Lesson 4, you might want to demonstrate another example of graphing a function in which rational numbers occur. An interesting one that is different from any in the exercises for Lesson 5 is $y = \frac{12}{x^2}$. Let x vary from –6 to 6.

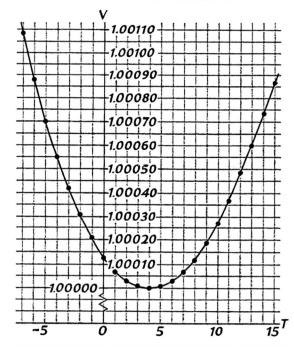

Consider at least one point for which x is between 0 and 1.

Review of Chapter 4

First day

Hand out duplicated copies of Transparency 4-15. *In doing this exercise, we will make algebraic trans-*

formations of the grade that I hope you will get on the next test. In the first transformation, we will add four to each coordinate of each point of the figure. First, have the students write the coordinates of the points as shown in the first figure below.

Next, show them how to transform each coordi-nate pair by adding four to each number. Finally, show them how to draw the image of A resulting from the transformation. Some other interesting transformations are illustrated in the other figures. The last is somewhat surprising in that the right side of the image A is curved.

1. Add 4 to each coordinate.

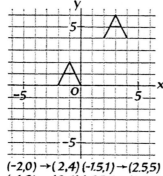

$(-2,0) \rightarrow (2,4)$ $(-1.5,1) \rightarrow (2.5,5)$
$(-1,2) \rightarrow (3,6)$ $(-0.5,1) \rightarrow (3.5,5)$
$(0,0) \rightarrow (4,4)$

2. Multiply each coordinate by 2.

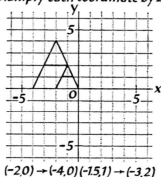

$(-2,0) \rightarrow (-4,0)$ $(-1.5,1) \rightarrow (-3,2)$
$(-1,2) \rightarrow (-2,4)$ $(0.5,1) \rightarrow (-1,2)$
$(0,0) \rightarrow (0,0)$

3. Multiply each coordinate by -3.

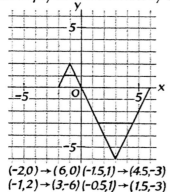

$(-2,0) \rightarrow (6,0)$ $(-1.5,1) \rightarrow (4.5,-3)$
$(-1,2) \rightarrow (3,-6)$ $(-0.5,1) \rightarrow (1.5,-3)$
$(0,0) \rightarrow (0,0)$

4. Square each coordinate.

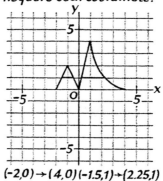

$(-2,0) \rightarrow (4,0)$ $(-1.5,1) \rightarrow (2.25,1)$
$(-1,2) \rightarrow (1,4)$ $(-0.5,1) \rightarrow (0.25,1)$
$(0,0) \rightarrow (0,0)$

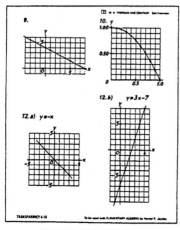

TRANSPARENCY 4-16

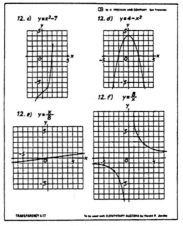

TRANSPARENCY 4-17

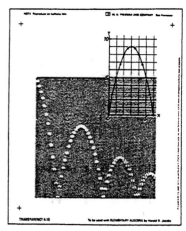

TRANSPARENCY 4-18

TRANSPARENCY 4-19

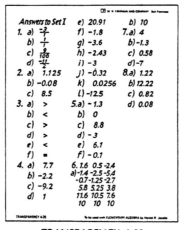

TRANSPARENCY 4-20

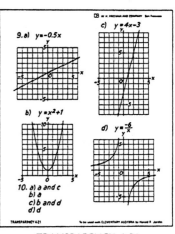

TRANSPARENCY 4-21

Graphs for discussing the Set III and IV exercises of Lesson 5 are given on Transparencies 4-16 and 4-17 and on the overlay for Transparency 4-18.

Second day

Show the cartoon at the top of Transparency 4-19. *People who are skilled at using the abacus can make calculations on it very quickly. Contests have been held between someone using an abacus and someone using an electronic calculator, in fact, in which the person using the abacus was the winner!*

A list of 15 numbers comparable to the sort used in such contests *is shown at the right of the cartoon. Keep the sum, shown at the bottom of the column, covered. Use your calculator to find their sum as quickly as you can but wait until I tell you to begin. Raise your hand when you are finished.* Note the time when the first hand is raised and again when half the hands are raised. Then reveal the correct answer. *Someone skilled at using an abacus can add these 15 numbers in just 15 seconds!* It took me about 45 seconds with my calculator.

Answers to the Set I exercises of the review lesson are given on Transparencies 4-20 and 4-21.

Chapter 5

EQUATIONS IN ONE VARIABLE

The logical approach to teaching beginning students how to solve equations in one variable seems to me to be psychologically wrong. Traditionally, we begin with linear equations, such as $x + 5 = 8$ and $3x = 12$, that can be solved in one step. Because we want our students to check their answers, we use equations whose solutions are integers. For an equation such as $x + 5 = 8$, the student might be expected to write

$$x + 5 = 8$$
$$x + 5 - 5 = 8 - 5$$
$$x = 3$$

Check: $\quad 3 + 5 = 8$
$$8 = 8$$

The trouble with this approach is that most students do not see the point of carrying out this routine when the answer is so obvious. They wonder why the problem should be done the "hard" way when there is such an easy way to solve it. It is easy to see why students often become irritated by a seemingly foolish routine and rebel at carrying it out.

The solution to this dilemma, it seems to me, is to encourage guessing at the start. The students have already had considerable experience at doing this before getting to Chapter 5. Look, for example, at exercise 6 of Set I in the Chapter 3 review. There they were solving equations such as $-9 + x = 2$ and $3x = -3$ by guessing. To tell the students that guessing is not allowed is self-defeating because, if this advice is taken seriously, they eventually become afraid to guess or may even become so demoralized that they no longer even *want* to guess.

How does one interest the student in learning the techniques for solving equations at this point in the course? By starting with a variety of equations, ranging from those having solutions that can be readily guessed to those having solutions that are difficult or impossible to guess, including not only equations having unique solutions but also those having no solutions or many solutions.

Lessons on inverse operations and equivalent expressions then prepare the student for learning

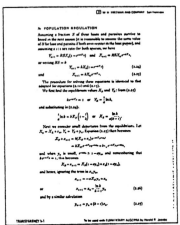

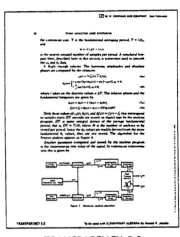

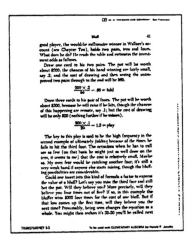

TRANSPARENCY 5-1   TRANSPARENCY 5-2   TRANSPARENCY 5-3

some of the formal procedures. A lesson on length and area serves as an introduction to the art of writing equations. Finally, a couple of lessons deal with "distance, rate, and time" problems. I feel that a serious weakness of many elementary algebra courses is the kind of "word" problem with which the student is confronted. Many of these problems have no practical value whatsoever and frequently cause the skeptical student to wonder about the usefulness of the subject. A weakness of the "word" problems themselves is that many of them can be solved more efficiently by simple arithmetic than by writing and solving an equation. Even worse, a great emphasis on such problems can consume valuable time that could be better spent making sure that the students are really mastering the fundamentals of the subject. To what avail has their study of algebra been if the students have spent many weeks on such problems and yet come back to school after summer vacation not even remembering how to add two fractions or find the square of a binomial? It is for these reasons that I have chosen to place less emphasis on the traditional "word" problems than is sometimes done.

Lesson 1
Equations

Show Transparencies 5-1, 5-2, and 5-3. *These pages from books on biology (*population regulation*), music*

(tone analysis and synthesis), and poker (bluffing) illustrate the use of equations in a variety of areas of study. The equations in the first two examples are so important that they are numbered for reference.

Some equations are obviously true: $1 + 1 = 2$, and some false: $3 \cdot 4 = 10$. These are the equations of arithmetic.

The equations of algebra are usually neither true nor false as they are written. Consider, for example, x + 2 = 7. *This equation becomes true or false depending on the number that replaces the variable. Numbers that make it true are called* solutions *of the equation. Some equations, like this one, have exactly one solution; some equations, such as* x² = 4, *have more than one solution, and some, such as* x = x − 3, *have no solutions.*

Some equations are easy to solve: the solution is either obvious or readily guessed. The equation, 2x = 10, *is an example. Others are not so easy. The equation,* x² − x = 8, *has two solutions, but someone would be very lucky to be able to find them by guessing. Others, such as* x⁵ + 3 = 11x, *are impossible to solve by guessing.*

Can you find a number that will make each of the following equations true? Show Transparency 5-4. The idea is for the students to guess *the solutions. It would be a mistake to explain any methods at this time. After the examples have been discussed, point out that a major goal of algebra is to learn efficient methods for solving equations that do not depend on guessing.*

TRANSPARENCY 5-4

1. $x - 2 = 16$ 6. $5^x = 125$

2. $3x = 45$ 7. $x + 7 = 6$

3. $\dfrac{40}{x} = 5$ 8. $x + x = 2x$

4. $x^2 = 9$ 9. $4 - x = 12$

5. $x = x + 3$ 10. $x^2 + 3x = 10$

TRANSPARENCY 5-5

Think of a number.
Add three.
Divide by four.
Subtract eight.
The result is two.

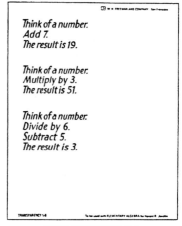

Think of a number.
Add 7.
The result is 19.

Think of a number.
Multiply by 3.
The result is 51.

Think of a number.
Divide by 6.
Subtract 5.
The result is 3.

TRANSPARENCY 5-6

Lesson 2
Inverse Operations

A number puzzle

Since beginning our study of algebra, we have done quite a few number tricks. Show Transparency 5-5. *Although these directions sound like those for a number trick, they are not. A number trick should work for* any *number that you might start with. These directions only work for* one *number; so this might be called a number* puzzle *rather than a number trick. Can you figure out what the number is?*

Think of a number.
Add three.
Divide by four.
Subtract eight.
The result is two.

Give your students time to think about this before any discussion of a possible procedure. Those who discover the original number, 37, will probably have worked backward, reasoning as follows. Two is the result of subtracting eight from what number? Ten. Ten is the result of dividing what number by four? Forty. Forty is the result of adding three to what number? 37. The number originally thought of was 37.

List these steps to the right of the first set, as shown below.

Think of a number.	Thirty-seven.
Add three.	Subtract three.
Divide by four.	Multiply by four.
Subtract eight.	Add eight.
The result is two	Two.

To get back to the original number, we performed a series of inverse *operations. Inverse operations are the subject of today's lesson.* The three number puzzles on Transparency 5-6 can be used to develop the rest of the lesson.

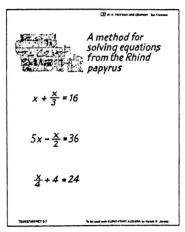

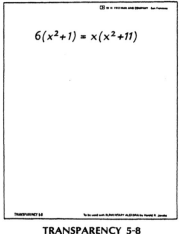

TRANSPARENCY 5-7   **TRANSPARENCY 5-8**   **TRANSPARENCY 5-9**

Transparency 5-7 is for use in discussing the Set IV exercise of Lesson 1.

Lesson 3
Equivalent Equations

You have had some practice at guessing solutions to equations. Show Transparency 5-8. *Can you find any solutions to this equation by guessing?*

$$6(x^2 + 1) = x(x^2 + 11)$$

(The equation has three solutions: 1, 2, and 3.) After the class has had time to discover these, point out that an equation is not considered solved until we have found *all* of the numbers that can replace the variable to make it true. *At this point in our study of equations, we have no idea of how many solutions an equation such as this might have. In today's lesson, we will learn a method for finding the solutions to some simple equations that does not require guessing.*

Transparency 5-9 is for use in discussing the Set IV exercise of Lesson 2.

Transparency 5-10 may be used to develop Lesson 3 in accord with the text. Examples different from those in the text are given on Transparency 5-11. I suggest alternating the solutions, solving

each odd-numbered example with your students and having them try to solve the even-numbered ones on their own.

Lesson 4
Equivalent Expressions

Show Transparency 5-12. *Everyone has known this equation "for ages" because it is true regardless of what number replaces* x. *Two expressions that are equal for all values of their variables, such as* $2x \cdot 2$ *and* $4x$, *are called* equivalent. Show Transparency 5-13. *Here are more equations. Which of these do you suppose "everyone has known for ages"?*

Use these examples to develop the ideas of Lesson 4. (Be careful not to create the impression that equations 2, 5, 6, or 9 are false. They are simply not true for every value of their variables.)

Transparency 5-14 can be used to discuss the Set IV exercise of Lesson 3.

Lesson 5
More on Solving Equations

The bears problem

The following exercise is by W. W. Sawyer, who

If the two bricks and three 1-pound weights exactly balance the nine 1-pound weights, how much does each brick weigh?

TRANSPARENCY 5-10

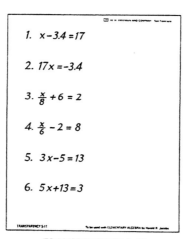

1. $x - 3.4 = 17$

2. $17x = -3.4$

3. $\frac{x}{8} + 6 = 2$

4. $\frac{x}{6} - 2 = 8$

5. $3x - 5 = 13$

6. $5x + 13 = 3$

TRANSPARENCY 5-11

"But Gershon, you can't call it Gershon's equation if everyone has known it for ages."

TRANSPARENCY 5-12

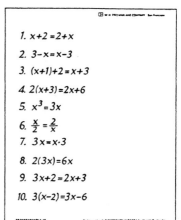

1. $x + 2 = 2 + x$

2. $3 - x = x - 3$

3. $(x+1) + 2 = x + 3$

4. $2(x+3) = 2x + 6$

5. $x^3 = 3x$

6. $\frac{x}{2} = \frac{2}{x}$

7. $3x = x \cdot 3$

8. $2(3x) = 6x$

9. $3x + 2 = 2x + 3$

10. $3(x-2) = 3x - 6$

TRANSPARENCY 5-13

calls it "a miniature problem in design." He introduces it with the following words:

When some complicated contrivance is being designed, each decision involves a long chain of consequences. If we alter the size or the weight or the strength of one part, we may have to re-design every other part. Often it is hard to see all the effects of each decision; we would like to postpone the moment when we commit ourselves. Algebra gives us a way of doing this. . . .

It is difficult to illustrate this procedure by examples drawn from actual engineering or scientific practice. The calculations are usually too involved, the technical considerations too many, for the work to serve as an introduction to elementary algebra. One is, therefore, forced to consider somewhat unreal situations. The problem discussed . . . is far from the normal routine of engineering. It is about a man who is troubled by bears, and builds a series of gates to keep them out.*

*W. W. Sawyer, *Vision in Elementary Mathematics* (Penguin, 1964), ch. 7.

TRANSPARENCY 5-14

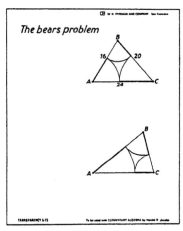

TRANSPARENCY 5-15

Hand out duplicated copies of Transparency 5-15. *This figure, an overhead view, shows three posts, labeled A, B, and C. The numbers represent the distances between them in feet. A man wants to mount one gate on each post. The gates are shown in heavy black and, as drawn, do not keep anything out. If bears appear from the northwest, however, the man can swing the gate at B to the left to close the gap between posts A and B. If the bears appear from the northeast, the man can leave the gate at B where it is and swing the gate at C up so that the side BC is closed. If the bears come from the south, he can swing gate A downward and close side AC. The problem is to figure out how wide to make the three gates so that they meet on each side without overlapping or leaving a gap.*

Let's make a guess. Suppose that the gate at A is 5 feet wide. Then gate B would have to be 11 feet wide and gate C would be 9 feet wide. But this would leave a gap along side AC of 10 feet, which the bears could get through.

Suppose that gate A is 6 feet wide instead. Have everyone record the results from the first guess and those from this one in a table such as the one shown here.

Gate A	5	6
Gate B	11	10
Gate C	9	10
Gap	10	8

Now let the students make their own guesses for gate A. Someone may reason from the results of guessing 5 and 6 that increasing gate A by 1 foot results in making the gap 2 feet smaller. If so, increasing gate A another 4 feet should make the gap disappear. That this reasoning does lead to the correct answer is easily verified.

One way to solve the problem, then, is to simply guess the answer; if we are lucky, we may even guess the correct answer on the first try. Another way would be to make some guesses and then discover the correct answer by reasoning from the results obtained. A third way is to "roll all of the guesses into one" and let gate A be x feet wide. This means that gate B is $16 - x$ feet wide and gate C is $20 - (16 - x)$ feet wide. We have encountered a pattern like $20 - (16 - x)$ before. (See Chapter 3, Lesson 4, exercises 7 and 13.) Because $20 - (16 - x) = 20 - 16 + x = 4 + x$, gate C must be 4 feet wider than gate A, as the numbers in our table confirm.

Filling in the lengths on each side as shown, we see

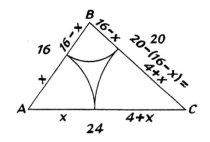

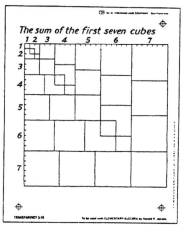

The sum of the first seven cubes

TRANSPARENCY 5-16

that the sum of x *and 4 +* x *must be equal to 24 if there is to be no gap on side AC. Solving the equation* x + (4 + x) = 24, *we get*

$$2x + 4 = 24$$
$$2x = 20$$
$$x = 10$$

Gate A should be 10 feet wide.

Have the class make up another problem by asking a student to choose different numbers for the distances between the posts. Use the second figure on Transparency 5-15 for this purpose. (Be sure that the numbers chosen could really be the sides of a triangle. Otherwise, one of the gate lengths will turn out to be negative!)

The Set IV exercise of Lesson 4 is about a rather surprising identity: the sum of the first *n* cubes is equal to the square of the sum of the first *n* integers. A clever way of illustrating this, devised by Solomon W. Golomb, was originally published in the May 1965 issue of *The Mathematical Gazette* and was also in Martin Gardner's column in the October 1973 issue of *Scientific American* titled "'Look-see' diagrams that offer visual proof of complex algebraic formulas." It is reproduced on Transparency 5-16. The figure is a square whose side is equal in length to the sum of the first seven integers; so its area is

$$(1 + 2 + 3 + 4 + 5 + 6 + 7)^2$$

It consists of 1 square of side 1, two squares of side 2, three squares of side 3, and so forth, up to seven squares of side 7. Some of the squares overlap (shown in dark color on the overlay), but each region of overlap corresponds to a hole of the same size. Because $1 \cdot 1^2 = 1^3$, $2 \cdot 2^2 = 2^3$, $3 \cdot 3^2 = 3^3$, and so forth, the total area is also

$$1^3 + 2^3 + 3^3 + 4^3 + 5^3 + 6^3 + 7^3$$

The most effective way to show this would be to cut out squares from heavyweight film, as indicated below, and place them on the pattern as you refer to them.

Color A: 1 square with side 1; 3 squares with sides 3 in L shape; 5 squares with sides 5 in L shape; and 7 squares with sides 7 in L shape.

Color B: 2 separate squares with sides 2; 4 squares with sides 4, 2 in each of 2 separate strips; and 6 squares with sides 6, 3 in each of 2 separate strips.

Simpler, although less effective, would be to use the overlay.

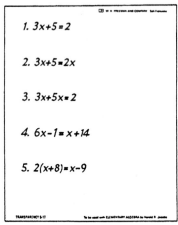

1. $3x+5=2$

2. $3x+5=2x$

3. $3x+5x=2$

4. $6x-1=x+14$

5. $2(x+8)=x-9$

TRANSPARENCY 5-17

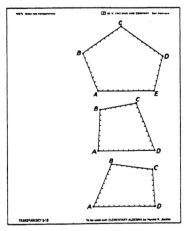

TRANSPARENCY 5-18

In the new lesson, the student learns how to solve linear equations in which the variable is on both sides. Examples different from those in the text are given on Transparency 5-17. Be sure to include checking each answer as part of each exercise.

Lesson 6
Length and Area

More on the bears problem

Hand out compasses and copies of Transparency 5-18 to your students. (Make two copies of the transparency and print it on both sides of the students' papers.) *Here are problems similar to the one we considered yesterday, but with more than three posts. The first figure shows an enclosure containing five posts. Write down the lengths of its sides in terms of the unit given.* Have a student guess a length for the gate on post A and show everyone how to draw an arc with center at A and that number as radius to represent gate A swinging from side AE to side AB. Draw arcs centered at B, C, D, and E so that the gates will close along sides AB, BC, and CD. Note the gap or overlap along side AE.

Now have your students turn their papers over and use algebra to figure out how long the gate on each post should be. Starting at post A and going clockwise around the figure, the lengths are labeled as shown in the next column.

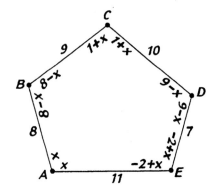

Writing an equation for side AE, we get

$$x + (-2 + x) = 11$$
$$2x - 2 = 11$$
$$2x = 13$$
$$x = 6.5$$

Draw arcs to show that this solution for x will cause the gates to close correctly on all five sides.

Have the students turn their papers back to the original side, make their own guesses for how long gate A should be in each of the other figures, and draw a set of arcs on each figure to check their guesses. After everyone has had time to do this for both figures, ask for a show of hands by those who guessed "the correct answer" for the second figure. The class will be astonished to discover that every-

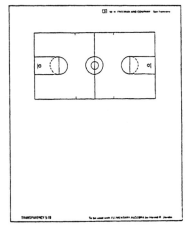

TRANSPARENCY 5-19

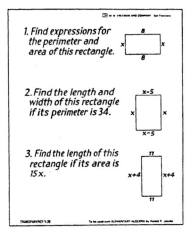

1. Find expressions for the perimeter and area of this rectangle.

2. Find the length and width of this rectangle if its perimeter is 34.

3. Find the length of this rectangle if its area is 15x.

TRANSPARENCY 5-20

TRANSPARENCY 5-21

one has. Then ask for a show of hands for the third figure. This time no one will have guessed the answer!

Turn the papers over once more and work out the algebra for each figure. In both cases, the variable disappears in the final equation, and so its value does not matter. The equation that results for the second figure, $11 = 11$, is true, and so any solution will work (although solutions in which one or more of the gates have negative lengths make little sense). The equation that results for the third figure, $8 = 12$, is false, and so the problem has no solution.

Some of your better students may enjoy exploring variations of the bears problem further. If so, they may discover that the nature of the solution depends on whether there is an odd or an even number of posts. Furthermore, for cases in which there is an even number of posts, the nature of the solution depends on a relationship between the lengths of the sides.

To introduce Lesson 6, you might ask your students what the figure on Transparency 5-19 represents. (The floor plan of a basketball court.) Ask if anyone knows its dimensions. (The U.S. court is 50 feet wide and 94 feet long.) Use it to review the meanings of the terms *perimeter* and *area*. Examples different from those in the text are given on Transparency 5-20.

Progressive Human Speed Records		
Speed in mph	*Vehicle*	*Date*
35?	Skis	3000 B.C.
57	Train, England	1839
131	Train, Germany	1903
150	Automobile, U.S.	1907
210	Airplane, France	1921
486	Airplane, Germany	1939
1,135	Rocketplane, U.S.	1951
2,905	Rocketplane, U.S.	1961
17,560	Spaceship, Russia	1962
24,791	Spaceship, U.S.	1969

TRANSPARENCY 5-22

Lesson 7
Distance, Rate, and Time

Almost everyone these days has heard of the Bee Gees. Here is a picture of a Gee Bee! Show Transparency 5-21. *This funny-looking little plane set the world speed record in 1932 by flying 296 miles per hour at the National Air Races in Cleveland. Because modern airplanes routinely fly at much greater speeds, it is hard to believe that less than fifty years ago the world record was only 296 miles per hour. In fact, a hundred years ago it was less than 90 miles per hour!*

Here is a table listing some progressive speed records in history. Show Transparency 5-22, *revealing one line at a time.*

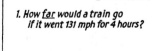

1. How <u>far</u> would a train go
   if it went 131 mph for 4 hours?

2. How <u>fast</u> is a car going
   if it goes 200 miles in 2.5 hours?

3. How <u>much time</u> would it take an
   airplane to go 84 miles
   if it goes 210 miles per hour?

TRANSPARENCY 5-23

TRANSPARENCY 5-24

TRANSPARENCY 5-25

Here is a short quiz to check how well you understand the relationship between distance, speed, and time (Transparency 5-23.) Have everyone copy the questions and try to answer them before any discussion. Introduce the formula $d = rt$, with the first question and show how it can be used to solve each problem algebraically.

Transparency 5-24 is for use in discussing the Set IV exercise of Lesson 6.

Lesson 8
Rate Problems

First day

For most classes, it is probably a good idea to spend two days on this lesson, assigning the Set II exercises on the first day and the Set III exercises on the second.

Show Transparency 5-25. *These photographs show Jupiter and three of its moons emerging from behind our own moon. One method by which astronomers have found the distance of the earth from the sun has as its basis the timing of the eclipses of the moons of Jupiter. When the earth is moving in its orbit away from Jupiter, the time intervals between these eclipses as observed from the earth are longer than when the earth is moving toward Jupiter. The* reason for this is that, as the distance from the earth to Jupiter increases, the time that it takes light to travel to the earth from Jupiter also increases.

Show Transparency 5-26. *From these observations, it has been determined that it takes light about 16 minutes and 40 seconds to cross the earth's orbit. Show how this information and the fact that light travels at a speed of 186,000 miles per second can be used to find the distance of the earth from the sun. (16 minutes, 40 seconds = 1,000 seconds; 186,000 miles per second × 1,000 seconds = 186,000,000 miles; $\frac{186,000,000 \text{ miles}}{2} = 93,000,000 \text{ miles}$.)*

Transparency 5-27 is for use in discussing the Set IV exercise of Lesson 7.

The example used in the text to develop Lesson 8 has as its basis the prize-winning flight of the Gossamer Condor. If you want to present it to your class, you might first show Transparency 5-28, which shows Otto Lilienthal, one of the pioneers of heavier-than-air flight, flying one of his hang gliders in 1894.

The photograph of the Cossamer Condor that is in the text is reproduced on Transparency 5-29. Some of the following information taken from "Science and the Citizen," *Scientific American*, October 1977, may be of interest to your class.

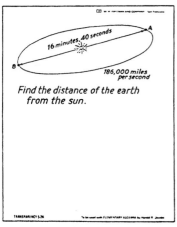

Find the distance of the earth from the sun.

TRANSPARENCY 5-26

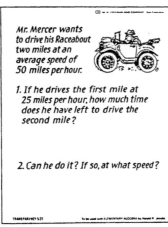

Mr. Mercer wants to drive his Raceabout two miles at an average speed of 50 miles per hour.

1. If he drives the first mile at 25 miles per hour, how much time does he have left to drive the second mile?

2. Can he do it? If so, at what speed?

TRANSPARENCY 5-27

TRANSPARENCY 5-28

The modern era of man-powered flight began in 1935, when two British engineers developed a propeller-driven "flying bicycle" that traveled through the air for short distances. In 1959 a British industrialist under the auspices of the Royal Aeronautical Society offered a prize now worth $85,000 for the first man-powered flight in which the craft took off from level ground without assistance, flew in a figure-eight pattern over a mile course, and passed over 10-foot hurdles at the start and finish. No stored power or buoyant gases were permitted.

The Gossamer Condor, which won the prize in 1977, was developed by Paul B. MacCready, the president of AeroVironment, Inc., of Pasadena, California. It is 30 feet long, 8 feet high, and has a wing span of 96 feet, yet weighs only 70 pounds. It is flown by pedaling a bicycle chain connected to a propeller behind the cockpit. The prize-winning flight took place at the Kern County Airport in Shafter, California.

The main difference between the Gossamer Condor and earlier man-powered aircraft is that it was based on hang-glider technology. The pilot, Bryan Allen, a 24-year-old championship bicycle racer, is also an experienced hang-glider pilot.

In finding the plane's airspeed from the data given, we assume for simplicity that it flew back and forth along a straight path rather than in a figure-eight. Show the students how to set up and solve the problem, as is done in the text.

The flight of the Gossamer Condor, August 1977.

4 minutes, 20 seconds against a 2 mph wind

3 minutes with the same wind

What was the plane's airspeed?

TRANSPARENCY 5-29

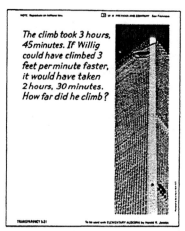

The climb took 3 hours, 45 minutes. If Willig could have climbed 3 feet per minute faster, it would have taken 2 hours, 30 minutes. How far did he climb?

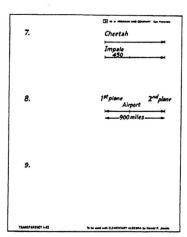

7. Cheetah

Impala
450

8. 1st plane   2nd plane
Airport
900 miles

9.

TRANSPARENCY 5-30	TRANSPARENCY 5-31

TRANSPARENCY 5-32

Second day

Show Transparency 5-30. *The world record for the highest land speed is held by Gary Gabelich, who drove his car, The Blue Flame, 650 miles per hour on the Bonneville Salt Flats in 1970.*

Suppose that Gary and (insert the name of one of your students here) *have a race. Suppose that* (your student) *can go 120 miles per hour. How many miles per minute is this? Suppose that Gary goes "only" 600 miles per hour to give* (your student) *a chance. How many miles per minute is this? If* (your student) *gets a headstart of 2 minutes, how long will it take Gary to catch up?* (30 seconds.)

Review of Chapter 5

First day

Show the photograph on Transparency 5-31, keeping the problem covered. *One morning in May 1977, an experienced mountain climber named George Willig started at the base of the south tower of the World Trade Center in New York City and climbed to the top. He was immediately arrested and threatened with a $250,000 lawsuit by the city, but the next day it was decided that all he would have to*

*pay was $1.10, a penny for each story he had climbed.**

Reveal the problem beside the photograph and give everyone a chance to solve it before any discussion.

The figures on Transparency 5-32 are for use in discussing the Set III exercises of Lesson 8. Transparency 5-33 can be used to discuss the Set IV exercise.

Second day

You might begin with the problem on Transparency 5-34, which is from an algebra book (*Text-Book of Algebra*, by Joseph V. Collins) published in 1893.

A train on the Northwestern line passes from London to Birmingham in 3 hours; a train on the Great Western line, which is 15 miles longer, traveling at a speed which is less by 1 mile per hour, passes from one place to the other in $3\frac{1}{2}$ hours. Find the length of each line.

The answers to the Set I exercises of the chapter review are given on Transparencies 5-35 and 5-36.

---

*Articles on Willig's climb appeared in *Time*, June 6, 1977, p. 27, and *Newsweek*, June 6, 1977, p. 33.

$r = 10 + 32t$

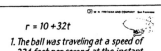

1. The ball was traveling at a speed of 234 feet per second at the instant it was caught. How many seconds had passed between the time that the ball was thrown and the time that it was caught?

$d = 16t^2$

2. How far above the ground was the blimp from which the baseball was thrown?

TRANSPARENCY 5-33

A train on the Northwestern line passes from London to Birmingham in 3 hours; a train on the Great Western line, which is 15 miles longer, traveling at a speed which is less by 1 mile per hour, passes from one place to the other in 3½ hours. Find the length of each line.

TRANSPARENCY 5-34

Answers to Set I
1. a) False
   b) True
   c) Neither
   d) True
2. a) -16
   b) 13
   c) 8
   d) No number is 2 less than itself.
   e) 8 and -8
   f) 6
3. a) 3 and -1
   b) 0 and -7
   c) 2,3,4,5,6
4. a) Add 7, divide by 4.
   b) Multiply by 3, subtract 8.

c) Add y, multiply by 6.
d) Divide by 2, subtract y.
5. a) -13
   b) 13
   c) -1.6
   d) 2
   e) 63
   f) 3
6. a) -2
   b) 30
   c) 3
   a) -21
   e) 18
7. a) 2(x-5)
   b) x^2+8x
   c) (3+7)x or 10 +x

d) (3.7)x or 21x
e) 4x-1
8. a) 11x
   b) 18x
   c) 11+x
   d) 3x
   e) $6x^3$
   f) 6+3x
9. a) -3
   b) -2.2
   c) 8
   a) -1
   e) 5
   f) 0
10. a) 3
    b) -1.5
    c) 2.4
    d) -215

TRANSPARENCY 5-35

11. a) per, 2x+16; area, 8x
    b) per, 12x; area, $9x^2$
    c) per, 2x+14; area, 9x-18
12. a) 14, 12, 24
    b) 4.5, 14.5, 8.5, 5.5
    c) 6, 8, 6, 8
    d) 13, 10, 13, 10
13. a) AE=20, EB=26
    b) AB=24, CE=10
    c) AE=69, EB=18, CD=87
14. a) 20
    b) 20x miles

15. a)
t	0	1	2	3	4
d	0	9	18	27	36
    b) directly
    c) d = 9t
16. a) 2x feet per minute
    b) Ollie ——3x——→ ——3(2x)——→ Alice, 1125
    c) 3x+3(2x)=1125
    d) 125
    e) Alice swam 750 feet; Ollie swam 375 feet.

TRANSPARENCY 5-36

Chapter 6
EQUATIONS IN TWO VARIABLES

Having learned how to solve linear equations in one variable, the students are now ready for a variety of equations in two variables. The first lesson is about solving such equations by choosing values of one variable and finding the corresponding values of the other. This is followed by a lesson on formulas, with emphasis on changing the subject of a formula from one of its variables to another.

The remaining lessons build on many of the ideas introduced in Chapter 2. After a lesson emphasizing the fact that the solutions of linear equations in two variables can be represented graphically, lessons dealing with the concepts of intercepts and slope prepare the way for the final lesson on the slope-intercept form. The knowledge gained in the lessons on graphing will help the students to understand more fully the work with simultaneous equations in the chapter to follow.

Lesson 1
Equations in Two Variables

Show Transparency 6-1. *This photograph shows the staircase of the Nathaniel Russell House in Charles-ton, South Carolina, built in 1806.* Show Transparency 6-2. *When architects design stairways, they use the equation*

$$2x + y = 23$$

in which x *represents the height of each step and* y *represents its length from front to back, each dimension being given in inches. Notice that, unlike the equations we have been solving, this equation contains two variables. Can you figure out how to solve it for* x *and* y?

Give your students time to discover that it has many possible solutions. Show how to write some of the solutions as ordered pairs. The diagram on Transparency 6-2 is a scale drawing of the stairway corresponding to the solution (6, 11). Have your students illustrate other solutions on graph paper, letting 1 unit represent 2 feet. Include a couple of "extreme" cases, such as (1, 21) and (11, 1). In practice, *x* is chosen to be between 6 and 7.

Be sure to emphasize that the equation has an *unlimited* number of solutions: *for any number that we choose for* x, *we can always find a number for* y *so that the resulting ordered pair will make the equation*

58

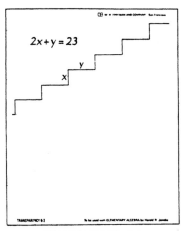

$$2x + y = 23$$

y

x

A formula for
the size of your shoes

Men's shoes
$n = 3\ell - 25$

Women's shoes
$n = 3\ell - 22$

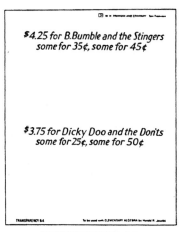

$4.25 for B. Bumble and the Stingers
some for 35¢, some for 45¢

$3.75 for Dicky Doo and the Don'ts
some for 25¢, some for 50¢

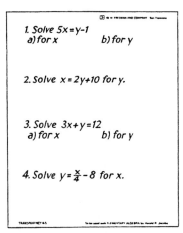

1. Solve $5x = y - 1$
   a) for x   b) for y

2. Solve $x = 2y + 10$ for y.

3. Solve $3x + y = 12$
   a) for x   b) for y

4. Solve $y = \frac{x}{4} - 8$ for x.

true. When the equation is applied to designing stairways, however, only solutions in which both x and y are positive numbers have any meaning.

Lesson 2
Formulas

A formula for the size of your shoes

Shoe stores have to carry a much larger stock of shoes than most people realize. It is not uncommon for a store to carry as many as sixty-five sizes of a single style and to carry as many as one hundred different types of shoes.

Show Transparency 6-3. Men's and women's shoe sizes are measured on different scales. In the formulas given here, n represents a person's shoe size and ℓ represents the length of his or her foot in inches. Find the shoe size of a man whose feet are 1 foot long (11.) Find the shoe size of a woman whose feet are 1 foot long. (14.)

Humphrey Bogart's shoe size was 10. How long were his feet? ($11\frac{2}{3}$ inches.) Greta Garbo's shoe size is 7. How long are her feet? ($9\frac{2}{3}$ inches.) If you know your shoe size, figure out how long your feet are.

From these exercises, it is clear that it is easier to find the shoe size given foot length than the other way around. This is because of the way in which the formulas are written. The shoe size, n, is by itself on the left side of the equal sign and is called the subject *of the formula. The length of the foot, ℓ, is in an*

expression on the right side of the equal sign. We can change the subject of the formula to ℓ by solving it for ℓ in terms of n. Show the students how this is done, emphasizing that we are carrying out the very same steps as those in the Bogart, Garbo, and student's own shoe-size problems. In effect, we are solving all such problems at the same time. Show how the new formula with ℓ as its subject can be more easily used to find the answers to these problems.

Transparency 6-4 can be used to discuss the Set IV exercise of Lesson 1.

Additional examples of how to solve a formula for a given variable are on Transparency 6-5.

Lesson 3
Graphing Linear Equations

Show Transparency 6-6. *In 1960, Captain Joseph Kittinger stepped out of a balloon over New Mexico at an altitude of 102,800 feet (almost 20 miles) and fell 16 miles before his parachute opened! As he fell through the air, his rate of speed increased. A formula that holds for the first few seconds is*

$$y = 32x$$

in which y *represents the speed in feet per second and* x *represents the time in seconds. What kind of function is this? (A direct variation.) What would you expect its graph to look like? (A straight line through the origin.) Make a table, letting* x = 0, 1, 2, 3, 4, *and 5, and graph the function, choosing an appropriate scale for the* y-*axis.*

Now rewrite the equation in the form

$$32x - y = 0$$

Because this equation is equivalent to the equation y = 32x, *it has the same graph, a straight line through the origin.*

Show Transparency 6-7. *Here are more equations in two variables. What do their graphs look like?* Show how to graph each one by solving it for *y*, making a table, and plotting points. Then transform each one to the form $ax + by = c$ and point out that every equation equivalent to an equation of this form has a graph that is a straight line if *a* and *b* are not both zero. Explain that such equations are called *linear equations* in two variables and that an equation written in the form $ax + by = 0$ is said to be written in *standard form.*

Transparency 6-8 is for use in discussing the Set IV exercise of Lesson 2.

Lesson 4
Intercepts

Show Transparency 6-9. *Back in the ninth century, before mechanical clocks had been invented, King Alfred the Great used candles to measure the passage of time. The candles, placed in a lantern to protect them from drafts, were made long enough to last exactly four hours.*

If the candles were eight inches tall, how tall were they after burning for one hour? Draw a graph, using axes as shown, to show what happened to the height of one of these candles as it burned. Can you write a formula for the height of the candle, y, *in terms of the time that has passed,* x? (y = 8 − 2x.) *Write the equation in standard form.* (2x + y = 8.) *How might we have told from this equation that its graph is a straight line if we had not drawn the graph? (It is a linear equation in two variables and the graphs of such equations are straight lines.) Where does the line cross the* x-*axis? (At 4.) What does this number mean? (That the candle lasted four hours.) Where does the line cross the* y-*axis? (At 8.) What does this number mean? (That the candle was originally eight inches tall.) The numbers corresponding to the points in which a graph crosses the coordinate axes are called its* intercepts. *What are the intercepts of the candle graph? (The* x-*intercept is 4 and the* y-*intercept is 8.)*

If we know that the graph of an equation is a line, then we can use its intercepts to draw the line. More examples of how to do this are on Transparency 6-10.

The graphs for the Set III exercises of Lesson 3 are included on Transparencies 6-11 and 6-12. Transparency 6-13 is for use in discussing the Set IV exercise.

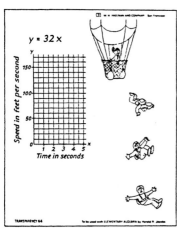

TRANSPARENCY 6-6

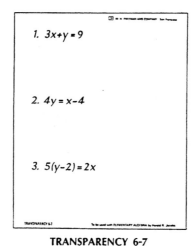

1. $3x+y=9$

2. $4y=x-4$

3. $5(y-2)=2x$

TRANSPARENCY 6-7

$w=5.5h-220$

$h=\dfrac{w+220}{5.5}$

Robert Earl Hughes once weighed 1,069 pounds.

TRANSPARENCY 6-8

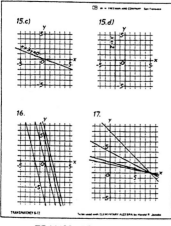

The candle clocks of King Alfred the Great

TRANSPARENCY 6-9

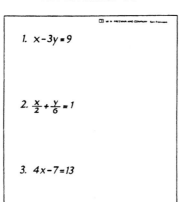

1. $x-3y=9$

2. $\dfrac{x}{2}+\dfrac{y}{6}=1$

3. $4x-7=13$

TRANSPARENCY 6-10

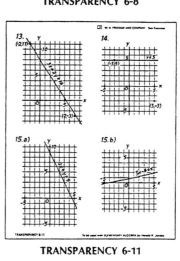

TRANSPARENCY 6-11

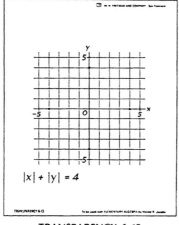

TRANSPARENCY 6-12

$|x|+|y|=4$

TRANSPARENCY 6-13

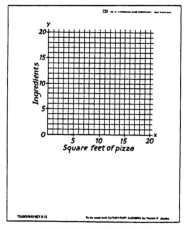

TRANSPARENCY 6-14   TRANSPARENCY 6-15

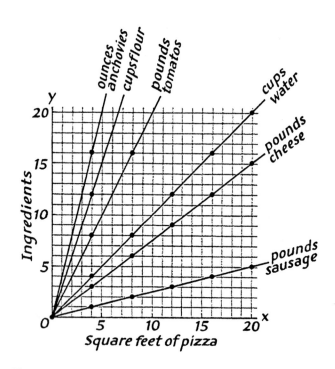

Lesson 5
Slope

Show Transparency 6-14. *The pizza in this photograph was, for a time, the largest one ever made. Baked at the Pizza Inn in Little Rock, Arkansas, it weighed 1,200 pounds and covered an area of 494 square feet.*

Some of the ingredients in a recipe for making 1 square foot of pizza are listed in the table below the photograph. Let's see what happens when the recipe is increased. Copy the list and then fill in the amounts of each ingredient needed to make 4, 8, 12, 16, and 20 square feet of pizza. Fill in the first column to get everyone started.

Now show Transparency 6-15, and have everyone draw and label a pair of axes as shown and plot those points from the table that will fit. *Because the amount of each ingredient needed varies directly with the number of square feet of pizza made, the points that we have plotted for each ingredient lie on a line.* Have your students draw and label the lines as shown at the left. *Which line is the "steepest"? Which is the least steep? The steepness of a line is measured by a number called its* slope. Use the lines to explain how slope is found by drawing some slope triangles and introducing the terms *rise* and *run.*

Note that the slopes of the lines are simply the numbers in the original recipe since they indicate the

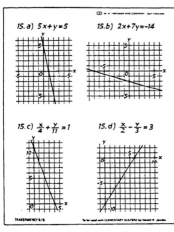

15.a) $5x+y=5$ 15.b) $2x+7y=-14$

15.c) $\frac{x}{4}+\frac{y}{11}=1$ 15.d) $\frac{x}{2}-\frac{y}{3}=3$

TRANSPARENCY 6-16

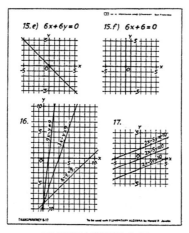

15.e) $6x+6y=0$ 15.f) $6x+6=0$

16.   17.

TRANSPARENCY 6-17

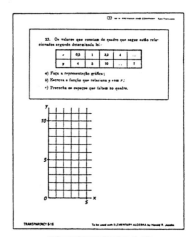

TRANSPARENCY 6-18

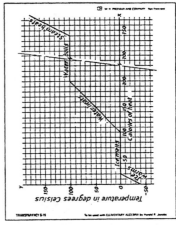

TRANSPARENCY 6-19

amount of each ingredient corresponding to 1 square foot of pizza.

Transparencies 6-16 and 6-17 contain figures for checking the Set III exercises of Lesson 4. Transparency 6-18 is for use in discussing the Set IV exercise.

Lesson 6
The Slope-Intercept Form

Anyone who has ever eaten a popsicle knows that, no matter how cold it is when taken out of the freezer, it doesn't take very long for it to begin melting. Show Transparency 6-19. *This graph shows the change in temperature of one gram of ice as it absorbs heat. (The line at the right indicates that the graph as been "folded.") What is the temperature of the ice at the beginning? (–40° C.) What is its temperature as it melts? (0° C.) What is its temperature as it boils? (100° C.) How many calories does it take to melt the ice? (80.) How many calories to heat the water produced from freezing to boiling? (100.) How many calories to boil the water? (540.) What is the slope of the graph as the ice warms? (2.) This means that its temperature rises two degrees for each calorie of heat that is absorbed. What is the slope of the graph as the ice melts? (0.) What does this mean? (That the tem-* perature does not rise during this time.) *What is the slope of the graph as the water heats? (1.) What does this mean? (That the temperature of the water rises one degree Celsius for each calorie of heat that is added.) This is, in fact, the basis for the definition of a calorie: the amount of heat necessary to raise the temperature of one gram of water one degree Celsius. What happens as the water boils? (The slope of the graph is again zero: the temperature of the water is not changing.) What happens as the steam heats? (The slope of the graph is 2: the temperature of the steam rises two degrees for each calorie of heat that is absorbed.)*

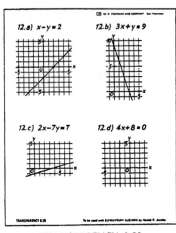

TRANSPARENCY 6-20

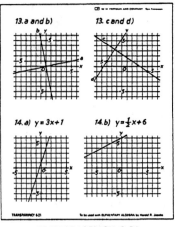

TRANSPARENCY 6-21

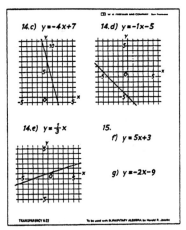

TRANSPARENCY 6-22

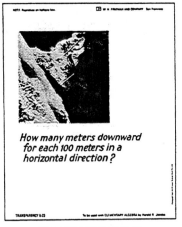

TRANSPARENCY 6-23

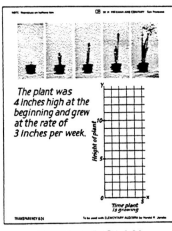

TRANSPARENCY 6-24

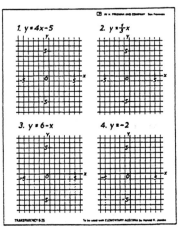

TRANSPARENCY 6-25

Transparencies 6-20, 6-21, and 6-22 contain figures for checking some of the Set III exercises of Lesson 5. Transparency 6-23 is for use in discussing the Set IV exercise.

Transparency 6-24 can be used to introduce the lesson on the slope-intercept form of an equation as presented in the text. Additional examples for use in explaining how to graph linear equations in slope-intercept form that are different from those in the text are on Transparency 6-25. You may want to print copies of this transparency as a worksheet for your students.

Review of Chapter 6

First day

Show Transparency 6-26. *This page is from a Korean mathematics book. Can you figure out what it is about?* (At the top are equations for three linear functions, together with a table of some of their values. The figure shows graphs of all three functions on one pair of axes.) *What do you think the symbols in the box in the lower right corner of the page indicate?* (Those in the box pointing to "a" must mean slope and those in the box pointing to "+b" must mean y-intercept.)

TRANSPARENCY 6-26

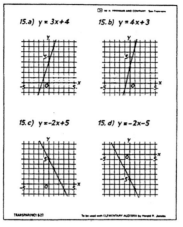

TRANSPARENCY 6-27

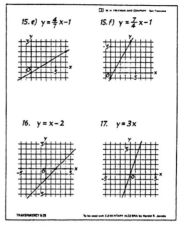

TRANSPARENCY 6-28

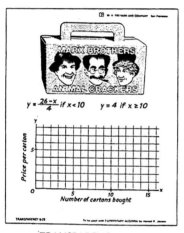

TRANSPARENCY 6-29

TRANSPARENCY 6-30

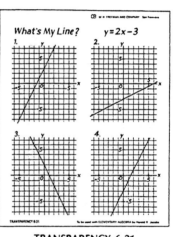

TRANSPARENCY 6-31

The graphs of the three functions are parallel lines. How could we have told this from their equations? (All three have the same slope: 2.) Where does each line cross the y-axis? (At 0, 3, and –4 respectively.)

Transparencies 6-27 and 6-28 contain figures for checking the Set III exercises of Lesson 6. Transparency 6-29 is for use in discussing the Set IV exercise.

Second day: "What's My Line?"

Show Transparency 6-30. *One of the longest-running shows on television was the program "What's My Line?" It began in 1950 and did not go off the air until 1973. This photograph shows the master of ceremonies of the program, John Daly, with panelists Arlene Francis, Ernie Kovacs, Dorothy Kilgallen, and Bennett Cerf. People with unusual occupations appeared on each program and the panelists tried to figure out what they did by asking questions that could be answered yes or no. Today we are going to play a game of "What's My Line?" but with a somewhat different meaning from that of the television show.* Show Transparency 6-31. *The "contestant" is the equation* $y = 2x - 3$, *and the idea is to find out which one of the four lines is its graph. Use your knowledge of the slope-intercept form of an equation to figure out which one it is. In addition to having the students identify the correct answer, the*

Set I Answers

1. a) 12
   b) -23
   c) 12
   d) 49
   e) 17
   f) 17
2. a) Yes
   b) No
   c) Yes
   d) No
   e) Yes
   f) Yes
3. a) (1,8), (2,4),(4,2),(8,1)
   b) Unlimited
   c) (1,10), (2,6),(3,2)
   d) None
4. a) $x = \frac{y+7}{3}$
   b) $y = 3x-7$
   c) $x = \frac{1-y}{5}$
   d) $y = 1-6x$
5. a) $h = \frac{2a}{b}$
   b) $b = \frac{2a}{h}$
   c) 12
   d) $4 = \frac{2(12)}{6}$
   e) $6 = \frac{2(12)}{4}$
6. a) $100
   b) $1.50
   c) $d = \frac{c-15m}{100}$
   d) 4
7. a) $8x-y=10$
   $a=8, b=-1, c=10$
   b) $3x+6y=4$
   $a=3, b=6, c=4$
   c) $x-7y=9$
   $a=1, b=-7, c=9$
   d) $5x=-1$
   $a=5, b=0, c=-1$
8. a) $y=12-6x$
   b) $y=\frac{x-8}{5}$
   c) $y=\frac{1-2x}{7}$
   d) $y=4x-9$
9. a) 15 and 9
   b) 3.5 and -7
   c) 0 and 0
   d) 4 and 9

TRANSPARENCY 6-32

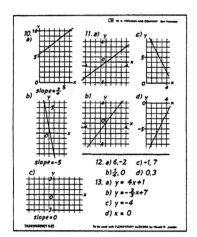

TRANSPARENCY 6-33

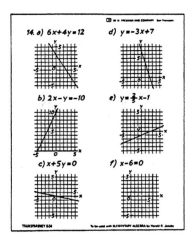

TRANSPARENCY 6-34

fourth graph, you might ask them to figure out the equations of the other three lines by means of their slopes and intercepts.

The answers to the Set I exercises of the chapter review are given on Transparencies 6-32, 6-33, and 6-34.

Chapter 7
SIMULTANEOUS EQUATIONS

Following the study of linear equations in two variables, attention is now turned to pairs of such equations. To help the students understand both the meaning of simultaneous equations and the rationale behind the various methods for solving them, emphasis is on picturing the equations as line-segment diagrams. We begin by considering only equations in which the coefficients of one of the variables are the same, with individual lessons on solving by addition and subtraction and a third lesson dealing with the more general case.

Next, the students learn how to solve simultaneous equations by graphing, not because this is a practical method, but because it helps them understand that such equations can have a unique solution, infinitely many solutions, or no solutions at all. This work also serves as a good review of graphing linear equations by finding their intercepts and by using the slope-intercept form.

After a lesson on solving simultaneous equations by substitution, the chapter finishes with a lesson on mixture problems. Although such problems can be solved by means of a single variable, writing a pair of simultaneous equations in two variables is often a more straightforward method. It is for this reason that mixture problems have not been introduced earlier. Like those of the distance-rate-time problems considered in Chapter 5, the examples of mixture problems in the text lead the students through the solutions step by step. Such problems are included in the rest of the Set I exercises in the book to provide the students with additional practice in solving them.

TRANSPARENCY 7-1

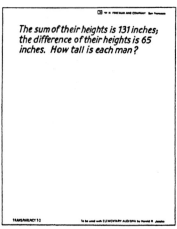

The sum of their heights is 131 inches; the difference of their heights is 65 inches. How tall is each man?

TRANSPARENCY 7-2

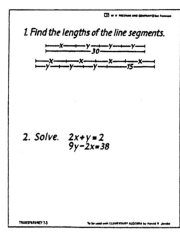

1. Find the lengths of the line segments.

2. Solve. $2x + y = 2$
$9y - 2x = 38$

TRANSPARENCY 7-3

Lesson 1
Simultaneous Equations

Transparencies 7-1 and 7-2 can be used to present the puzzle about Don Koehler and Mihaly Mesyaros that is in the text. After showing Transparency 7-1, show Transparency 7-2 without either overlay. Point out that the problem is to figure out two numbers about which we have been given two clues. Add Overlay A, which illustrates the first clue, and write the addition equation beginning at the point marked with the dot. Note that this equation has many solutions.

Remove Overlay A and add Overlay B, which illustrates the second clue. Write the subtraction equation beginning at the point marked with the dot. Note that this equation also has many solutions.

Now, without removing Overlay B, put Overlay A back in place, point out that the two equations are called *simultaneous equations,* and develop the rest of the lesson as is done in the text. Two more examples are included on Transparency 7-3.

Lesson 2
Solving by Subtraction

Show Transparency 7-4. *Yesterday we considered a puzzle about the world's tallest and shortest men. Here is a similar puzzle about the longest and shortest days of the year. The length of time between sunrise and sunset changes from one day to another. The longest day of the year, the summer solstice, is in late June and the shortest day, the winter solstice, in late December. In Los Angeles, the sum of the lengths of these two days is 1459 minutes and their difference is 271 minutes. If you use this exercise, you should be able to obtain the corresponding data for your own community from a local newspaper office because the times of sunrise and sunset as well as the length of daylight in hours and minutes are usually published in the weather section. Can you figure out from these clues how long each day is?*

Letting x *and* y *represent the number of minutes in the longest and shortest days respectively, we can write the simultaneous equations,*

$$x + y = 1459$$
$$x - y = \ \ 271$$

TRANSPARENCY 7-4

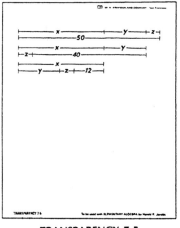

TRANSPARENCY 7-5

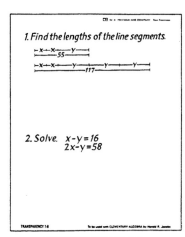

TRANSPARENCY 7-6

Solving these equations, we get (865, 594). The longest day in Los Angeles is 14 hours and 25 minutes and the shortest day is 9 hours and 54 minutes.

Transparency 7-5 is for use in discussing the Set IV exercise of Lesson 1.

Strips for introducing the lesson on solving simultaneous equations by subtraction are on the last page of the transparency book. You might cover the two strips labeled x with adhesive film of one color and the five strips labeled y with adhesive film of another color before cutting them out. Place them on a clear piece of film as shown in the drawing below. Give everyone time to try to figure out x

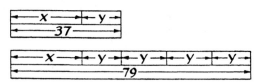

and y and then have someone explain his or her method. *Because the second figure contains three more y's than the first and is 42 units longer, each y must be 14 units. Because the first figure contains one x and y and is 37 units long, x must be 23 units.*

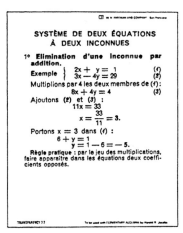

TRANSPARENCY 7-7

Write the simultaneous equations illustrated by the figures and show how this reasoning is equivalent to subtracting one equation from the other. The additional examples on Transparency 7-6 can now be considered before the students read the lesson.

Lesson 3
More on Solving by Addition and Subtraction

Show Transparency 7-7. *Here is part of a page from a French mathematics book. Can you figure out what it is about?*

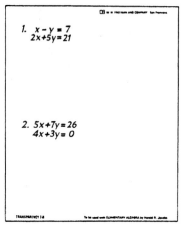

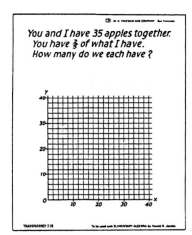

| TRANSPARENCY 7-8 | TRANSPARENCY 7-9 | TRANSPARENCY 7-10 |

This example serves as a nice introduction to solving a pair of simultaneous equations in two variables that cannot be immediately solved by either adding or subtracting them. After the students have seen that the third equation was obtained by multiplying both sides of the first equation by 4 and that the second and third equations were then added to eliminate y, ask them to copy the two original equations and then try adding them directly. Note that the resulting equation, $5x - 3y = 30$, still contains both x and y. Now consider what would happen if we tried subtracting, say, the first equation from the second instead. The resulting equation, $x - 5y = 28$, still contains both x and y.

Why does multiplying the second equation by 4 and then adding the resulting equation to the first one produce an equation in which y does not appear? Use the space at the bottom of the transparency to show how x could have been eliminated instead. The following format makes the strategy especially clear:

$$\textcircled{2}x + y = 1 \xrightarrow{\times 3} \textcircled{6}x + 3y = 3$$
$$\textcircled{3}x - 4y = 29 \xrightarrow{\times 2} -\textcircled{6}x - 8y = 58$$
$$11y = -55$$
$$y = -5$$

Additional examples different from those in the text are given on Transparency 7-8.

Transparency 7-9 is for use in discussing the Set IV exercise of Lesson 2. Because the students had not learned a method for solving a pair of simultaneous equations such as

$$x + 2y = 361$$
$$2x + y = 362$$

in Lessons 1 or 2, be sure to let someone who has solved the problem explain the procedure that he or she used.

Lesson 4
Graphing Simultaneous Equations

Show the puzzle at the top of Transparency 7-10. *This puzzle was given on a radio quiz program several years ago ("Testing One Two Three," KNX radio, Los Angeles). The first person calling in who was able to figure it out would have won five dollars, yet no one among the three people given the puzzle could give the correct answer.*

One way to solve it is with a graph. Letting x *represent the number of apples that I have and* y *represent the number that you have, we can translate*

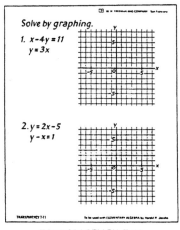

Solve by graphing.

1. $x - 4y = 11$
   $y = 3x$

2. $y = 2x - 5$
   $y - x = 1$

TRANSPARENCY 7-11

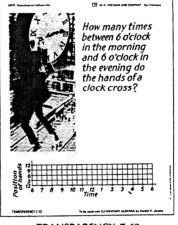

How many times between 6 o'clock in the morning and 6 o'clock in the evening do the hands of a clock cross?

TRANSPARENCY 7-12

---

the clues into the simultaneous equations,

$$x + y = 35$$

$$y = \frac{2}{5}x$$

Have your students draw and label a pair of axes as shown on Transparency 7-10. Note that one unit on each axis represents two apples.

After graphing the two equations, observe that each line represents the solutions of one of the equations and that the point in which the lines intersect represents the solution of both equations. *To solve a pair of simultaneous equations by graphing, we graph both equations on one pair of axes and find the coordinates of the point or points in which the graphs intersect.*

Additional exercises providing practice in solving simultaneous equations by graphing are given on Transparency 7-11.

Lesson 5
Inconsistent and Equivalent Equations

Show the photograph on Transparency 7-12. *This photograph shows Harold Lloyd, a popular film comedian of the twenties, hanging from one of the hands of a clock on a building many floors above the street. The scene, which appears in the movie* Safety Last, *was filmed without the use of trick photography!*

Here is a puzzle about a clock. Show the puzzle at the right of the photograph. *How many times between 6 o'clock in the morning and 6 o'clock in the evening do the hands of a clock cross?* One way to find the answer to this question is to draw a graph. Have your students copy the axes and help them in starting to draw the graph. The solid lines represent the positions of the minute hand and the dashed lines represent the positions of the hour hand. The answer, 11 times, can be seen by counting the crossing points in the finished graph, which is shown below.

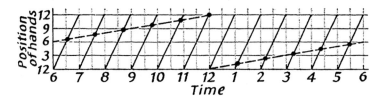

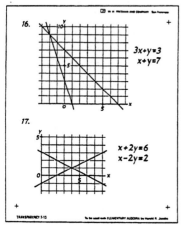

TRANSPARENCY 7-13

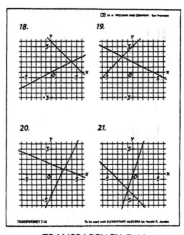

TRANSPARENCY 7-14

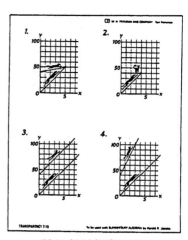

TRANSPARENCY 7-15

Although this puzzle can be solved without a graph, a graph makes it easy to see its solutions. The same is true of simultaneous equations. Although simultaneous equations can be solved without a graph, a graph makes it easy to see their solutions.

The graphs for the Set III exercises of Lesson 4 are reproduced on Transparencies 7-13 and 7-14. Transparency 7-15 is for use in discussing the Set IV exercise.

To introduce the lesson on inconsistent and equivalent equations, you might duplicate and hand out copies of Transparency 7-16. Solve the first pair of simultaneous equations first by graphing them and then by writing the second equation in standard form and using either addition or subtraction. Note that the graphs intersect in exactly one point and that the pair of equations has a unique solution.

Graph the second pair of equations and then try to solve them by writing the second equation in standard form and subtracting it from the first equation. Note that the equations contradict each other; *such equations are* inconsistent. *The contradiction leads to the false equation 0 = 3. That the simultaneous equations do not have a solution is indicated by the fact that their graphs do not intersect.*

Graph the third pair of equations. Observe that both graphs are the same line. Try to solve them by

either doubling both sides of the first equation or halving both sides of the second equation and then subtracting. *That the equations are* equivalent *and have infinitely many solutions is indicated by the fact that their graphs are a single line.*

Lesson 6
Substitution

Using a mirror to solve some simultaneous equations

Have your students copy the first equation on Transparency 7-17 while you give each one a mirror. *Note that the X and the Y in this equation are capital letters. Place the edge of the mirror on the line indicated and hold it vertically so that the reflecting side faces the equation. Copy the reflection of the equation below it to form the pair of simultaneous equations:*

$$X = Y - 18$$
$$81 - Y = X$$

So far we have learned two ways to solve a pair of equations such as these: by addition or subtraction and by graphing. Write each equation in standard form

$$X - Y = -18$$
$$X + Y = 81$$

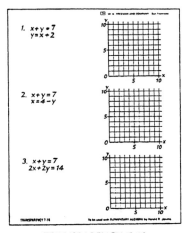

1. $x + y = 7$
   $y = x + 2$

2. $x + y = 7$
   $x = 4 - y$

3. $x + y = 7$
   $2x + 2y = 14$

TRANSPARENCY 7-16

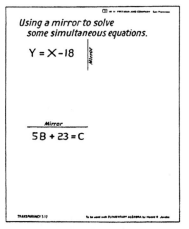

Using a mirror to solve
some simultaneous equations.

$Y = X - 18$ |Mirror

_____Mirror_____

$5B + 23 = C$

TRANSPARENCY 7-17

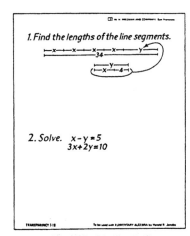

1. Find the lengths of the line segments.

2. Solve. $x - y = 5$
          $3x + 2y = 10$

TRANSPARENCY 7-18

and solve them by addition.

Another way to solve a pair of simultaneous equations is by substitution. Note that each equation tells us something about X: X is equal to Y − 18 and it is also equal to 81 − Y. Substituting Y − 18 for X in the second equation, we have

$$81 - Y = Y - 18, \text{ etc.}$$

As a second example, reveal the second equation on Transparency 7-17. After your students have copied it, have them place the edge of the mirror on the line indicated and hold it horizontally so that the reflecting side faces the equation. *Again copy the reflection of the equation below it to form the pair of simultaneous equations*

$$5B + 23 = C$$
$$2B + 53 = C$$

Give everyone a chance to try to solve these equations by substitution before discussing the solution.

Additional examples for demonstrating how to use the substitution method are given on Transparency 7-18.

The graphs for the Set III exercises of Lesson 5 are reproduced on Transparency 7-19. Transparency 7-20 is for use in discussing the Set IV exercise.

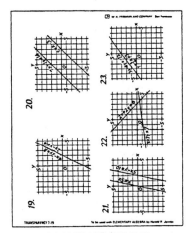

TRANSPARENCY 7-19

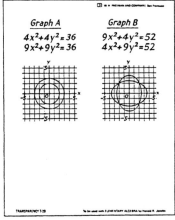

Graph A
$4x^2 + 4y^2 = 36$
$9x^2 + 9y^2 = 36$

Graph B
$9x^2 + 4y^2 = 52$
$4x^2 + 9y^2 = 52$

TRANSPARENCY 7-20

Lesson 7
Mixture Problems

First day

Show Transparency 7-21 with the overlay in place. *According to George Polya, variations of this problem date back several centuries.* The version presented here is adapted from Philip Kaplan's *Posers* (Harper & Row, 1963), page 34.

Because this problem affords an excellent way of introducing mixture problems, I suggest that you have your students copy it before helping them solve it by means of a pair of simultaneous equations. Let x and y represent the numbers of hens and rabbits, respectively, and use the numbers of heads and feet to write equations.

The problem of Archimedes and the king's crown presented in Lesson 7 of the text is considerably more subtle than the hens-and-rabbits problem and you may wish to present it to the class after discussing the exercises of Lesson 6. If so, Transparency 7-22 can be used for this purpose.

Transparency 7-23 is for use in discussing the Set IV exercise of Lesson 6.

Second day

Times have certainly changed since the problem on Transparency 7-24 was written. It is from Wentworth's *Algebra*, published in 1891.

Help the class in getting started by suggesting that x represent the number of days that the man worked and y represent the number of days that he did not work.

---

*George Polya, *Mathematical Discovery*, volume 1 (Wiley, 1962), p. 23.

Review of Chapter 7

First day: "Peppermint Patty Cries for HELP!"

In the Peanuts strip of February 26, 1974, Peppermint Patty is sitting at her school desk reading the problem quoted on Transparency 7-25. In the last panel, she cries for help. The strip is reproduced in the Peanuts book titled *How Long, Great Pumpkin, How Long?* (Peanuts Parade Paperback No. 16) and might be shown to your class before showing Transparency 7-25. Help the class in setting up the problem, letting x represent the number of dimes and y represent the number of quarters:

$$x + y = 20$$
$$10x + 25y + 90 = 25x + 10y$$

After the solution has been found, your students should enjoy seeing Transparencies 7-26 and 7-27. The letter on Transparency 7-26 was published in the *Los Angeles Times* several days after the Peppermint Patty strip had appeared in the paper. It contains the solution in terms of two variables, worked out in a rather clumsy fashion. A few days later, the *Times* published eight more letters about the problem and several of them are reproduced on Transparency 7-27. Mary Woodruff's algebraic solution is interesting, considering that it might have been 70 years since she had studied algebra. Several other writers offer nice solutions using simple arithmetic.

Transparency 7-28 is for use in discussing the Set IV exercise of Lesson 7.

A farmer has hens and rabbits. Between the two there are 30 heads and 86 feet. How many of each animal does the farmer have?

TRANSPARENCY 7-21

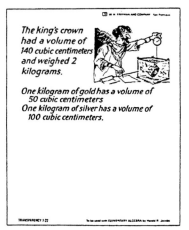

The king's crown had a volume of 140 cubic centimeters and weighed 2 kilograms.

One kilogram of gold has a volume of 50 cubic centimeters
One kilogram of silver has a volume of 100 cubic centimeters.

TRANSPARENCY 7-22

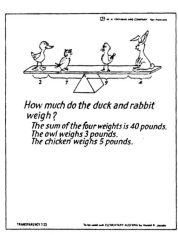

How much do the duck and rabbit weigh?
The sum of the four weights is 40 pounds.
The owl weighs 3 pounds.
The chicken weighs 5 pounds.

TRANSPARENCY 7-23

A workman was hired for 40 days, at $1 for every day he worked, but with the condition that for every day he did not work he was to pay 45¢ for his board. At the end of the time he received $22.60. How many days did he work?

TRANSPARENCY 7-24

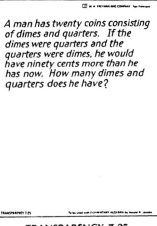

A man has twenty coins consisting of dimes and quarters. If the dimes were quarters and the quarters were dimes, he would have ninety cents more than he has now. How many dimes and quarters does he have?

TRANSPARENCY 7-25

TRANSPARENCY 7-26

TRANSPARENCY 7-27

The price of 9 citrons and 7 wood apples is 107; the price of 7 citrons and 9 wood apples is 101. Tell me quickly the price of a citron and of a wood apple. *Mahavira, c.850 A.D.*

TRANSPARENCY 7-28

Two wine merchants enter Paris, one of them with 64 casks of wine, the other with 20. Since they have not enough money to pay the customs duties, the first pays 5 casks of wine and 40 francs, and the second pays 2 casks of wine and receives 40 francs in change. What is the price of each flask of wine and the duty on it?

Chuquet, 1484

TRANSPARENCY 7-29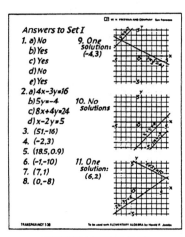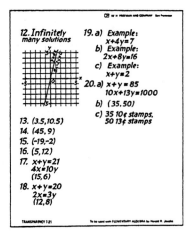

TRANSPARENCY 7-29   TRANSPARENCY 7-30   TRANSPARENCY 7-31

Second day

The problem of the wine merchants on Transparency 7-29 was written by the fifteenth-century French mathematician Nicolas Chuquet. Your students will need some direction in how to write a pair of simultaneous equations to represent the information in it. *Let x represent the price of a cask of wine in francs and y represent the number of francs of duty on it.*

How much duty is each merchant supposed to pay? (64y *francs and* 20y *francs*). *What is the value of what each merchant pays the customs inspector?* (5x + 40 *francs and* 2x − 40 *francs.*) *What equations can we write from all of this information?* (64y = 5x + 40 *and* 20y = 2x − 40.) It might be instructive to have your students solve these equations both by addition or subtraction and by substitution.

Transparencies 7-30 and 7-31 can be used to check the answers to the Set I review exercises.

Chapter **8**

EXPONENTS

Although Archimedes used the first law of exponents in *The Sand Reckoner,* the complete theory of exponents was not developed until the seventeenth century. Descartes is credited with introducing the symbolism that we now use and it was the English mathematician John Wallis who explained the meaning of negative numbers and rational numbers as exponents for the first time.

Up to this point in the course, the idea that the students have had of an exponent is that it means to multiply a number by itself one or more times. This concept of exponentiation as repeated multiplication is meaningful only if we restrict exponents to be integers larger than one. It is important, in extending one's knowledge of exponents to the negative integers and zero, to realize that such exponents are not to be interpreted in the same way. Sheila Tobias, in her book *Overcoming Math Anxiety* (Norton, 1978), tells of having difficulty with negative exponents because of not understanding this.

The chapter begins with a lesson on large numbers in which scientific notation is discussed. The first three laws of exponents are considered in subsequent lessons before a pattern with powers of two

is used to suggest meanings for zero and the negative integers as exponents. The practicality of negative exponents is reinforced with a lesson on writing small numbers in scientific notation.

A lesson dealing with powers of products and quotients introduces the remaining laws of exponents. The chapter closes with a lesson on exponential functions that includes a variety of interesting applications dealing with growth and decay.

Lesson 1
Large Numbers

An ideal way to open this lesson is to show the film *A Rough Sketch for a Proposed Film Dealing with the Powers of Ten and the Relative Size of Things in the Universe.* It was made for the Commission on College Physics by the office of Charles Eames and is available for rental or purchase from Pyramid Films, P.O. Box 1048, Santa Monica, California, 90406. The film begins with a view of a man on a Miami golf course and takes the viewer outward to the edge of the universe and then inward to the nucleus of an atom, all in just eight minutes. It was

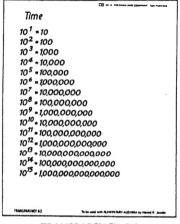

TRANSPARENCY 8-1   TRANSPARENCY 8-2

inspired by the book *Cosmic View: The Universe in 40 Jumps* by Kees Boeke (The John Day Company, 1957).

After showing the film or some transparencies made from the book, you might show Transparency 8-1, which is a photograph of a part of the constellation Monoceros. The overlay gives the distances in miles to the sun, the nearest star beyond the sun, and the nearest galaxy beyond the Milky Way. These numbers can be used to introduce and develop the lesson on powers of ten and scientific notation.

Lesson 2
A Fundamental Property of Exponents

Show the column of powers of ten on Transparency 8-2, keeping the rest of the transparency covered. Have your students copy it, writing each number in decimal form to the right of its form as a power of ten.

If we think of the numbers in this table as representing seconds of time, an astonishing range of times is included. Beginning at the top, we have 10 seconds, followed by 100 seconds. One minute of time falls between these numbers, because 1 minute equals 60 seconds. Indicate this on the transparency as shown below. *Between which two numbers does one hour fall? One day? Because one year out of every four is a leap year, we will consider a year to contain 365.25 days. Between which two numbers does one year fall? One century? One million years?*

$10^1 = 10$

$\quad\quad \longleftarrow 1$ minute (60)

$10^2 = 100$

$10^3 = 1,000$

$\quad\quad \longleftarrow 1$ hour (3,600)

$10^4 = 10,000$

$\quad\quad \longleftarrow 1$ day (86,400)

$10^5 = 100,000$

$10^6 = 1,000,000$

$10^7 = 10,000,000$

$\quad\quad \longleftarrow 1$ year (31,557,600)

$10^8 = 100,000,000$

$10^9 = 1,000,000,000$

$\quad\quad \longleftarrow 1$ century (3,155,760,000)

$10^{10} = 10,000,000,000$

$10^{11} = 100,000,000,000$

$10^{12} = 1,000,000,000,000$

$10^{13} = 10,000,000,000,000$

$\quad\quad \longleftarrow 1$ million years

$10^{14} = 100,000,000,000,000$

$10^{15} = 1,000,000,000,000,000$

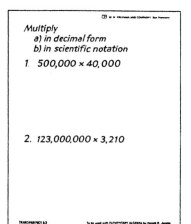

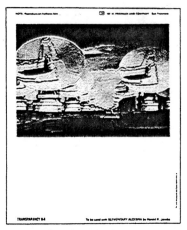

TRANSPARENCY 8-3   TRANSPARENCY 8-4

In fitting a person's life into this scale, we see that it doesn't take long for a newborn baby to become more than 10,000 seconds old because 10,000 seconds is less than a day. The baby passes the 100,000 second mark in his second day of life, and the 1,000,000 second mark in his twelfth day of life. Students who have calculators should check these and the following figures. *The 10,000,000 second mark is passed long before the baby reaches the age of one and the 100,000,000 second mark is passed not long after the child becomes three. A person does not pass the 1,000,000,000 second mark until his or her thirty-first year of life and no one makes it to 10,000,000,000.*

After the exercises for Lesson 1 have been discussed, you might use numbers from the table of powers of 10 to introduce the first law of exponents. Some examples comparing multiplying large numbers in decimal form with multiplying them in scientific notation are given on Transparency 8-3.

The following format, showing the solution of the first example, makes a comparison of the two methods especially straightforward.

$$
\begin{array}{r}
500{,}000 \\
\times\quad 40{,}000 \\
\hline
000000 \\
000000 \\
000000 \\
000000 \\
2000000 \\
\hline
20{,}000{,}000{,}000
\end{array}
\qquad
\begin{array}{r}
5\times10^5 \\
\times\ 4\times10^4 \\
\hline
20\times10^9 = 2\times10^{10}
\end{array}
$$

Lesson 3
Two More Properties of Exponents

Show Transparency 8-4. *In an article in the May 1975 issue of* Scientific American *titled "The Search for Extraterrestrial Intelligence," it is stated that "there can be little doubt that civilizations more advanced than the earth's exist elsewhere in the universe." The authors, Carl Sagan and Frank Drake, professors of astronomy at Cornell University, describe a proposed system of antennas for discovering and communicating with one of these civilizations. The system, called Cyclops, would consist of 1,500 antennas each 100 meters in diameter. This picture shows an artist's conception of what part of it would look like.*

Some scientists estimate that there may be 100 billion planets within our own galaxy capable of supporting life and that one planet out of 100 thousand might have a civilization at or exceeding the earth's present level of technological development. If these guesses are correct, how many such civilizations would there be in our galaxy? After your students have solved this problem by writing each of these numbers in decimal form and dividing, have them rewrite both the problem and solution in exponential form. The result is a simple illustration of the second law of exponents.

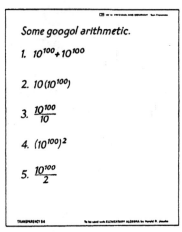

TRANSPARENCY 8-5   TRANSPARENCY 8-6   TRANSPARENCY 8-7

Lesson 4
Zero and Negative Exponents

Show Transparency 8-5. Because those students who did the Set IV exercise of Lesson 3 have already learned what a googol is, you might ask one of them to explain its meaning to the rest of the class. (The statement in the ad claiming that "there are googols of little creatures" on the earth is an extreme exaggeration. The number of *atoms* in the earth is less than 10^{50} and the number of atoms in the entire universe has been estimated at less than 10^{80}.*)

Show Transparency 8-6. *Can you apply your knowledge of large numbers to express the answers to the following problems either as powers of ten or in scientific notation?* In the following discussion, be sure to make clear why the first three laws of exponents can be applied to problems 2, 3, and 4, but not to problems 1 and 5.

Transparency 8-7 is for possible use in developing the new lesson in accord with the text.

---

*Isaac Asimov, "The Proton-Reckoner," an essay in *From Earth to Heaven* (Doubleday, 1966) and *Asimov on Numbers* (Doubleday, 1977).

Lesson 5
Small Numbers

Show Transparency 8-8 with the lines above *kilo* and the lines below *milli* covered. *The metric system was established in the last decade of the eighteenth century by a committee of French mathematicians and scientists. This table lists the prefixes decided upon at that time.*

The prefixes corresponding to positive powers of ten come from Greek words meaning ten, hundred, and thousand. Those corresponding to negative powers come from the Latin words having the same meaning.

Uncover the rest of the lines on Transparency 8-8 except the last two. *In 1958, new prefixes were added to cover powers of ten ranging from 10^{12} to 10^{-12}. All of the new prefixes came from the Greek except for* pico: tera *from the word for "monster,"* giga *from "giant,"* mega *from "great,"* micro *from "small," and* nano *from "dwarf." Pico comes from the Spanish word for "small."*

In 1962, two more prefixes were added to the list. Uncover the last two lines of the transparency. *They come from the Danish words for "fifteen" and "eighteen," respectively.*

---

*The information in this lesson is from Isaac Asimov's essay "Pre-fixing It Up," included in *Adding a Dimension* (Doubleday, 1964) and *Asimov on Numbers* (Doubleday, 1977).

10^{12}	tera
10^{9}	giga
10^{6}	mega
10^{3}	kilo
10^{2}	hecto
10^{1}	deka
10^{-1}	deci
10^{-2}	centi
10^{-3}	milli
10^{-6}	micro
10^{-9}	nano
10^{-12}	pico
10^{-15}	femto
10^{-18}	atto

A drop of water, 50 milligrams

An amoeba, 5 micrograms

A liver cell, 2 nanograms

A tobacco mosaic virus, 65 attograms

A hemoglobin molecule, 0.1 attogram

A carbon atom, 0.00002 attogram

An electron, 0.0000000009 attogram

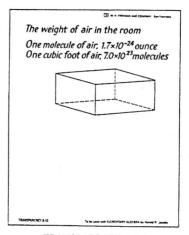

The weight of air in the room
One molecule of air, 1.7×10^{-24} ounce
One cubic foot of air, 7.0×10^{23} molecules

Note that each of the prefixes added to the original list indicates a number one thousand times the one below it. How does the number indicated by the largest prefix compare in size with the one indicated by the smallest one? (It is 10^{30} times as large.)

Show Transparency 8-9. *The metric unit of mass is the gram. Here are some very small masses. Use them to show how to write small numbers in scientific notation by expressing each mass in grams.*

Lesson 6
Powers of Products and Quotients

The weight of air in the room

Hold up a small paper bag filled with air. *Here is a paper bag filled with air. Suppose that all of the air in this classroom could be removed from the room and put into the bag. How much do you suppose the bag would weigh?* You might have several students tell their guesses.

Show Transparency 8-10. *Here is information that we can use to find out. Air is a mixture of gases, primarily nitrogen and oxygen, whose molecules have an average weight of 1.7×10^{-24} ounce. One cubic foot of air contains approximately 7.0×10^{23} molecules. Use this information to find out how much one cubic foot of air weighs.*

$$(7.0 \times 10^{23})(1.7 \times 10^{-24}) =$$
$$11.9 \times 10^{-1} \approx$$
$$1.2 \text{ ounces}$$

What other information do we need? (How many cubic feet of air there are in the room.) How can we determine this number? My classroom is 31 feet long, 27 feet wide, and 14 feet high; $31 \cdot 27 \cdot 14 = 11,718$ cubic feet. If we suppose that 10,000 cubic feet of this is occupied by air, the air would weigh 12,000 ounces, or about 750 pounds!

In the first step in solving this problem, we used one of the laws of exponents:

$$x^a \cdot x^b = x^{a+b}$$

to write $(10^{23})(10^{-24})$ *as* 10^{-1}. *In addition to this law, we have learned two other laws of exponents. Can you write them in symbols?*

$$\frac{x^a}{x^b} = x^{a-b} \quad and \quad (x^a)^b = x^{ab}$$

Having reviewed the first three laws of exponents, you might now present the new lesson. Because many students tend to confuse the "product of two powers" law with the "power of a product" law and the "quotient of two powers" law with the "power of a quotient" law, examples such as the following are useful in helping them to make the correct distinctions.

$$2^3 \cdot 2^4 = 2^{3+4} = 2^7 \quad \text{vs.} \quad 3^2 \cdot 4^2 = (3 \cdot 4)^2 = 12^2$$

$$\frac{5^6}{5^2} = 5^{6-2} = 5^4 \quad \text{vs.} \quad \frac{6^5}{2^5} = \left(\frac{6}{2}\right)^5 = 3^5$$

Transparency 8-11 is for use in discussing the Set IV exercise of Lesson 5.

Lesson 7
Exponential Functions

Show Transparency 8-12. *The first census was taken in the United States in 1790. This chart, published by the census bureau at the time, shows the number of people living in each of the 16 states and the "south western territory." The total came to slightly less than 4 million.*

Show Transparency 8-13. (You may wish to also hand out duplicated copies of it to your students to save time.) *Here is a table showing what has happened to the population of the United States since that first census.* (Note that there is space for adding the number for 1980 when it becomes available. The population in 1970 was 203 million.) Have your students plot points corresponding to the numbers in the table and connect them with a smooth curve. *The population in 1700 has been estimated at 250,000 and the population in 1750 at 1,200,000. Add these points to the graph.*

Benjamin Franklin once said that the population of the United States doubles itself every twenty-five years. Let's see if this claim is reasonable for the period from 1800 to 1900. Rounding the population in 1800 to 5 million, we can construct the following table:

YEAR	POPULATION IN MILLIONS
1800	5
1825	10
1850	20
1875	40
1900	80

Plotting points corresponding to these numbers on our graphs, we get the result shown here. (The points are indicated as open circles.) We see that they do come

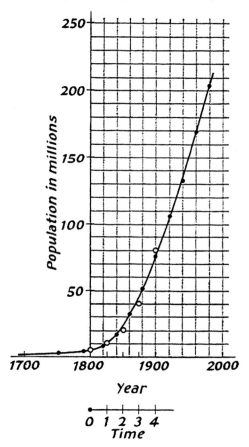

Object	Mass in kg
Sun	2×10^{30}
Earth	6×10^{24}
Moon	7×10^{22}
Whale	1×10^{5}
Elephant	4×10^{3}
Human	6×10^{1}
Flea	3×10^{-4}
Dust particle	1×10^{-6}
Hydrogen atom	2×10^{-27}
Electron	9×10^{-31}

TRANSPARENCY 8-11

TRANSPARENCY 8-12

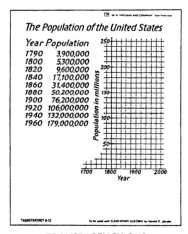

The Population of the United States

Year	Population
1790	3,900,000
1800	5,300,000
1820	9,600,000
1840	17,100,000
1860	31,400,000
1880	50,200,000
1900	76,200,000
1920	106,000,000
1940	132,000,000
1960	179,000,000

TRANSPARENCY 8-13

close to lying on the population curve. Because our table begins with the year 1800, it is reasonable to renumber it as 0. Taking twenty-five years as our unit of time and showing the doubling pattern of the population, we can rewrite the table as follows:

	POPULATION
TIME	IN MILLIONS
0	5
1	$5 \cdot 2$
2	$5 \cdot 2 \cdot 2$
3	$5 \cdot 2 \cdot 2 \cdot 2$
4	$5 \cdot 2 \cdot 2 \cdot 2 \cdot 2$

According to this pattern, what do you suppose should correspond to a time of 10? (*Five multiplied by the product of 10 twos.*) What should correspond to a time of x? (*Five multiplied by the product of x twos.*) We can express this pattern in terms of the formula

$$y = 5 \cdot 2^x$$

in which x represents the time and y represents the population in millions.

The population is a function of the time and, because the variable representing the time is an exponent, it is called an exponential function. The students are now prepared to understand the definition of an exponential function given on page 377 of the text.

TRANSPARENCY 8-14

Transparency 8-14 is for use in discussing the Set IV exercise of Lesson 6.

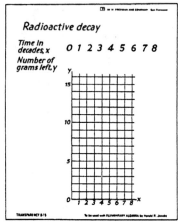

TRANSPARENCY 8-15

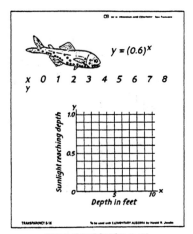

TRANSPARENCY 8-16

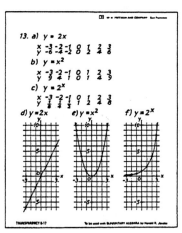

TRANSPARENCY 8-17

After the exercises of Lesson 6 have been discussed, Transparency 8-15 might be used to present the example of radioactive decay as a decreasing exponential function given on pages 377 and 378 of the text.

Review of Chapter 8

First day: Fish with headlights

Show the picture of the fish on Transparency 8-16. *Although most fish live near the surface of the sea, some live deep in the ocean where it is forever dark. Many of these fish are able to produce their own light. The fish shown in this drawing, for example, is called a lantern fish because it can cast a beam of light as far as two feet in front of it.*

The sunlight reaching a given depth in the ocean is an exponential function of the depth. It is given by the formula

$$y = (0.6)^x$$

in which x represents the depth in feet and y represents the fraction of the sunlight that reaches that depth from the surface. Graph this function by letting x = 0, 1, 2, 3, 4, 5, 6, 7, and 8 and connecting the points with a smooth curve. Rounding each value of y to the nearest hundredth, the table is

x	0	1	2	3	4	5	6	7	8
y	1	0.60	0.36	0.22	0.13	0.08	0.05	0.03	0.02

The graph shows how the amount of light decreases from the surface to a depth of 8 feet. Fish have been discovered at depths of more than 27,000 feet. What do you think the graph would look like if it were extended that far? (The curve would be extremely close to the x-axis. The fraction of light from the surface reaching a depth of 27,000 feet, $(0.6)^{27,000}$, is an extremely small number.)

The answers to exercise 13 of Lesson 7 are given on Transparency 8-17. Transparency 8-18 is for use in discussing the Set IV exercise.

Second day

If you can obtain a copy of *Growth*, a LIFE Science Library book published in 1965 by Time-Life Books, show your students the photographs on page 76. *These pictures were taken by Mr. H. B. Osgood of his daughter Grace as she grew from infancy to adolescence. The first photograph was taken when Grace was eight weeks old and the rest at yearly intervals.*

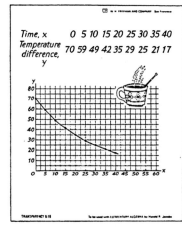

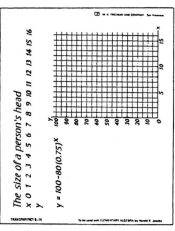

TRANSPARENCY 8-18

TRANSPARENCY 8-19

1. a) 10^5
   b) 13.5
   c) 10^{-7}
   d) 0.0007
   e) 2×10^5
2. a) 2.2×10^6
   b) 2.2×10^1
   c) 10,000,000
   d) It becomes 7 less.
3. a) 30,000
   b) 0.0003
   c) 0.62
   d) 85,000,000
4. a) 7×10^{11}
   b) 4.12×10^7
   c) 2×10^{-3}
   d) 1.0002×10^1
5. a) $\frac{1}{8}$
   b) $\frac{1}{625}$
   c) 1
   d) 36
   e) $\frac{1}{36}$
   f) 64
6. a) x^{-6}
   b) x^{-16}
   c) x^{-5}
   d) x^7
   e) x^{-11}
   f) x^2
7. a) >
   b) <
   c) >
   d) =
8. a) $64x^3$
   b) $\frac{1}{7y}$
   c) $25x^{10}$
   d) $\frac{x^4}{y^8}$
9. a) 5.4×10^{16}
   b) 1.4×10^{-2}
   c) 8×10^{-6}
   d) 5.5×10^{-8}
   e) 8.1×10^9
   f) 2×10^6
10. a) <
    b) >
    c) =
    d) >
    e) >
    f) =
    g) <

TRANSPARENCY 8-20

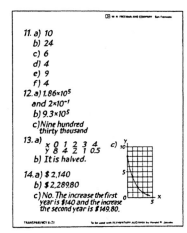

11. a) 10
    b) 24
    c) 6
    d) 4
    e) 9
    f) 4
12. a) 1.86×10^5 and 2×10^{-1}
    b) 9.3×10^5
    c) Nine hundred thirty thousand
13. a)

x	0	1	2	3	4
y	8	4	2	1	0.5

    b) It is halved.
    c)
14. a) $2,140
    b) $2,289.80
    c) No. The increase the first year is $140 and the increase the second year is $149.80.

TRANSPARENCY 8-21

Show Transparency 8-19. (You may wish to hand out duplicated copies as well.) *As a person grows, his head changes in size according to the formula*

$$y = 100 - 80(0.75)^x$$

in which x represents age in years and y represents the size of the head as a percentage of its size when the person becomes an adult. Use this formula to find the values of y when x is equal to 0, 1, and 2. After your students have determined these, fill in the rest of the table as shown below and have them graph the function.

x	0	1	2	3	4	5	6	7	8
y	20	40	55	66	75	81	86	89	92

x	9	10	11	12	13	14	15	16
y	94	95	96	97	98	99	99	99

Answers to the Set I exercises of the review lesson are given on Transparencies 8-20 and 8-21.

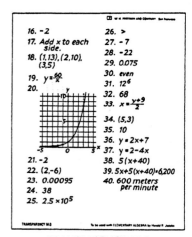

TRANSPARENCY M-1

Midterm Review: Set I

1. 8x−24
2. True
3. 3y
4. True
5. −18
6. $\frac{80}{x}$
7. False
8. 4^{-1}
9. $-x\cdot{}^{-7}$
10. True
11. −49
12. $\frac{100}{x}$
13. −42
14. True
15. x+−y
16. False
17. 5^{-2}
18. 16
19. x^{-10}
20. −23
21. $p=\frac{1}{rt}$
22. False
23. <
24. True
25. An even number
26. $\frac{x^2}{y}$
27. 3x
28. False
29. Add 7 to each side.
30. $\frac{1}{36}$
31. At (0,−9)
32. =
33. −1
34. $8x^6$
35. 10^{-3}
36. 5+x
37. y^5
38. False
39. y=4x−3
40. x^9

TRANSPARENCY M-2

Midterm Review: Set III

1. 1.2×10^4
2. $x^2<3x$
3. ⠿ ⠿ ⠿ ⠿ ⠿
4. x+2y=−4
5. −6
6. −2.25
7. −2
8. $-12x^2$
9. 120
10.
11. −2.2
12. 7
13. (2,5)
14. 2^6
15.

TRANSPARENCY M-3

16. −2
17. Add x to each side.
18. (1,13),(2,10),(3,5)
19. $y=\frac{60}{x}$
20.
21. −2
22. (2,−6)
23. 0.00095
24. 38
25. 2.5×10^5
26. >
27. −7
28. −22
29. 0.075
30. even
31. 12^6
32. 68
33. $x=\frac{y+9}{2}$
34. (5,3)
35. 10
36. y=2x+7
37. y=2−4x
38. 5(x+40)
39. 5x+5(x+40)=6,200
40. 600 meters per minute

MIDTERM REVIEW

The midterm review consists of three exercise sets, each containing forty items. Set I can be done orally. Sets II and III can be taken as practice tests. Answers to Sets I and III are given on Transparencies M-1, M-2, and M-3.

Chapter 9
POLYNOMIALS

Although the vocabulary for talking about polynomials in this chapter is new to the students, they have already encountered polynomial expressions many times throughout the course. Polynomial expressions appeared for the first time, in fact, in the first lesson of Chapter 1. Finding the value of a polynomial for a given value of its variable was introduced in Chapter 1, Lesson 7. Polynomial functions were studied in Chapter 2 and polynomial equations in one and two variables in Chapters 5 and 6. Some of the algebra of first-degree polynomials appeared in Chapter 7 when the students learned how to solve simultaneous equations by addition and subtraction. Chapter 8, on exponents, prepared the way for the work in this chapter with polynomials of higher degree.

The chapter begins with a lesson on the vocabulary and algebra of monomials. This is followed by a lesson on the basic properties of polynomials. The rest of the chapter deals with the algebra of polynomials. The patterns for the square of a binomial and the product of the sum and difference of two terms are so important that an entire lesson is devoted to them. Rectangles are used extensively to help the students visualize the operations of multiplication and division. They are especially helpful in making the algorithm of division meaningful. Division problems in which there is a remainder have been deliberately omitted from the course partly for the same reason that elementary school children do not encounter division problems with remainders in arithmetic until they have first mastered the procedure and partly because such problems make little sense until algebraic fractions have been studied.

Lesson 1
Monomials

Show Transparency 9-1. *This diagram from the California Driver's Handbook shows the stopping distance of a car as a function of its speed. The part of each bar in white represents the reaction distance: the distance traveled by the car between the time the driver sees a reason for stopping and the time that he applies the brakes. The part in black represents the braking distance: the distance traveled by the car between the time the brakes are applied and the time that it comes to a complete stop. The distances shown are for a car with perfect four-wheel brakes being driven under ideal conditions.*

If we let x *represent the speed in miles per hour, then an expression that approximates the stopping distances in feet given in this diagram is*

$$0.055x^2 + 1.1x - 2$$

Write this below the diagram and have your students verify it by finding its value when $x = 45$. Explain that it is an example of a *polynomial*, the subject of the next unit of the course. Having just found the value of this polynomial for one value of *x*, your students will readily recognize that it consists of three "parts": $0.055x^2$, $1.1x$, and 2. Explain that these parts are called *monomials* and use them as examples to develop the rest of the lesson in accord with the text.

Examples of how monomials are added, subtracted, and multiplied are given on Transparency 9-2.

Lesson 2
Polynomials

A "pyramid club" is usually a scheme to get rich by means of a chain letter. The idea is to send some money to someone and, through a series of letters, to eventually receive many times that amount of money in return. Show Transparency 9-3, reading the cartoon to your class. *The letters in the pyramid club in this cartoon were evidently used to actually build a pyramid!*

Show Transparency 9-4. *The number of blocks in a pyramid of the type shown in the cartoon is given by the expression*

$$\frac{1}{3}x^3 + \frac{1}{2}x^2 + \frac{1}{6}x$$

in which x *represents the number of layers in the pyramid. Use it to find how many blocks are in a pyramid having three layers.* $\left[\frac{1}{3}(3)^3 + \frac{1}{2}(3)^2 + \frac{1}{6}(3) = 14.\right]$ *Check your answer by counting the number of blocks in the diagram.* (1 + 4 + 9 = 14.) Add the overlay to the transparency. *Now use the expression to find how many blocks are in a pyramid having twelve layers.* $\left[\frac{1}{3}(12)^3 + \frac{1}{2}(12)^2 + \frac{1}{6}(12) = 650.\right]$ *Explain that the expression* $\frac{1}{3}x^3 + \frac{1}{2}x^2 + \frac{1}{6}x$ *is an example of a polynomial because it indicates the addition of several monomials,* and develop the rest of the lesson in accord with the text.*

Transparency 9-5 is for use in discussing the Set IV exercise of Lesson 1.

---

*The pyramiding in a chain letter in fact leads to a geometric sequence, not a polynomial—the B.C. cartoon and our block counting not to the contrary. Geometric sequences are introduced in Chapter 17.

TRANSPARENCY 9-1

If possible, write each of the following expressions as a monomial.

1. $2x^4 + 3x^4$

2. $(2x^4)(3x^4)$

3. $4x^5 - x^5$

4. $(4x^5)(-x^5)$

5. $5x^3 + x^2$

6. $(5x^3)(x^2)$

7. $(3x^2)^5$

TRANSPARENCY 9-2

TRANSPARENCY 9-3

The number of blocks in a pyramid:
$\frac{1}{3}x^3 + \frac{1}{2}x^2 + \frac{1}{6}x$

TRANSPARENCY 9-4

Which monomial is larger?

x	$10x^2$	$2x^{10}$
0	0	0
1	10	2
2	40	2,048
3	90	118,098
4	160	2,097,152
5	250	19,531,250
⋮	⋮	⋮

TRANSPARENCY 9-5

Lesson 3
Adding and Subtracting Polynomials

A remarkable card trick

This trick is quite astonishing, yet very easy to do.* Shuffle a deck of playing cards and deal nine of them face down on a student's desk. Ask the student to choose one of the cards and hold it up for the class and you to see. Put the other eight cards together in a pile and place the card chosen by the student face down on top of the pile. Put the rest of the deck face down on top of these cards.

Now start dealing cards one at a time from the top of the deck, placing them face-up in a pile and counting backward from 10 to 1 at the same time to see if the number on any card is the same as the number being counted. (Aces count as ones and face cards as tens.) If there is a match (for example, a seven turns up as you say seven), stop dealing on that pile and begin another one, counting again from ten. If there is no match, "kill" the pile by placing a card face down on top of it. Do this until you have made four piles altogether. Now add the numbers showing on the piles and count out that additional number of cards from the deck. The last one will be the card originally chosen.

---

*The trick is described in *Mathematics, Magic and Mystery* by Martin Gardner (Dover, 1956), pp. 7–9.

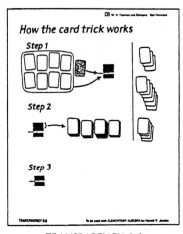

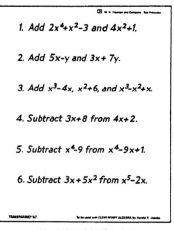

1. Add $2x^4+x^2-3$ and $4x^2+1$.

2. Add $5x-y$ and $3x+7y$.

3. Add x^3-4x, x^2+6, and x^3-x^2+x.

4. Subtract $3x+8$ from $4x+2$.

5. Subtract x^4-9 from x^4-9x+1.

6. Subtract $3x+5x^2$ from x^5-2x.

The speed of an elevator
in feet per second:
$-0.5x^4+4x^3-12x^2+16x$

x	Speed
0	
1	
2	
3	
4	

TRANSPARENCY 9-6	**TRANSPARENCY 9-7**	**TRANSPARENCY 9-8**

Transparency 9-6 is for use in explaining how the trick works. The figure below shows what you might add to it as you discuss the various steps.

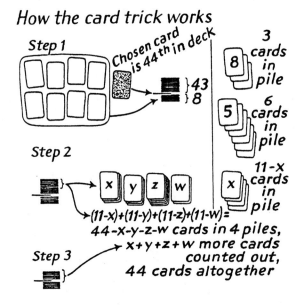

The result of the first step is that the chosen card ends up as the 44th in the deck. The figures on the right side of the transparency help the students discover that, if there is a match in counting out a pile in step 2 and the matched number is x, then the number of cards in the pile is $11 - x$. This is true even if there is no match. Because we count backward from 10 as we place cards on the pile, the card that "kills" the pile counts as 0; the number of cards in the pile is $11 - 0 = 11$.

In forming the four piles, then, we use $(11 - x) + (11 - y) + (11 - z) + (11 - w) = 44 - x - y - z - w$ cards, in which x, y, z, and w represent the numbers of the cards on the tops of the piles (zero for each pile that is killed). We now count out $x + y + z + w$ more cards from the deck. Because $(44 - x - y - z - w) + (x + y + z + w) = 44$, this means that 44 cards are counted out altogether, regardless of the number of cards in each of the four piles. Because the chosen card is the 44th card in the deck, it turns up again as the last card counted out.

Point out that the expressions $44 - x - y - z - w$ and $x + y + z + w$ are examples of polynomials in four variables and that, to understand how the card trick works, we had to add them. *In today's lesson, we will learn how to add and subtract polynomials in one or more variables.*

Examples different from those in the text are provided on Transparency 9-7.

Transparency 9-8 is for use in discussing the Set IV exercise of Lesson 2.

Lesson 4
Multiplying Polynomials

The Fibonacci number trick

Before your students come into the room, put a large card above the front board. The more noticeable the card is, the better. I use one that is bright yellow.

Throughout our study of algebra, we have done a variety of number tricks. Here is one that is especially surprising. Write the numbers from 1 through 10 in a column on your paper. Now ask a student to come to the board and write two different numbers on it while you are facing the class. I suggest specifying that each number contain two digits to prevent your students from being able to carry out the calculations too quickly. Calculators should not be used for the same reason.

The rest of the class should copy the numbers from the board on the first two lines of their numbered column. The student who chose the numbers should now erase them from the board, return to his desk, and do the same.

Add the two numbers and write the sum on the third line. Then add the second and third numbers and write their sum on the fourth line. Continue to do this until numbers have been written on all ten lines, each number being found by adding the two numbers before it. While the class is absorbed in doing this, you must find out what the seventh number is. To do this, ask a student to tell what he has gotten for the seventh number, saying that number tricks are frequently spoiled by mistakes and that this is a check to see that everyone is on the right track. It is best to do this before everyone has had time to find all ten numbers.

Now tell your students to find the sum of all ten numbers. While they are doing this, multiply the seventh number by eleven. Quietly take down the card from the board, write the result on it with a felt-tip marker (one that does not squeak!), and put the card back with the number facing the wall. The

class should be sufficiently absorbed in adding the ten numbers that you can do this without anyone noticing.

After having everyone check their work, ask them if the final answer depends on the two numbers originally chosen. Because it does not for many of the tricks that we have considered, it is important for everyone to think about this. Let someone who recognizes that the final answer cannot be predicted before the trick begins explain his reasoning to the rest of the class. Then tell the class that you do not know either of the numbers originally chosen and turn the card over to show the number on the back. Those students who have gotten the right answer and who had noticed that the card was above the board before class will be astonished at seeing it on the card.

To find out how the trick works, we will represent the two numbers originally chosen as x *and* y. *Write these symbols to the right of the original numbers. The rest of the numbers are determined by these two. How would you represent the third number?* (x + y). *The fourth number?* (y + (x + y) = x + 2y). Have everyone write expressions for the rest of the numbers in the same way. Point out that these expressions are polynomials in two variables.

1. x

2. y

3. $x + y$

4. $x + 2y$

5. $2x + 3y$

6. $3x + 5y$

7. $5x + 8y$

8. $8x + 13y$

9. $13x + 21y$

10. $21x + 34y$

A Number Trick
Think of a two-digit number in which the units digit is more than the tens digit. Multiply the difference between the digits by nine and add the result to your original number.
What do you notice about the result?

TRANSPARENCY 9-9

TRANSPARENCY 9-10

1. Multiply $x+5$ and $2x+3$.

2. Multiply $4x+1$ and $3x-8$.

3. Multiply $6x$ and x^2-4x+2.

4. Multiply $x-3$ and $5x^3+x-7$.

TRANSPARENCY 9-11

When the list is complete, observe that the coefficient of x in each successive polynomial is identical with the coefficient of y in the preceding polynomial. *Now add all ten polynomials. The sum, $55x + 88y$, has a simple relationship to one of the ten polynomials in our list. Can you figure out which one it is?* (It is 11 times the seventh polynomial: $11(5x + 8y) = 55x + 88y$.)

If no one has recalled by now that you found out what the seventh number was while the class was doing the original arithmetic (I have had many classes in which this had been forgotten), the light should begin to dawn on some of your students at this point. Many, thinking that the card was above the board all the time, will still be mystified and so this is the time to confess everything.

Transparency 9-9 is for use in discussing the Set IV exercise of Lesson 3.

Transparency 9-10 can be used to present the lesson on multiplying polynomials in accord with the text. A facsimile of the Marmaduke book, published under the title *Marmaduke Multiply's*, is available in both clothbound and paperbound form from Dover Publications, Inc., 180 Varick Street, New York, N.Y., 10014. It includes an interesting commentary by E. F. Bleiler.

You might begin by multiplying 37 by 42 and comparing the process with multiplying $3x + 7$ and $4x + 2$ as shown below.

$$
\begin{array}{r} 37 \\ \times\,42 \\ \hline 74 \\ 148 \\ \hline 1554 \end{array}
\qquad
\begin{array}{r} 30 + 7 \\ 40 + 2 \\ \hline 60 + 14 \\ 1200 + 280 \end{array}
\qquad
\begin{array}{r} 74 \\ 1480 \\ \hline 1554 \end{array}
$$

	30	7
40	1200	280
2	60	14

$$
\begin{array}{r} 3x + 7 \\ \times\ 4x + 2 \\ \hline 6x + 14 \\ 12x^2 + 28x \\ \hline 12x^2 + 34x + 14 \end{array}
$$

	3x	7
4x	$12x^2$	$28x$
2	$6x$	14

Additional examples different from those in the text are provided on Transparency 9-11. (Note that in this lesson the column format for multiplication is used exclusively. The row format is introduced in Lesson 5.)

Lesson 5
More on Multiplying Polynomials

Wallace Judd has written a clever book titled *Patterns to Play on a Hundred Chart* (Creative Publications, 1975) suitable for use in both arithmetic and beginning algebra classes. The following activity is adapted from this book.

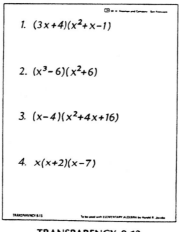

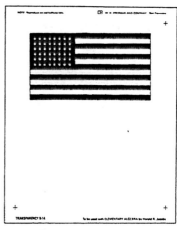

1. $(3x+4)(x^2+x-1)$

2. $(x^3-6)(x^2+6)$

3. $(x-4)(x^2+4x+16)$

4. $x(x+2)(x-7)$

TRANSPARENCY 9-12   **TRANSPARENCY 9-13**   **TRANSPARENCY 9-14**

Hand out duplicated copies of Transparency 9-12 and ask your students to choose a number in the pattern that is not in the border and draw an X through it. Demonstrate with an example on the overhead projector as shown below.

$$-50\ -49\ -48\ -47\ -46\ -45\ -44\ -43\ -42\ -41$$
$$-40\ -39\ -38\ -37\ -36\ -35\ -34\ -33\ -32\ -31$$
$$-30\ -29\ -28\ -27\ -26\ -25\ -24\ -23\ -22\ -21$$
$$-20\ -19\ \textcircled{-18}\ -17\ \textcircled{-16}\ -15\ -14\ -13\ -12\ -11$$
$$-10\ -9\ -8\ -7\ -6\ -5\ -4\ -3\ -2\ -1$$
$$0\ 1\ \textcircled{2}\ 3\ \textcircled{4}\ 5\ 6\ 7\ 8\ 9$$
$$10\ 11\ 12\ 13\ 14\ 15\ 16\ 17\ 18\ 19$$
$$20\ 21\ 22\ 23\ 24\ 25\ 26\ 27\ 28\ 29$$
$$30\ 31\ 32\ 33\ 34\ 35\ 36\ 37\ 38\ 39$$
$$40\ 41\ 42\ 43\ 44\ 45\ 46\ 47\ 48\ 49$$

Circle the four numbers at the corners of the X and multiply each pair of opposite numbers. Subtract the smaller result from the larger. Repeat the procedure with another number. What do you notice? (If no mistakes are made, the answer will always be 40.)

After your students have discovered this, help them develop the following proof, writing it in the space below the pattern. PROOF: *Let* x *represent the number at the upper left of the crossed-out number. How would the other three numbers be represented in terms of* x? *Multiplying, we get* $x(x + 22) = x^2 + 22x$ *and* $(x + 2)(x + 20) = x^2 + 22x + 40$. Do the second multiplication in the column format.

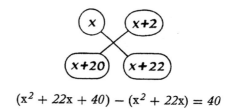

$$(x^2 + 22x + 40) - (x^2 + 22x) = 40$$

After the proof is complete, show how to do the second multiplication using the row format. More examples for use in illustrating the row format are given on Transparency 9-13.

Lesson 6
Squaring Binomials

Show Transparency 9-14 with overlay A in place. *This picture shows the flag of the United States as it appeared for the longest period of time in our nation's*

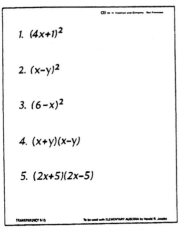

1. $(4x+1)^2$

2. $(x-y)^2$

3. $(6-x)^2$

4. $(x+y)(x-y)$

5. $(2x+5)(2x-5)$

TRANSPARENCY 9-15

Choose three consecutive integers. Square the second number and multiply the first by the last. What do you notice?

TRANSPARENCY 9-16

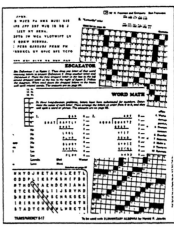

TRANSPARENCY 9-17

history: from 1912, when Arizona was admitted to the union, to 1959, when Alaska became a state. It contained forty-eight stars, arranged in six rows with eight stars in each row. For just one year, July 4, 1959, to July 4, 1960, the flag contained forty-nine stars. How do you suppose they were arranged? After someone has guessed, remove overlay A and add overlay B to show the arrangement: seven rows with seven stars in each row. Leaving overlay B in place, add overlay C. *How many of you thought the arrangement looked like this? We will use this pattern to introduce today's lesson.*

The array contains 7^2 stars. Because $7 = 3 + 4$, it is also correct to say that it contains $(3 + 4)^2$ stars. Is it correct to say that $(3 + 4)^2 = 3^2 + 4^2$? No, as we can see by drawing squares around 3^2 and 4^2

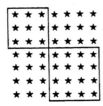

stars. There are two more sets of stars, each containing $3 \cdot 4$ stars; so $(3 + 4)^2 = 3^2 + 2(3 \cdot 4) + 4^2$.

Now replace 3 and 4 with x and y respectively, rewriting the pattern as $(x + y)^2 = x^2 + 2xy + y^2$.

Explain that $x + y$ is called a *binomial* because it is a polynomial that contains two terms and that $x^2 + 2xy + y^2$ is called a *trinomial* because it is a polynomial that contains three terms. The square of the binomial $x + y$, then, is the trinomial $x^2 + 2xy + y^2$.

Examples for developing the rest of the lesson are given on Transparency 9-15.

Transparency 9-16 is for use in discussing the Set IV exercise of Lesson 5.

Lesson 7
Dividing Polynomials

Show Transparency 9-17. *A popular type of puzzle that is frequently found in crossword puzzle books has as its basis long division. In this type of puzzle, letters or other symbols are substituted for some or all of the digits and the idea is to figure out what the digits are.*

Hand out duplicated copies of Transparency 9-18. *Here is a puzzle of this sort in which many of the digits have been replaced by boxes. Can you figure out what the missing digits are?* This is probably best done as a class exercise with everyone working together. By figuring out the missing digits, your students will have a good review of the long-

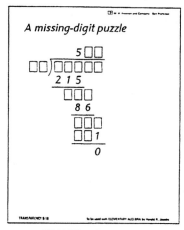

TRANSPARENCY 9-18

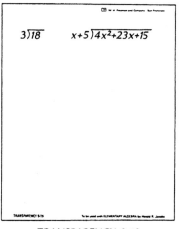

TRANSPARENCY 9-19

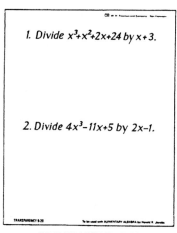

TRANSPARENCY 9-20

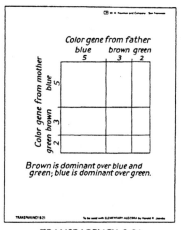

TRANSPARENCY 9-21

division algorithm and will be prepared to learn how to divide polynomials.

Show Transparency 9-19. *Division can be illustrated by means of rectangles in essentially the same way that we have learned to picture multiplication. Each of these division problems can be interpreted as finding the other dimension of a rectangle from one of its dimensions and its area.* Draw figures as shown below on the lower part of the transparency to illustrate this.

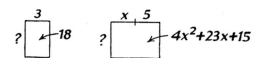

Show how to do the second division by filling in the second figure as shown below. Then show how to do it by long division.

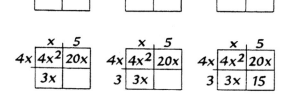

Additional examples for use in demonstrating how polynomials are divided are given on Transparency 9-20.

Transparency 9-21 is for use in discussing the Set IV exercise of Lesson 6.

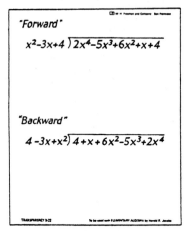

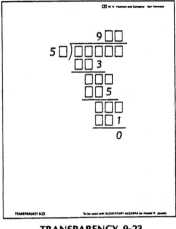

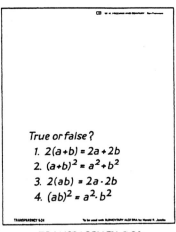

TRANSPARENCY 9-22   TRANSPARENCY 9-23   TRANSPARENCY 9-24

Review of Chapter 9

First day

Show the upper half of Transparency 9-22. *When dividing one polynomial into another, it is customary to first arrange each polynomial in descending powers of the variable as shown in this problem.* Reveal the rest of the transparency. *What would happen if both polynomials were arranged in* ascending *powers of the variable instead? Copy both versions of the problem and see if you can carry out the division in each case.* If every division came out even, the arrangement in ascending order would work just as well for dividing polynomials as the traditional arrangement. For divisions that do not come out even, however, the ascending arrangement does not produce satisfactory results.

Transparency 9-23 is for use in discussing the Set IV exercise of Lesson 7.

Second day: A short true or false test

In the Peanuts strip of October 3, 1968, Linus is trying to guess the pattern of answers to a true or false test. In the last panel, he says "If you're smart, you can pass a true or false test without being smart!" The strip is reproduced in the Peanuts book titled "My Anxieties Have Anxieties" (Peanuts Parade Paperback No. 18) and might be shown to your class before showing Transparency 9-24.

The questions on the test are about four similar equations that students sometimes think are true for all real numbers. If you show the cartoon to your students before they take the test, they might enjoy observing afterward that anyone who has followed Linus's reasoning (the first question is almost always true, the second is false to balance the true one, the third is false to break the pattern, the fourth is true) has answered all of the questions correctly!

Answers to Set I

1. $(a+b)^2 = a^2+2ab+b^2$; how to square a binomial.

2.a) False   e) N.p.
 b) False   f) $162x^4$
 c) False   g) $-3x^5$
 d) True   h) N.p.
 e) True

3.a) 240
 b) 240
 c) 28
 d) 81
 e) 0
 f) 9,801

4.a) $2x^4$
 b) N.p.
 c) $-6x^4$
 d) $9x^8$

c)

	x	-4
$8x$	$8x^2$	$-32x$
1	x	-4

$8x^2-31x-4$

d)

	y	-11
x	xy	$-11x$
-9	$-9y$	99

$xy-11x-9y+99$

5.a)

	$3x$	5
$3x$	$9x^2$	$15x$
5	$15x$	25

$9x^2+30x+25$

b)

	$7x$	y
$7x$	$49x^2$	$7xy$
$-y$	$-7xy$	$-y^2$

$49x^2-y^2$

6.a) x^4+5x^3+9x
 b) $3x^6-6x^2+6x$
 c) $-x^5+5$
 d) $-3x^3+7x+5$

7.a) x^2+4x
 b) $x^2+7x+12$
 c) $8x^2-16x$
 d) $8x^2-17x+2$
 e) x^2+x-30

TRANSPARENCY 9-25

---

f) $30x^2+x-1$
g) $x^2+14x+49$
h) $49x^2$
i) $4x^2-12x+9$
j) x^2-100
k) x^6+6x^3+9
l) x^6-9

8.a) $-4y$
 b) x^4-x^3+1
 c) x^3+7x^2+8x-4
 d) $15x+15y$
 e) $35x^2+36x-20$
 f) $2x^5-2x^4+7x^3+x^2-4x$
 g) x^2+3x-1
 h) x^3+4x^2+14x

9.a) per., $20x^4$; area, $25x^8$
 b) per., $6x+16$;

area, $27x-9$
c) per., $10x-12$; area, $4x^2-30x-16$
d) per., $24x$; area, $36x^2-25$

10.a) 12
 b) 225
 c) $6xy$
 d) 25

11.a) $9x+4$
 b) $20x^2+23x-21$
 c) $x^3+x^2-11x-3$
 d) x^2+4x+2
 e) x^4+x
 f) x^4-x-2
 g) x^5+x^4-x-1
 h) x^3-x^2+x-1

TRANSPARENCY 9-26

Answers to the Set I exercises of the review lesson are given on Transparencies 9-25 and 9-26.

Chapter 10

FACTORING

In this chapter, the students learn how to "unmultiply." The first two lessons, on factoring integers and monomials, prepare them for the remaining lessons on factoring polynomials. Starting with a lesson on factoring out the greatest common factor to every term of a polynomial, they next learn how to factor polynomials of the form $x^2 + bx + c$. Although the trial-and-error technique developed in this lesson can be used to factor both the difference of two squares and trinomial squares, these patterns are so important that attention is directed to them in the next two lessons. A lesson on factoring polynomials of the form $ax^2 + bx + c$ in which a is not 1 and a lesson on factoring higher-degree polynomials conclude the chapter.

The rectangle diagrams that were used as an aid to understanding polynomial multiplication and division continue to serve as a valuable device not only in visualizing factoring but also as a way to solve problems as challenging as factoring the sum of two cubes. Although few applications are found

in this chapter, a variety of anecdotes and exercises of mathematical and historical interest are included in both the text and lesson plans.

Factoring is not only difficult, but also subtle in theory. The problem is that one must be careful about what counts as a factor and what sort of factors are allowed. We factor $3x^2 + 3$ as $3(x^2 + 1)$, but we do not generally factor $3x^2 + 1$ as $3\left(x^2 + \dfrac{1}{3}\right)$, although that is a factorization. The reason is that we usually stop with integer factors so that the process will come to a definite end.

Similarly, we factor $x^2 - 4$ as $(x + 2)(x - 2)$, but we do not always ask students to factor $x^2 - 2$ as $(x + \sqrt{2})(x - \sqrt{2})$. This is because we do not always want factors with irrational terms. Finally, we do not factor $x^2 + 1$ in this course, though it may be factored as $(x + \sqrt{-1})(x - \sqrt{-1})$, because we do not allow imaginary terms here.

The remarkable fact is that every polynomial in x can be factored as a product of linear and quadratic terms with real coefficients and the only non-factorable polynomials in x are sums of squares, such as $x^2 + 1$ or $(x - 9)^2 + 25$ or $x^2 + 2$ (remembering that $x^2 + 2 = x^2 + (\sqrt{2})^2$).

Lesson 1
Prime and Composite Numbers

The first of the following exercises has as its basis an idea that Martin Gardner described in his Mathematical Games column "Some new and dramatic demonstrations of number theorems with playing cards" in the November 1974 issue of *Scientific American*.

If you have been to a college football game or have watched one on television, you have probably seen some card stunts played at halftime. As numbers are called, people sitting in a certain section of the stands hold up and turn over cards to form a picture or pattern of some sort. We are going to try a couple of mathematical card stunts that, if carried out correctly, will form some interesting patterns.

Half of your class (or at least 16 students if you have a small class) should stand in a row at the front of the room. Give each student a large card with a number printed on it so that the consecutive positive integers appear from left to right. Each student should look at his card so that he knows the number he is holding.

Stunt 1

Tell everyone to turn around and face the front wall with his back to the class. Now tell everyone holding a number evenly divisible by 1 to turn around. (Everyone should now be facing the class.) Tell everyone holding a number evenly divisible by 2 to turn around. (Just the students holding odd numbers should now be facing the class.) Continue in the same fashion, telling everyone holding a number evenly divisible by 3 to turn around, everyone holding a number evenly divisible by 4, and so forth until you have told everyone holding a number evenly divisible by the last number to turn around. Note that in general as you call off the number n, every nth student counting from the left should turn around. Go slowly so that everyone has time to think and so that you can correct anyone who makes a mistake.

When these directions have been carried out, the students holding the numbers 1, 4, 9, and 16 should be facing the class. Ask the students who are seated what these numbers have in common. (They are square numbers.) *So now we know who the squares are!*

Stunt 2

Tell the students who did the first stunt to be seated and have the rest of the class (again at least 16 students) come to the front for the second stunt. Hand out the cards in order from left to right as before.

This time each person is not to turn around more than twice; if asked to turn around a third time, the student should put down his card and return to his seat. The stunt starts this time with everyone facing the class. Repeat the directions as before, telling everyone holding a number evenly divisible by 1 to turn around and so forth to the last number held. This time, the student holding the number 1 will end up facing the wall, the students holding the prime numbers will be facing the class, and the students holding the composite numbers will have gone to their seats.

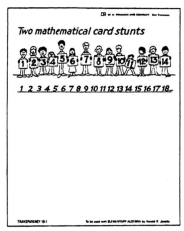

TRANSPARENCY 10-1

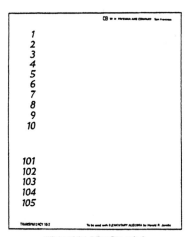

TRANSPARENCY 10-2

To explain how the stunts work, show Transparency 10-1 and either hand out duplicated copies of it or have everyone copy the row of positive integers at the top of a sheet of paper. Now write a 1 below each number evenly divisible by 1, a 2 below each number evenly divisible by 2, and so forth, up to an 18 under each number evenly divisible by 18. The result is shown below.

1	2	3	4	5	6	7	8	9	10	11	12	13	14	15	16	17	18
1	1	1	1	1	1	1	1	1	1	1	1	1	1	1	1	1	1
	2	3	2	5	2	7	2	3	2	11	2	13	2	3	2	17	2
			4		3		4	9	5		3		7	5	4		3
					6		8		10		4		14	15	8		6
											6				16		9
											12						18

In the first stunt, everyone turned around for every number by which they were evenly divisible. Which people turned around the most? (Those holding the 12 and 18.) Who turned around the least? (The person holding the 1.) Everyone started out facing the wall. How many times could you turn around and end up facing the class? (1, 3, 5, . . . ; an odd number of times.) The integers by which a number is evenly divisible are called its factors. *Which numbers have an odd number of positive factors? (The square numbers.)* To show why, pair the factors of each number as shown below.

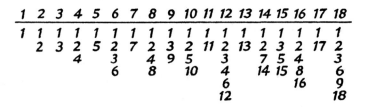

Doing this reveals that each square number has an odd number of positive factors because one factor is paired with itself; it is *because* of this, in fact, that these numbers are squares.

In the second stunt, everyone started out facing the class. Who turned around exactly once and ended up facing the wall? (The person holding the 1.) Who turned around exactly twice and ended up facing the class? (The people holding the numbers having exactly two positive factors: the number itself and 1.) Who sat down? (The people holding numbers having more than two positive factors.)

At this point, show the upper part of Transparency 10-2 and have everyone make a list of the first ten integers as shown on the next page. *Every prime*

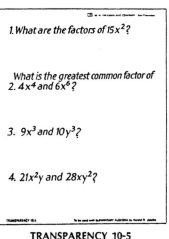

TRANSPARENCY 10-3 **TRANSPARENCY 10-4** **TRANSPARENCY 10-5**

number can be factored into positive factors in only one way: as the product of itself and 1. Composite numbers can be factored into positive factors in more than one way.

$$
\begin{array}{ll}
1 \\
2 & prime \\
3 & prime \\
4 & 2\cdot2 = 2^2 \\
5 & prime \\
6 & 2\cdot3 \\
7 & prime \\
8 & 2\cdot2\cdot2 = 2^3 \\
9 & 3\cdot3 \\
10 & 2\cdot5
\end{array}
$$

Now show the lower part of the transparency and develop the rest of the lesson by trying to factor these five numbers into primes.

$$
\begin{array}{ll}
101 & prime \\
102 & 2\cdot3\cdot17 \\
103 & prime \\
104 & 2^3\cdot13 \\
105 & 3\cdot5\cdot7
\end{array}
$$

The terms *greatest common factor* and *relatively prime* can also be introduced through examples from these lists.

Lesson 2
Monomials and Their Factors

Show Transparency 10-3. *An article in the February 1978 issue of* Scientific American (Science and the Citizen, p. 89) *tells of the discovery of an interesting prime number. The number contains 317 digits, all of which are 1's. It is represented by the symbol R_{317}, the R indicating that it is written by repeating the digit 1. What makes the number interesting is that, before it was discovered, only three other prime numbers of this sort were known.* Show Transparency 10-4. *One of them is 11, represented by the symbol R_2, and the other two are R_{19} and R_{23}. To prove that a number as large as R_{317} is prime is extremely difficult; so difficult, in fact, that it could not be done without the aid of a computer.**

Because the Set IV exercise of Lesson 1 is about composite *R*-numbers, this would be an appropriate time to discuss it. The three questions might be reconsidered in light of the longer table on the transparency.

To introduce the new lesson, you might use the number trick discussed in the text. Additional examples are given on Transparency 10-5.

---

*The eleventh chapter of *Recreations in the Theory of Numbers*, by Albert H. Beiler (Dover, 1966), contains additional information on prime and composite *R*-numbers.

268,301	268,326	268,351	268,376
268,302	268,327	268,352	268,377
268,303	268,328	268,353	268,378
268,304	268,329	268,354	268,379
268,305	268,330	268,355	268,380
268,306	268,331	268,356	268,381
268,307	268,332	268,357	268,382
268,308	268,333	268,358	268,383
268,309	268,334	268,359	268,384
268,310	268,335	268,360	268,385
268,311	268,336	268,361	268,386
268,312	268,337	268,362	268,387
268,313	268,338	268,363	268,388
268,314	268,339	268,364	268,389
268,315	268,340	268,365	268,390
268,316	268,341	268,366	268,391
268,317	268,342	268,367	268,392
268,318	268,343	268,368	268,393
268,319	268,344	268,369	268,394
268,320	268,345	268,370	268,395
268,321	268,346	268,371	268,396
268,322	268,347	268,372	268,397
268,323	268,348	268,373	268,398
268,324	268,349	268,374	268,399
268,325	268,350	268,375	268,400

TRANSPARENCY 10-4

TRANSPARENCY 10-6

Which number is prime?

1311

1312

1313

1314

1315

1316

1317

1318

1319

1320

TRANSPARENCY 10-7

TRANSPARENCY 10-7

Write down a two-digit number three times to form a six-digit number. The result can be divided evenly by 3, 7, 13, and 37.

TRANSPARENCY 10-8

Lesson 3
Polynomials and Their Factors

Show Transparency 10-6. *As the integers get larger and larger, the primes among them become less and less frequent. In the list of one hundred consecutive numbers shown here, there is only one prime!* Although many of the numbers can be quickly eliminated in looking for the prime number (the even ones, for example), it is not an easy task to narrow the list of possibilities down to one. If you had a lot of time and patience, you might eventually be able to do this and find that the prime number is 268,343.*

Show Transparency 10-7. *Here is another list of consecutive integers that contains only one prime. In this case, it is much easier to narrow the possibilities down to one. Can you do so and figure out which number is prime?* (We can immediately eliminate the five even numbers and 1315, because it is divisible by 5. Of the four numbers left, 1311 and 1317 are both divisible by 3 and 1313 is divisible by 13. So the prime number must be 1319.)

Transparency 10-8 is for use in discussing the Set IV exercise of Lesson 2.

Transparency 10-9 may be used to introduce the new lesson in accord with the text. Additional examples are given on Transparency 10-10.

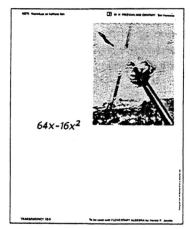

$64x - 16x^2$

TRANSPARENCY 10-9

Factor the following polynomials

1. $14x - 21$

2. $2x^3 + 8x$

3. $10x^2 - 3x^2$

4. $5x^5 + x^3 - x^2$

TRANSPARENCY 10-10

---

*This table is from *Famous Problems of Mathematics* by Heinrich Tietze (Graylock Press, 1965).

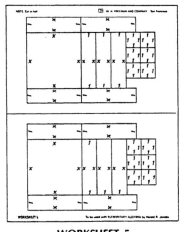

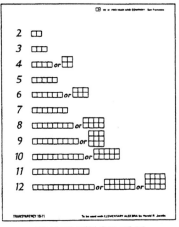

WORKSHEET 5   **TRANSPARENCY 10-11**

Lesson 4
Factoring Second-Degree Polynomials

To introduce this lesson, you might print enough copies of Worksheet 5 on heavy cardstock (file folders work very well) so that each student will receive a half sheet. Hand out scissors and letter-size envelopes. Tell your students to cut the pattern apart along the lines to separate the large square, the seven rectangular strips, and the six small squares from each other. After the experiment is finished, the pieces can be put into the envelopes and collected for future use by other classes. Also reproduce the pattern on a sheet of heavy-weight transparency film and cut out the pieces for use on your overhead projector. (You might cover them with a piece of colored adhesive film before cutting them out to make them easier to see on the screen.)

After the pieces have been prepared or handed out, begin by showing Transparency 10-11. *One way to tell whether a number is prime or composite is shown here. Can you tell what it is?* (*A prime number of squares can be arranged to make a rectangle in only one way. A composite number of squares can be arranged to make a rectangle in more than one way.*) *What do the dimensions of each rectangle represent?* (*The factors of the number when it is written as a product of two factors.*) To illustrate this, you might write below the three figures for 12: 12 = (1)(12) or 12 = (2)(6) or 12 = (3)(4).

The factors of polynomials can also be represented by the dimensions of rectangles. Take your large

square, *whose area represents* x^2, *and five of the rectangular strips, each of whose areas represents* x. *What polynomial is represented by their total area?* ($x^2 + 5x$.) *Put them together to form a rectangle.*

Illustrate this on the screen. *What are the dimensions of the rectangle?* (x *and* x + 5.) *So* $x^2 + 5x$ *can be factored as* x(x + 5). *How could we have told this without looking at a figure?* (*By observing that* x *is a common factor of both terms of the polynomial.*)

Now have everyone separate the pieces and add the six small squares, explaining that the area of each small square represents 1. *What polynomial is now represented by the total area?* ($x^2 + 5x + 6$.) *The terms of this polynomial have no common factor other than 1. Nevertheless, the pieces can be made into a rectangle. Can you figure out how to do it?* After your students have tried this at their desks, have someone show the solution on the screen. *What are the dimensions of the rectangle?* (x + 2 *and* x + 3.) *This means that* $x^2 + 5x + 6$ *can be written in factored form as* (x + 2)(x + 3).

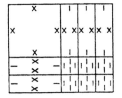

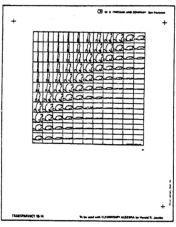

TRANSPARENCY 10-12 · TRANSPARENCY 10-13 · TRANSPARENCY 10-14

Factor if possible.
1. $x^2+8x+12$ 6. x^2+10x
2. $x^2-8x+15$ 7. $10x+25$
3. $x^2+3x-10$ 8. x^2+25
4. $x^2-3x-10$ 9. x^2-25
5. $x^2+10x+25$ 10. $x^2-10x-25$

A pattern for squaring numbers that end in 5

$1/5^2 = 2/25$
$2/5^2 = 6/25$
$3/5^2 = 12/25$
$4/5^2 = 20/25$
$5/5^2 = 30/25$
$6/5^2 = 42/25$
$7/5^2 = 56/25$
$8/5^2 = 72/25$
$9/5^2 = 90/25$

Now have everyone separate the pieces and add the other two rectangular strips. *Can you figure out a way to put these pieces together with the others to form a rectangle? What does the result indicate?* [*That* $x^2 + 7x + 6$ *can be factored as* $(x + 1)(x + 6)$.]

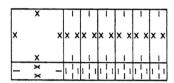

Take away four of the rectangular strips. Can you form a rectangle from the remaining pieces? (*No.*) *What does this suggest about the polynomial* $x^2 + 3x + 6$? (*That it cannot be factored.*)

Have everyone put the pieces into the envelope and then show how the polynomials $x^2 + 5x + 6$ and $x^2 + 7x + 6$ can be factored without looking at rectangles. Additional examples are given on Transparency 10-12.

Transparency 10-13 is for use in discussing the Set IV exercise of Lesson 3.

Lesson 5
Factoring the Difference of Two Squares

Show Transparency 10-14. *This picture, titled Beavers, was created by Leslie Mezei of the University of Toronto. Produced with the help of a computer, it shows successive transformations of a drawing of a beaver that are the result of changing its length and width.*

How many "beavers" are in the picture? If we agree to count even those that have been squashed flat, there are as many beavers as there are small squares: 121.

Now suppose that we make the picture smaller by removing the squares around the border. Add Overlay A. *How many beavers are left?* (*81.*) *How many were removed?* (*40.*) Write the equation $11^2 - 9^2 = 40$ in the space below the picture.

Suppose that the picture is made even smaller by removing a wider border of squares. Remove Overlay A and add Overlay B. *How many beavers are left now?* (*49.*) *How many have been removed from the original picture?* (*72.*) Write the equation $11^2 - 7^2 = 72$ below the first equation.

Remove Overlay B, add Overlay C, ask the same questions, and add the equation $11^2 - 5^2 = 96$ to the list. *We found the numbers of squares removed in each case by finding the difference of two squares. There is another way to find these numbers, which is by factoring. To discover what it is, write the polynomial* $x^2 - y^2$ *in factored form.* [$x^2 - y^2 = (x + y)(x - y)$.] Now write the first equation, $11^2 - 9^2 = 40$, in the form $(11 + 9)(11 - 9) = 20 \cdot 2 = 40$ and ask your students to do the same with the other two equations:

Factor if possible.

1. $x^2 - 36y^2$

2. $4 - x^6$

3. $x^2 + 9$

4. $3x^2 - 75$

5. $8x^2 + 8y^2$

TRANSPARENCY 10-15

+   *A polynomial for primes: $x^2 + x + 41$*   +

x	p	x	p	x	p	x	p
1	43	26	743	51	2693	76	▶5693
2	47	27	797	52	2797	77	6047
3	53	28	853	53	2903	78	6203
4	61	29	911	54	3011	79	6361
5	71	30	971	55	3121	80	6521
6	83	31	1033	56	▶3233	81	▶6683
7	97	32	1097	57	3347	82	▶6847
8	113	33	1163	58	3463	83	7013
9	131	34	1231	59	3581	84	▶7181
10	151	35	1301	60	3701	85	7351
11	173	36	1373	61	3823	86	7523
12	197	37	1447	62	3947	87	▶7697
13	223	38	1523	63	4073	88	7873
14	251	39	1601	64	4201	89	▶8051
15	281	40	▶1681	65	▶4331	90	8231
16	313	41	▶1763	66	4463	91	▶8413
17	347	42	1847	67	4597	92	8597
18	383	43	1933	68	4733	93	8783
19	421	44	▶2021	69	4871	94	8971
20	461	45	2111	70	5011	95	9161
21	503	46	2203	71	5153	96	▶9353
22	547	47	2297	72	5297	97	9547
23	593	48	2393	73	5443	98	9743
24	641	49	▶2491	74	5591	99	9941
25	691	50	2591	75	5741	100	10141

+

TRANSPARENCY 10-16

Find two numbers such that
their sum is 20 and the difference
of their squares is 80.

TRANSPARENCY 10-17

$$11^2 - 7^2 = (11 + 7)(11 - 7) = 18 \cdot 4 = 72$$
$$11^2 - 5^2 = (11 + 5)(11 - 5) = 16 \cdot 6 = 96$$

Additional examples that can be used to develop the rest of the new lesson are given on Transparency 10-15.

Transparency 10-16 and its overlay are for use in discussing the Set IV exercise of Lesson 4. The table is from Albert H. Beiler's *Recreations in the Theory of Numbers,* 2nd edition (Dover, 1966).* The polynomial produces prime numbers for each value of x from 1 through 39. The arrows on the overlay indicate for which values of x between 1 and 100 composite numbers are produced.

Use the space at the bottom to show how we can tell that composite numbers are produced for $x = 40$ and $x = 41$.

$$(40)^2 + (40) + 41 = 40(40 + 1) + 41$$
$$= 40(41) + 41$$
$$= 41(40 + 1)$$
$$= 41 \cdot 41$$

$$(41)^2 + (41) + 41 = 41(41 + 1 + 1)$$
$$= 41 \cdot 43$$

---

*Chapter 20 of this book contains a wealth of extremely interesting information about prime numbers.

Lesson 6
Factoring Trinomial Squares

Show the figure at the top of Transparency 10-17. *This is an algebra problem from the third century* A.D. *It was written by the Greek mathematician Diophantus and is a puzzle about two numbers.*

Reveal the problem below the figure. *To solve this puzzle, we might let* x *and* y *represent the two numbers. What equations can we write from the information given?* (x + y = 20 *and* $x^2 - y^2 = 80$.)

We have written two simultaneous equations in two variables. See if you can solve the equations by doing the following. First, notice that the left side of the second equation is the difference of two squares. Factor it. [(x + y)(x − y) = 80.] *Now substitute from the first equation into the second.* [20(x − y) = 80.] *Divide both sides of the resulting equation by 20.* (x − y = 4.) *Now finish the solution by pairing this equation with the equation* x + y = 20 *and solving the resulting pair of simultaneous equations.*

$$x - y = 4$$
$$x + y = 20$$

Adding, we get

$$2x = 24$$

so

$$x = 12$$

and

$$y = 8$$

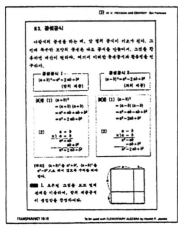

TRANSPARENCY 10-19

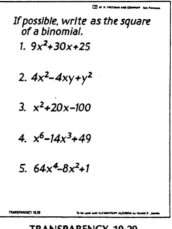

If possible, write as the square of a binomial.

1. $9x^2+30x+25$

2. $4x^2-4xy+y^2$

3. $x^2+20x-100$

4. x^6-14x^3+49

5. $64x^4-8x^2+1$

TRANSPARENCY 10-20

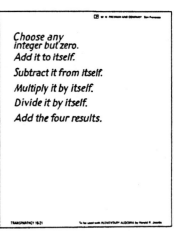

Choose any integer but zero.

Add it to itself.

Subtract it from itself.

Multiply it by itself.

Divide it by itself.

Add the four results.

TRANSPARENCY 10-21

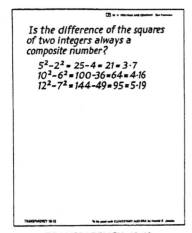

Is the difference of the squares of two integers always a composite number?

$5^2-2^2 = 25-4 = 21 = 3\cdot7$
$10^2-6^2 = 100-36 = 64 = 4\cdot16$
$12^2-7^2 = 144-49 = 95 = 5\cdot19$

TRANSPARENCY 10-18

Lesson 7
More on Factoring Second-Degree Polynomials

The number trick on Transparency 10-21 is from an article by David E. Williams in *The Mathematics Teacher* (September 1978, page 498) and is a nice application of factoring a trinomial square.

Try the trick out with several different integers, both positive and negative. What do you notice? (The result is always a square number, the square of the number that is 1 more than the original number.) See if you can explain why by letting x *represent the original number and carrying out the trick with it.*

Checking these numbers in the puzzle, we see that their sum is 12 + 8 = 20 and the difference of their squares is $12^2 - 8^2 = 144 - 64 = 80$.

Transparency 10-18 is for use in discussing the Set IV exercise of Lesson 5.

The new lesson is a fairly simple one and might be assigned for individual study. Transparencies 10-19 and 10-20 are for use if you wish to present the lesson to your class.

Choose any integer but zero.	x
Add it to itself.	$x + x = 2x$
Subtract it from itself.	$x - x = 0$
Multiply it by itself.	$x \cdot x = x^2$
Divide it by itself.	$\dfrac{x}{x} = 1$
Add the four results.	$2x + 0 + x^2 + 1 =$ $x^2 + 2x + 1 =$ $(x + 1)^2$

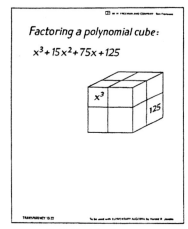

Factoring a polynomial cube:

$x^3 + 15x^2 + 75x + 125$

TRANSPARENCY 10-22

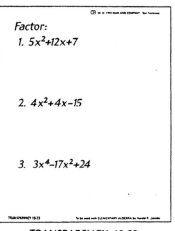

Factor:

1. $5x^2 + 12x + 7$

2. $4x^2 + 4x - 15$

3. $3x^4 - 17x^2 + 24$

TRANSPARENCY 10-23

TRANSPARENCY 10-24

Transparency 10-22 is for use in discussing the Set IV exercise of Lesson 6. *Note that the cube consists of eight pieces: one having a volume of* x^3, *three each having volumes of* $5x^2$, *three each having volumes of* 5^2x, *and one having a volume of* 5^3.

Examples for developing the new lesson are given on Transparency 10-23.

Lesson 8
Factoring Higher-Degree Polynomials

Show Transparency 10-24. *This is a page from a book in advanced algebra. Let's see if we can solve a couple of the problems.*

Show Transparency 10-25 and give everyone time to copy problem 8. *What do you notice about the first three terms of this polynomial? (They form a trinomial square.) Rewrite the polynomial, expressing the first three terms as the square of a binomial.* $[(x + 3y)^2 - 4.]$ *Notice that 4 is also a square; so this expression can be written as the difference of two squares.* $[(x + 3y)^2 - 2^2.]$ *Now factor the resulting expression, using the pattern for the difference of two squares.* $\big([(x + 3y) + 2][(x + 3y) - 2] = (x + 3y + 2)(x + 3y - 2).\big)$ *Show that this answer is correct by completing the following diagram.*

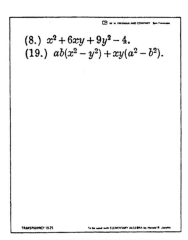

(8.) $x^2 + 6xy + 9y^2 - 4.$
(19.) $ab(x^2 - y^2) + xy(a^2 - b^2).$

TRANSPARENCY 10-25

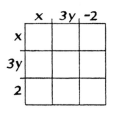 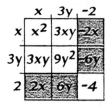

The diagram shows that $(x + 3y + 2)(x + 3y - 2) = x^2 + 6xy + 9y^2 - 4.$

Now copy problem 19 and use the distributive rule to remove the parentheses. ($abx^2 - aby^2 +$

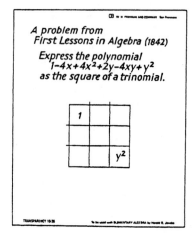

TRANSPARENCY 10-26

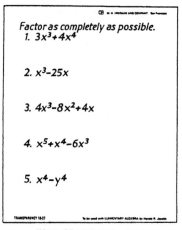

TRANSPARENCY 10-27

TRANSPARENCY 10-28

$xya^2 - xyb^2$.) *See if you can factor the resulting polynomial by completing the following diagram.*

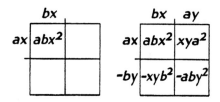

The diagram shows that $abx^2 - aby^2 + xya^2 - xyb^2 = (ax - by)(bx + ay)$.

Transparency 10-26 is for use in discussing the Set IV exercise of Lesson 7.

Examples for introducing Lesson 8 are given on Transparency 10-27.

Review of Chapter 10

First day

Most mathematics problems are written so that there is exactly enough information to solve them. Although it doesn't seem so, the following problem is no exception!*

---

*This problem is based on one that appeared in Martin Gardner's "Mathematical Games" column in the November 1970 issue of *Scientific American.*

Show Transparency 10-28. *To solve the problem, we will assume that the ages are positive integers. Factor 36 into three positive integers in as many ways as you can.* (Help your students get started doing this. There are eight possibilities, as shown below.) After checking this list with your students, add the overlay and ask them to find the sum of the three ages for each possibility and then look again at the clues. *Can you tell now how old the three daughters are?*

Possible ages	Sum
1, 1, 36*	38
1, 2, 18	21
1, 3, 12	16
1, 4, 9	14
1, 6, 6	13
2, 2, 9	13
2, 3, 6	11
3, 3, 4	10

*Not very likely.

If the sum of their ages doesn't give away the answer, it must be 13. The fact that there is an oldest daughter rules out the 1, 6, 6 case; so the ages must be 2, 2, and 9.

Transparency 10-29 is for use in discussing the Set IV exercise of Lesson 8.

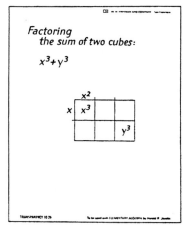

TRANSPARENCY 10-29

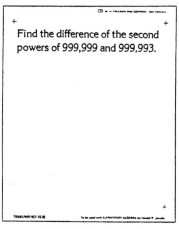

TRANSPARENCY 10-30

TRANSPARENCY 10-31

Second day

Show Transparency 10-30. *Here is a problem from a nineteenth-century algebra book* that can be solved in two different ways. The obvious way to solve it is rather difficult to carry out. An electronic calculator, in fact, would not be able to do it by this method because of the number of digits entailed. From what you have learned about factoring, you know another way that is considerably easier to carry out. Solve the problem by this method.*

$$999,999^2 - 999,993^2$$
$$= (999,999 + 999,993)(999,999 - 999,993)$$
$$= (1,999,992)(6)$$
$$= 11,999,952$$

After the method has been discussed (write it on the upper half of the transparency), add the overlay to show the longer solution.

Answers to the Set I exercises of the review lesson are given on Transparencies 10-31 and 10-32.

TRANSPARENCY 10-32

**First Lessons in Algebra* by Ebenezer Bailey (Jenks and Palmer, 1842).

Chapter 11
FRACTIONS

In order to be successful with algebraic fractions, a student has to understand arithmetic fractions. Unfortunately, many students do not. It is for this reason that each lesson in this chapter begins with a discussion of arithmetic fractions before any examples of fractions containing variables are considered.

The first lesson reviews two interpretations of arithmetic fractions and the meaning of equivalent fractions. The process of "cancelling" is purposely avoided because many students quickly gain the impression that anything may be crossed out of the numerator and denominator of a fraction without making any difference. Several of the exercises in this lesson are designed to help prevent this impression.

In the second lesson, the transition from arithmetic fractions to algebraic fractions is made, with emphasis on the fact that the algebraic properties of both types of fractions are the same.

The remaining lessons deal with the basic operations with fractions. Diagrams are used to help the students understand their meanings. Although complex fractions are not of great importance in an elementary course, the chapter closes with a lesson on them because the procedures used to simplify them provide a nice review of many of the principles introduced earlier.

Lesson 1
Fractions

The puzzle on Transparency 11-1 once appeared in Ripley's "Believe It or Not!" The answer would be rather obvious if four cuts were allowed. How to do it with three takes some thought. Give your students time to think about it before revealing the solution by drawing the cuts as shown below.

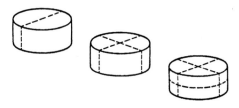

Add the overlay and use the fact that the shaded part of the cake may be said to be $\frac{1}{2}$ or $\frac{2}{4}$ or $\frac{4}{8}$ of it to review the interpretation of the fraction $\frac{a}{b}$ as meaning $a\left(\frac{1}{b}\right)$. Then consider the more general interpretation of a fraction as an indicated quotient before returning to the cake example to discuss equivalent fractions.

Lesson 2
Algebraic Fractions

A fraction orchard*

Duplicate Transparency 11-2 and give a copy to each student, together with a ruler. *This picture represents an overhead view of the southwest corner of an infinite orchard in which each circle represents a tree. The fractions are related to the positions of the trees. For example, where are the trees labeled with fractions whose numerators are zero? (Along the bottom row.) Where are the trees labeled with fractions whose denominators are three? (Along the fourth vertical row from the left.)*

Imagine that you are standing at the spot marked with an X. If you look into the orchard, you will be able to see many of the trees, but not all of them. For example, you will see the tree $\frac{0}{1}$, but not the trees $\frac{0}{2}$, $\frac{0}{3}$, $\frac{0}{4}$, $\frac{0}{5}$, $\frac{0}{6}$, $\frac{0}{7}$, and $\frac{0}{8}$ because they are behind it. Line up your ruler with the X and this row of trees and draw a line through the trees behind the $\frac{0}{1}$ tree. Now slowly turn the ruler counterclockwise about the

X until the view of one or more trees is blocked by a tree in front of it or them. (*The first case is the $\frac{2}{8}$ tree behind the $\frac{1}{4}$ tree.*) Show how to draw a line beyond the $\frac{1}{4}$ tree as in the diagram below. Illustrate one more case, the $\frac{2}{6}$ tree in line with the $\frac{1}{3}$ tree, and then give your students time to continue on their own, drawing lines through the rest of the trees that cannot be seen from the position marked X. When all the lines have been drawn, shade in the circles representing all of the trees that can be seen. The finished figure is shown below.

A fraction orchard

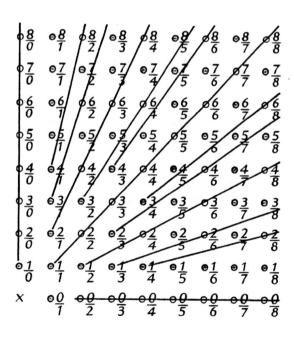

---

*This activity is adapted from Martin Gardner's column titled "The lattice of integers considered as an orchard or a billiard table" in the May 1965 issue of *Scientific American,* reprinted in *Martin Gardner's Sixth Book of Mathematical Games from Scientific American* (Scribners, 1971).

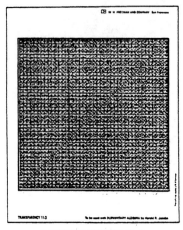

TRANSPARENCY 11-3

TRANSPARENCY 11-4

If possible, reduce each of the following fractions.

1. $\dfrac{2x}{10x}$

2. $\dfrac{x+2}{x+10}$

3. $\dfrac{10}{20x+5}$

4. $\dfrac{x+4}{x^2-16}$

5. $\dfrac{x^3-3x}{3x-9}$

TRANSPARENCY 11-5

By the time the work is finished, everyone should be able to answer to following questions. *What do you notice about the fractions within each set of trees that are in line with the X?* (*They are equivalent to each other, including even the first vertical row, in which the fractions are undefined.*) *What do you notice about the fractions corresponding to the trees that can be seen from the X?* (*They cannot be reduced. The fractions in the second vertical row, however, can be written as integers.*) *Where are the trees having fractions whose values are equal to 1?* (*Along the diagonal from the X to the upper right.*) *Where are the trees having fractions whose values are less than 1?* (*Below this diagonal.*) *Which tree or trees have fractions with the smallest value?* (*Those on the bottom row.*) *Which tree has the fraction with the largest value?* $\left(The\ \dfrac{8}{1}\ tree.\right)$

When your students are finished with this exercise, they will be interested in seeing the picture on Transparency 11-3. *Created by M. R. Schroeder of the Bell Telephone Laboratories with the help of a computer, this picture shows the fraction orchard extending 256 rows in each direction. The trees that can be seen from the lower left corner are represented by white squares and those that cannot be seen are represented by black squares. Note how the diagonal row of trees having fractions equivalent to 1 stands out.*

Transparency 11-4 is for use in discussing the Set IV exercise of Lesson 1.

Examples for use in presenting the new lesson are given on Transparencies 11-5 and 11-6.

Lesson 3
Adding and Subtracting Fractions

Some patterns with fractions: Pattern 1

Show the first two problems on Transparency 11-7 (keep the others covered) and ask your students to copy and simplify each as indicated below.

1. $\dfrac{1+2+3}{4+5+6} = \dfrac{6}{15} = \dfrac{2}{5}$

2. $\dfrac{7+8+9}{10+11+12} = \dfrac{24}{33} = \dfrac{8}{11}$

After checking these solutions, ask your students to guess what the next problem is, write it down, and figure out the answer.

3. $\dfrac{13+14+15}{16+17+18} = \dfrac{42}{51} = \dfrac{14}{17}$

Change to fractions that have the same denominator.

6. $\frac{x}{4}$ and $\frac{1}{x}$

7. $\frac{2}{x+3}$ and $\frac{x}{x-1}$

Find the value of $\frac{4x+7}{x-2}$ if

a) $x=5$

b) $x=0$

c) $x=-3$

d) $x=2$

TRANSPARENCY 11-6

Some patterns with fractions
Pattern 1.

1. $\frac{1+2+3}{4+5+6}$

2. $\frac{7+8+9}{10+11+12}$

3. $\frac{13+14+15}{16+17+18}$

4. $\frac{100+101+102}{103+104+105}$

TRANSPARENCY 11-7

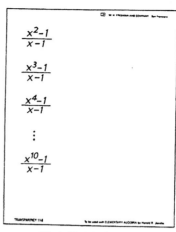

$\frac{x^2-1}{x-1}$

$\frac{x^3-1}{x-1}$

$\frac{x^4-1}{x-1}$

$\vdots$

$\frac{x^{10}-1}{x-1}$

TRANSPARENCY 11-8

After they have done this, reveal the problem and check its solution.

Look again at the three problems we have solved. Reveal problem 4. Can you guess the answer to this problem without doing any adding?

4. $\dfrac{100 + 101 + 102}{103 + 104 + 105} = \dfrac{101}{104}$

After someone has pointed out that the answer to each problem is "in the middle" of the problem, for example,

$$\frac{1 + \boxed{2} + 3}{4 + \boxed{5} + 6} = \frac{2}{5}$$

give your students time to prove it, letting x represent the first term in the numerator. (They may need some direction in representing the rest of the fraction in terms of x.)

$$\frac{x + (x + 1) + (x + 2)}{(x + 3) + (x + 4) + (x + 5)} =$$

$$\frac{3x + 3}{3x + 12} = \frac{3(x + 1)}{3(x + 4)} = \frac{x + 1}{x + 4}$$

Transparency 11-8 is for use in discussing the Set IV exercise of Lesson 2.

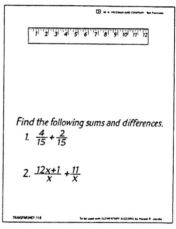

Find the following sums and differences.

1. $\frac{4}{15} + \frac{2}{15}$

2. $\frac{12x+1}{x} + \frac{11}{x}$

TRANSPARENCY 11-9

To introduce the new lesson, you might either hand out rulers or show Transparency 11-9 and have your class consider the problem of adding $\frac{1}{3}$ and $\frac{1}{4}$ by thinking of them as fractions of a foot. *Because $\frac{1}{3}$ of a foot is 4 inches and $\frac{1}{4}$ of a foot is 3 inches, together they add up to 7 inches, which is $\frac{7}{12}$ of a foot. So $\frac{1}{3} + \frac{1}{4} = \frac{7}{12}$.* Examples for possible

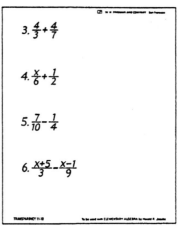

use in developing the rest of the lesson are given on Transparencies 11-9 and 11-10.

Lesson 4
More on Addition and Subtraction

Hold up a bottle of Coca-Cola. *The recipe for Coca-Cola has always been a carefully kept secret. It is never written down and only two people in the entire world know it. To help insure that the recipe will not be lost, they never travel on the same airplane.*

Even though the recipe is a well-guarded secret, many of the ingredients of Coca-Cola are known. Show Transparency 11-11. *The label on the bottle lists some of them: carbonated water, sugar, caramel color, phosphoric acid, and caffeine. The other ingredients are referred to only as "natural flavorings." They include cinnamon, nutmeg, lavender, lime juice, citrus oils, glycerine, and something known as "7X."*

Coca-Cola is $\frac{1}{11}$ sugar by weight.* What fraction of the drink do all of the rest of the ingredients make up? $\left(\frac{10}{11}.\right)$ *How did you get this answer?* $\left(1 - \frac{1}{11} = \frac{10}{11}.\right)$ *Suppose that Coca-Cola is $\frac{1}{a}$*

ingredient 7X by weight. *What fraction of the drink would all of the rest of the ingredients make up?* Help your students to figure this out, expressing the answer as a single fraction.

$$1 - \frac{1}{a} = \frac{a}{a} - \frac{1}{a} = \frac{a-1}{a}$$

These questions lead naturally into the new lesson. Additional examples are given on Transparencies 11-12 and 11-13. $\Big($Note that the answers to examples 5 through 7 are not unique. For example, some correct answers to example 6 are $\frac{1}{x+y} + \frac{3}{x+y}$, $\frac{2}{x+y} + \frac{2}{x+y}$, and $\frac{4-x}{x+y} + \frac{x}{x+y}.\Big)$

Transparency 11-14 is for use in discussing the Set IV exercise of Lesson 3.

Lesson 5
Multiplying Fractions

In 1858, a young man from Scotland named Alexander Rhind bought an old leather scroll while taking a trip through Egypt.* The scroll, which turned

---

*This amounts to an alarming five teaspoons of sugar per eight-ounce bottle!

*The information in this lesson is from Chapter 9 of *Mathematics in the Time of the Pharoahs*, by Richard J. Gillings (The M.I.T. Press, 1972.)

Write as a fraction.

1. $2\frac{5}{8}$

2. $x+\frac{1}{7}$

3. $3x-\frac{4}{x}$

4. $x-2+\frac{1}{x+2}$

TRANSPARENCY 11-12

Write as the sum of two fractions.

5. $\frac{x+y}{4}$

6. $\frac{4}{x+y}$

Write as the difference of a fraction and an integer.

7. $\frac{x^2+x-6}{3}$

TRANSPARENCY 11-13

Ollie's method | Correct method

TRANSPARENCY 11-14

out to have been written in about 1650 B.C., was so brittle that Rhind could not unroll it in order to find out what it said. After his death the scroll was acquired by the British Museum, but its contents still remained a mystery. It was thought that it might contain the words of a famous pharoah or tell of the construction of the pyramids. Finally, in 1927, someone figured out how to unroll the scroll without damaging it, and the mystery was solved. Show Transparency 11-15. *This photograph of the scroll shows what was inside. Instead of telling something about ancient Egypt, it turned out to be a list of 26 problems on adding fractions. Add the overlay. Four of the problems are shown here.*

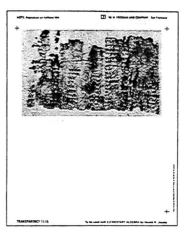

TRANSPARENCY 11-15

$$\frac{1}{14}+\frac{1}{21}+\frac{1}{42}$$

$$\frac{1}{18}+\frac{1}{27}+\frac{1}{54}$$

$$\frac{1}{22}+\frac{1}{33}+\frac{1}{66}$$

$$\frac{1}{30}+\frac{1}{45}+\frac{1}{90}$$

Solve the first one by expressing it as a single fraction.

$$\frac{3}{42}+\frac{2}{42}+\frac{1}{42}=\frac{6}{42}=\frac{1}{7}$$

Do the same with the second one.

$$\frac{3}{54}+\frac{2}{54}+\frac{1}{54}=\frac{6}{54}=\frac{1}{9}$$

All four problems are alike in a very basic way. Can you tell what it is? $\left(\text{All are of the form } \frac{1}{2x}+\frac{1}{3x}+\frac{1}{6x}.\right)$ If necessary, help your students to discover this. *Solve all four problems at once by expressing the sum* $\frac{1}{2x}+\frac{1}{3x}+\frac{1}{6x}$ *as a single fraction.* $\left(\frac{1}{2x}+\frac{1}{3x}+\frac{1}{6x}=\frac{3}{6x}+\frac{2}{6x}+\frac{1}{6x}=\frac{6}{6x}=\frac{1}{x}.\right)$ Use

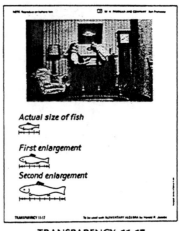

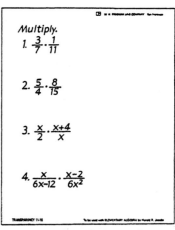

| TRANSPARENCY 11-16 | TRANSPARENCY 11-17 | TRANSPARENCY 11-18 |

your result to find the answers to the last two problems. $\left(\dfrac{1}{11}\ and\ \dfrac{1}{15}.\right)$

Transparency 11-16 is for use in discussing the Set IV exercise of Lesson 4.

Transparencies 11-17 and 11-18 might be used to present the new lesson in accord with the text.

Lesson 6
More on Multiplication

Some patterns with fractions: Pattern 2

Show the problem at the top of Transparency 11-19 (keep the others covered) and ask someone to explain what the three dots indicate. *How many products are to be added altogether? (99.) Rather than actually adding them, we may be able to guess the correct answer by considering some simpler problems.* Reveal the three problems below, asking your students to solve them and look for a pattern.

$$\frac{1}{1}\cdot\frac{1}{2}=\frac{1}{2}$$

$$\frac{1}{1}\cdot\frac{1}{2}+\frac{1}{2}\cdot\frac{1}{3}=\frac{2}{3}$$

$$\frac{1}{1}\cdot\frac{1}{2}+\frac{1}{2}\cdot\frac{1}{3}+\frac{1}{3}\cdot\frac{1}{4}=\frac{3}{4}$$

Those students who recognize the pattern will now be able to guess the answer to the original problem, $\dfrac{99}{100}$. *By writing these first three sums as*

$$\frac{1}{2}=1-\frac{1}{2},\quad \frac{2}{3}=1-\frac{1}{3},\quad and\quad \frac{3}{4}=1-\frac{1}{4}$$

and by noticing that

$$\frac{1}{1}\cdot\frac{1}{2}=\frac{1}{1}-\frac{1}{2},\quad \frac{1}{2}\cdot\frac{1}{3}=\frac{1}{2}-\frac{1}{3},$$

and in general

$$\frac{1}{n}\cdot\frac{1}{n+1}=\frac{1}{n}-\frac{1}{n+1},$$

we can see why the general formula

$$\frac{1}{1\cdot 2}+\frac{1}{2\cdot 3}+\frac{1}{3\cdot 4}+\cdots+\frac{1}{n(n+1)}=1-\frac{1}{n+1}$$

is valid. You may wish to show your students this approach to this sum or you may treat it as an exercise in pattern discovery and see if they can guess the form

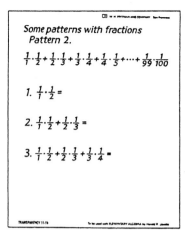

TRANSPARENCY 11-19

During the time the picture was taken, about 3 out of every 4,000 of the 10^{26} cobalt-60 atoms used broke apart. How many was that?

TRANSPARENCY 11-20

TRANSPARENCY 11-21

$$\frac{1}{1 \cdot 2} + \frac{1}{2 \cdot 3} + \cdots + \frac{1}{n(n + 1)} = \frac{n}{n + 1}$$

Transparency 11-20 is for use in discussing the Set IV exercise of Lesson 5.

Transparencies 11-21 and 11-22 might be used to present the new lesson as is done in the text. (Show how to do example 3 by using the pattern $(a + b)(c + d) = ac + ad + bc + bd$ and by first changing each factor to a fraction.)

Lesson 7
Dividing Fractions

Music from a meter stick

Mark a meter stick made from hard wood as shown in the diagram on the next page and ask your wood-shop teacher to cut it into the nine pieces indicated. Throw the shortest piece away. If the other eight are dropped in order from longest to shortest on a hard surface such as the floor of your classroom, the sounds produced will be the notes of a musical scale. After demonstrating this to your class, show Transparency 11-23 with the overlay in

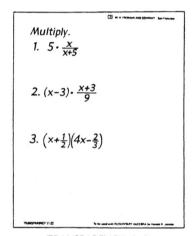

Multiply.
1. $5 \cdot \dfrac{x}{x+5}$

2. $(x-3) \cdot \dfrac{x+3}{9}$

3. $\left(x+\dfrac{1}{2}\right)\left(4x-\dfrac{2}{3}\right)$

TRANSPARENCY 11-22

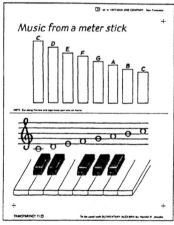

TRANSPARENCY 11-23

C	D	E	F	G	A	B	C
	16.2	30.6	43.6	55.8	66.6	76.3 84.9 93.0

place. *If we express the lengths of the sticks as fractions of the length of the longest stick, they are:*

C	D	E	F	G	A	B	C
1	$\frac{8}{9}$	$\frac{4}{5}$	$\frac{3}{4}$	$\frac{2}{3}$	$\frac{3}{5}$	$\frac{8}{15}$	$\frac{1}{2}$

(You might also give each student a duplicated copy of this transparency.) *Write these fractions on the sticks. They were chosen by the Pythagoreans in the fifth century B.C. and are arranged according to a pattern that can be discovered by division. Because the lengths are expressed as fractions of the length of the stick labeled C, it is clear that the stick labeled D is $\frac{8}{9}$ as long. Write this in the space below these sticks.*

How do the lengths of the sticks labeled D and E compare with each other? One way to compare them is by division. Remove the overlay to allow room for the following presentation on the lower part of the transparency.

To indicate the division of $\frac{4}{5}$ by $\frac{8}{9}$, we can write

$$\frac{\frac{4}{5}}{\frac{8}{9}}$$

One way to do the division is to multiply the numerator and denominator of this fraction by $\frac{9}{8}$:

$$\frac{\frac{4}{5} \cdot \frac{9}{8}}{\frac{8}{9} \cdot \frac{9}{8}} = \frac{\frac{4}{5} \cdot \frac{9}{8}}{1} = \frac{4}{5} \cdot \frac{9}{8} = \frac{4 \cdot 9}{5 \cdot 8} = \frac{9}{5 \cdot 2} = \frac{9}{10}$$

Explain that $\frac{9}{8}$ is called the reciprocal of $\frac{8}{9}$ and that, in general, the reciprocal of the fraction $\frac{x}{y}$ is the fraction $\frac{y}{x}$. Show why dividing by a fraction is equivalent to multiplying by its reciprocal:

$$\frac{\frac{a}{b}}{\frac{c}{d}} = \frac{\frac{a}{b} \cdot \frac{d}{c}}{\frac{c}{d} \cdot \frac{d}{c}} = \frac{\frac{a}{b} \cdot \frac{d}{c}}{1} = \frac{a}{b} \cdot \frac{d}{c}$$

Now have your students compare each of the remaining pairs of successive lengths by dividing the first fraction in each pair into the second: $\frac{3}{4} \div \frac{4}{5}$, $\frac{2}{3} \div \frac{3}{4}$, etc. *Writing the results as shown here, expressed both as common fractions and in decimal form rounded to the nearest hundredth, a pattern emerges.*

C	D	E	F	G	A	B	C
1	$\frac{8}{9}$	$\frac{4}{5}$	$\frac{3}{4}$	$\frac{2}{3}$	$\frac{3}{5}$	$\frac{8}{15}$	$\frac{1}{2}$
	$\frac{8}{9}$	$\frac{9}{10}$	$\frac{15}{16}$	$\frac{8}{9}$	$\frac{9}{10}$	$\frac{8}{9}$	$\frac{15}{16}$
	0.89	0.90	0.94	0.89	0.90	0.89	0.94

Comparing them to the arrangement of the keys on a keyboard, we see that the length quotient of two white keys that are separated by a black key (a whole step) is either 0.89 or 0.90, whereas the length quotient of two white keys that are not separated by a black key (a half step) is 0.94.

---

*In "even-tempered tuning," each whole step ratio is exactly the same: $\left(\frac{1}{\sqrt[12]{2}}\right)^2 \approx 0.8909\ldots$

Simplify.

1. $\left(1-\frac{1}{1}\right)\left(1-\frac{1}{2}\right)\left(1-\frac{1}{3}\right)\left(1-\frac{1}{4}\right)\cdots\left(1-\frac{1}{100}\right)$

2. $\left(1-\frac{1}{2}\right)\left(1-\frac{1}{3}\right)\left(1-\frac{1}{4}\right)\left(1-\frac{1}{5}\right)\cdots\left(1-\frac{1}{99}\right)$

3. $\left(1-\frac{1}{3}\right)\left(1-\frac{1}{4}\right)\left(1-\frac{1}{5}\right)\left(1-\frac{1}{6}\right)\cdots\left(1-\frac{1}{98}\right)$

TRANSPARENCY 11-24

Divide.

1. $9\frac{1}{2} \div \frac{1}{2}$

2. $\frac{3}{5} \div 6$

3. $\frac{x-2}{4} \div \frac{x}{8}$

4. $(x^2-25) \div \frac{x+5}{10}$

TRANSPARENCY 11-25

Transparency 11-24 is for use in discussing the Set IV exercise of Lesson 6. The third problem, which is not included in the text, might be considered after the other two have been checked.

Further examples for use in teaching the division of fractions are given on Transparency 11-25.

Lesson 8
Complex Fractions

Show Transparency 11-26. *Here are a couple of problems about fractions from a Swedish algebra book. The Swedish word for fraction, "bråk," is closely related to our own. Do you see why? (It seems to be related to the word "break." The word "fraction" is related to the word "fracture," which also means "break.") Look at the two problems. What do you think the last word of the instruction means? ("Dubbelbråk" looks like "double break." Our term for a "double break" is "complex fraction.")*

Use the two problems to illustrate how to transform complex fractions into simple fractions. Additional examples are given on Transparency 11-27.

Bestäm ekvivalenta enkla bråk till följande dubbelbråk.

1. $\dfrac{x+1}{1+\frac{1}{x}}$ 2. $\dfrac{1-\frac{2}{x}+\frac{1}{x^2}}{x-\frac{1}{x}}$

TRANSPARENCY 11-26

Simplify.

1. $\dfrac{x-\frac{1}{8}}{2x-\frac{1}{4}}$

2. $1+\dfrac{\frac{2}{3}}{1+\frac{4}{5}}$

3. $\dfrac{\frac{x}{y}}{\frac{1}{x}+\frac{1}{y}}$

TRANSPARENCY 11-27

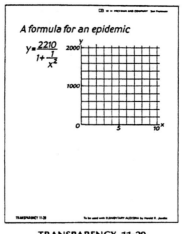

A formula for an epidemic

$$y = \frac{2210}{1 + \frac{1}{x^2}}$$

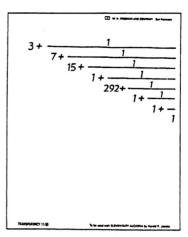

$$3 + \cfrac{1}{7 + \cfrac{1}{15 + \cfrac{1}{1 + \cfrac{1}{292 + \cfrac{1}{1 + \cfrac{1}{1 + \cfrac{1}{1}}}}}}}$$

| TRANSPARENCY 11-28 | TRANSPARENCY 11-29 | TRANSPARENCY 11-30 |

Review of Chapter 11

First day

Show Transparency 11-28. *This photograph, taken in 1918, shows a policeman directing traffic in New York City. His nose and mouth are covered by a mask because of the terrible flu epidemic of that year. The epidemic started in Kansas in March and swept across the world, killing more than 20 million people before it came to an end eight months later.* An interesting description of the epidemic is given on pages 547–548 of *The People's Almanac,* by David Wallechinsky and Irving Wallace (Doubleday, 1975).

Show Transparency 11-29. *A formula such as this might represent the number of cases in a certain city as a function of time.*

$$y = \frac{2210}{1 + \frac{1}{x^2}}$$

In the formula, x *represents the number of weeks from the beginning of the epidemic and* y *represents the number of cases. Use the formula to find* y *when* x = 1, 2, 3, 4, 5, *and 10.*

x	1	2	3	4	5	10
y	1105	1768	1989	2080	2125	≈ 2188

Graph this function by plotting the six points corresponding to this table and connecting them with a smooth curve.

Why can't we find y *when* x = 0? (*We cannot divide by 0.*) *It can be shown that when* x *is very close to 0,* y *is also very close to 0; so we can extend the curve toward the origin. What do you think happens to* y *when* x *becomes very large?* (*The fraction* $\frac{1}{x^2}$ *gets very close to 0; so* y *gets very close to 2210.*) *The table and graph show that the rate at which the number of cases increases becomes less and less as time goes by.*

Transparency 11-30 is for use in discussing the Set IV exercise of Lesson 8. More of Lambert's expression is given on the transparency than in the text; so you should circle the part that the students worked with before discussing the problem. It appears on the cover of C. D. Olds's *Continued Fractions,* volume 9 of the New Mathematical Library, originally published by Random House and now available from the Mathematical Association of America, 1529 Eighteenth Street, NW, Washington, D.C. 20036.

Second day
Some patterns with fractions: Pattern 3

Hand out duplicated copies of Transparency 11-31 and ask your students to solve the first two problems, expressing the answers as single fractions.

$$1 - \frac{1}{2^2} = \frac{3}{4}$$

$$\left(1 - \frac{1}{2^2}\right)\left(1 - \frac{1}{3^2}\right) = \frac{3}{4} \cdot \frac{8}{9} = \frac{2}{3}$$

After everyone has done this, ask if anyone can guess the answer to the third problem. *The first two answers suggest that the third one might be $\frac{1}{2}$, but it isn't.*

$$\left(1 - \frac{1}{2^2}\right)\left(1 - \frac{1}{3^2}\right)\left(1 - \frac{1}{4^2}\right) = \frac{2}{3} \cdot \frac{15}{16} = \frac{5}{8}$$

Give everyone time to solve the remaining problems.

$$\left(1 - \frac{1}{2^2}\right)\left(1 - \frac{1}{3^2}\right)\left(1 - \frac{1}{4^2}\right)\left(1 - \frac{1}{5^2}\right) = \frac{5}{8} \cdot \frac{24}{25} = \frac{3}{5}$$

$$\left(1 - \frac{1}{2^2}\right)\left(1 - \frac{1}{3^2}\right)\left(1 - \frac{1}{4^2}\right)\left(1 - \frac{1}{5^2}\right)\left(1 - \frac{1}{6^2}\right) = \frac{3}{5} \cdot \frac{35}{36} = \frac{7}{12}$$

$$\left(1 - \frac{1}{2^2}\right)\left(1 - \frac{1}{3^2}\right)\left(1 - \frac{1}{4^2}\right)\left(1 - \frac{1}{5^2}\right)\left(1 - \frac{1}{6^2}\right)\left(1 - \frac{1}{7^2}\right) = \frac{7}{12} \cdot \frac{48}{49} = \frac{4}{7}$$

$$\left(1 - \frac{1}{2^2}\right)\left(1 - \frac{1}{3^2}\right)\left(1 - \frac{1}{4^2}\right)\left(1 - \frac{1}{5^2}\right)\left(1 - \frac{1}{6^2}\right)\left(1 - \frac{1}{7^2}\right)\left(1 - \frac{1}{8^2}\right) = \frac{4}{7} \cdot \frac{63}{64} = \frac{9}{16}$$

After checking the answers, make a table as shown below.

n	2	3	4	5	6	7	8	k	
$P(n)$*	$\frac{3}{4}$	$\frac{2}{3}$	$\frac{5}{8}$	$\frac{3}{5}$	$\frac{7}{12}$	$\frac{4}{7}$	$\frac{9}{16}$		Reduced form
$P(n)$	$\frac{3}{4}$	$\frac{4}{6}$	$\frac{5}{8}$	$\frac{6}{10}$	$\frac{7}{12}$	$\frac{8}{14}$	$\frac{9}{16}$	$\frac{k+1}{2k}$	Unreduced form

*$\left(1 - \frac{1}{2^2}\right)\left(1 - \frac{1}{3^2}\right) \cdots \left(1 - \frac{1}{n^2}\right) = P(n)$, or the product of the first n terms.

Some patterns with fractions
 Pattern 3.

$1 - \frac{1}{2^2} =$

$\left(1 - \frac{1}{2^2}\right)\left(1 - \frac{1}{3^2}\right) =$

$\left(1 - \frac{1}{2^2}\right)\left(1 - \frac{1}{3^2}\right)\left(1 - \frac{1}{4^2}\right) =$

$\left(1 - \frac{1}{2^2}\right)\left(1 - \frac{1}{3^2}\right)\left(1 - \frac{1}{4^2}\right)\left(1 - \frac{1}{5^2}\right) =$

$\left(1 - \frac{1}{2^2}\right)\left(1 - \frac{1}{3^2}\right)\left(1 - \frac{1}{4^2}\right)\left(1 - \frac{1}{5^2}\right)\left(1 - \frac{1}{6^2}\right) =$

$\left(1 - \frac{1}{2^2}\right)\left(1 - \frac{1}{3^2}\right)\left(1 - \frac{1}{4^2}\right)\left(1 - \frac{1}{5^2}\right)\left(1 - \frac{1}{6^2}\right)\left(1 - \frac{1}{7^2}\right) =$

$\left(1 - \frac{1}{2^2}\right)\left(1 - \frac{1}{3^2}\right)\left(1 - \frac{1}{4^2}\right)\left(1 - \frac{1}{5^2}\right)\left(1 - \frac{1}{6^2}\right)\left(1 - \frac{1}{7^2}\right)\left(1 - \frac{1}{8^2}\right) =$

TRANSPARENCY 11-31

TRANSPARENCY 11-31

Answers to Set I

1. a) $\frac{1}{5}$ c) Not possible
   b) $\frac{1}{3}$ d) $\frac{3}{x}$
   c) $\frac{2}{15}$ e) Not possible
        f) $\frac{x+7}{x-7}$

2. a) False   5. a) $\frac{4}{9}$
   b) True   b) $\frac{10}{4}$
   c) True   c) $\frac{7}{11}$
   d) False   d) $\frac{43}{94}$
   e) True
   f) True

3. a) -5 6. a) $\frac{4}{5}$
   b) None   b) $\frac{1}{2}$
   c) 0 and 3   c) $\frac{2x-1}{x^2}$

4. a) $\frac{1}{6}$ d) $\frac{x^2-y^2}{xy}$
   b) $\frac{x^2}{4}$

TRANSPARENCY 11-32

---

   e) $\frac{5x^2+x}{5}$ 9/10. a) $x^2-x+\frac{6}{25}$
   f) $\frac{4x^2+1}{2x+1}$ b) $63+\frac{2}{x}-\frac{1}{x^2}$

7. a) $\frac{x}{8}+\frac{1}{2}$ 11. a) 24
   b) $\frac{3x}{3x+7}-\frac{7}{3x+7}$ b) $\frac{2(x+14)}{x+7}$
   c) $1+\frac{2}{x^2-x}$ c) $\frac{x+y}{3}$
   d) $2x-\frac{2}{x}$ d) $\frac{1}{2x^3}$

8. a) $\frac{x^5}{6}$ e) $3x^4+x$
   b) 1   f) 1
   c) $-\frac{4}{x}$ 12. a) $\frac{x+1}{y^2}$
   d) $\frac{x(x-2)}{2}$ b) x^2+x
   e) $\frac{x^2-y^2}{x^2}$ c) $\frac{4}{7}$
   f) $\frac{1}{4}$ d) $\frac{2x^2+5x}{2x+4}$

TRANSPARENCY 11-33

You may wish to conceal the row showing the unreduced form from your students and see what patterns they can discover. Some of them are shown below.

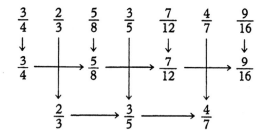

However, the unreduced form of the answers reveals the general pattern. In more advanced courses, this formula would be proved algebraically by induction.

Answers to the Set I exercises of the review lesson are given on Transparencies 11-32 and 11-33.

Chapter **12**

SQUARE ROOTS

Having learned how to work with fractional expressions, the students can now turn attention to the properties of radical expressions.

In the first lesson, square root is defined and the radical sign introduced, as is the fact that the number of square roots that a number has depends on whether it is positive, negative, or zero. That many square roots are irrational is hinted at in this lesson but not discussed in depth until Chapter 14.

A table of squares and approximate square roots is given at the end of the first lesson for use in solving some of the exercises throughout this chapter and the following ones. Although I feel that students should be permitted to use calculators whenever they wish, they may need to be reminded in the course of studying this chapter that a major goal in learning algebra is not merely to obtain answers but to learn techniques for manipulating various types of expressions. There is no reason to learn how to rationalize the denominator of an expression such as $\dfrac{6}{\sqrt{2}}$, for example, if we are merely interested in a decimal approximation of it. It is just as easy to divide 6 by $\sqrt{2}$ on a calculator as it is to multiply $\sqrt{2}$ by 3. It is important, how-

ever, to know that $\dfrac{6}{\sqrt{2}}$ and $3\sqrt{2}$ are equivalent expressions and this is something that we cannot learn from pushing keys on a calculator.

The use of the exponent $\dfrac{1}{2}$ to represent square roots is introduced in the second lesson so that the square-root laws for products and quotients can be obtained from the corresponding power laws. The square-root laws are then applied to expressing square roots of integers, monomials, and fractions in simple radical form in this lesson and the following one.

The basic operations with square roots are discussed in lessons 4, 5, and 6 and the chapter closes with a lesson on radical equations.

Lesson 1
Squares and Square Roots

A puzzle with two squares

Ask your students to draw two squares at the top of a sheet of graph paper as shown on Transparency

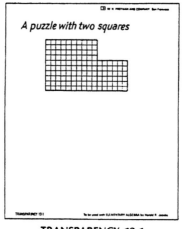

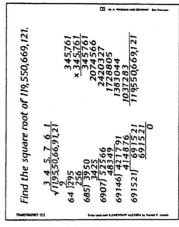

TRANSPARENCY 12-1

TRANSPARENCY 12-2

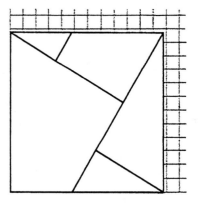

12-1. Everyone should observe that each side of the larger square is 10 units long, that each side of the smaller square is 6 units long, and that the areas of the squares are 100 and 36 square units respectively. Give each student a pair of scissors while the squares are being drawn.

Point out that 100 is called *the square* of 10, whereas 10 is called *a square root* of 100: $100 = 10^2$ and $10 = \sqrt{100}$. Have your students write the corresponding equations for the other square: $36 = 6^2$ and $6 = \sqrt{36}$. Now add the two lines indicated below to the figure on the transparency, asking everyone to copy them and then cut the figure into the five pieces that result.

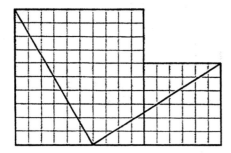

The puzzle is to rearrange the pieces to form a single square. This can be done without turning any of them over. After everyone has had time to try to do this, have a student who has discovered the solution (pictured in the next column) show it on the projector.

What is the area of the square formed? (Although we cannot easily find it by counting the small squares in the figure, we can reason that it is 136 since it was formed from two squares having areas of 100 and 36.) What is the length of its side? ($\sqrt{136}$.) By placing the square on the grid of the graph paper as shown, we see that this number is between 11 and 12.

The rest of the lesson might now be developed in accord with the text.

Lesson 2
Square Roots of Products

Show the problem at the top of Transparency 12-2, keeping the solution covered. *Here is a problem that has an interesting story behind it. It was once given to a ten-year-old boy to solve and the boy figured it out*

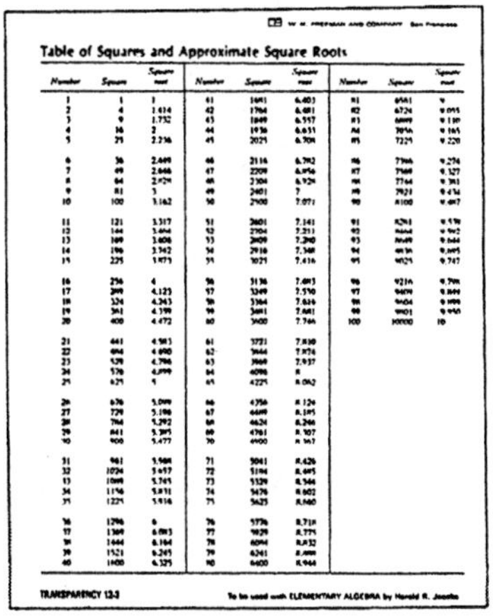

TRANSPARENCY 12-3

The method can be discovered by comparing the processes of addition and multiplication. Have your students make the table shown below, skipping a couple of lines between the headings of two columns and the first row.

Addition	Multiplication
(Leave space here.)	
$x + x = 2x$	$x \cdot x = x^2$
$2x + 2x = 4x$	$x^2 \cdot x^2 = x^4$
$3x + 3x = 6x$	$x^3 \cdot x^3 = x^6$
$4x + 4x = 8x$	$x^4 \cdot x^4 = x^8$
$\vdots$	$\vdots$

Add coefficients and exponents to the first row so that it reads:

$$1x + 1x = 2x \qquad x^1 \cdot x^1 = x^2$$

Then fill in the following entries in the left column, asking your students to copy them and fill in what they think should go in the right column.

$$0x + 0x = 0x$$

$$\tfrac{1}{2}x + \tfrac{1}{2}x = 1x$$

According to the pattern, the entries in the right column should be $x^0 \cdot x^0 = x^0$ *and* $x^{1/2} \cdot x^{1/2} = x^1$. *Do these results make any sense?* (If x is not 0, $x^0 = 1$; $1 \cdot 1 = 1$; so the first entry makes sense if x is not 0.) *The second entry makes sense if* $x^{1/2}$ *means* $\sqrt{x}$. *For this reason, the expression* $x^{1/2}$ *is defined to mean* $\sqrt{x}$ *whenever x is a nonnegative number. A useful result of considering square roots to be powers is that the laws of exponents that we aleady know continue to be true. For example, if we let* $a = \dfrac{1}{2}$ *in the power of a product law,*

$$(xy)^a = x^a y^a$$

we get

$$(xy)^{1/2} = x^{1/2} y^{1/3} \quad \text{or} \quad \sqrt{xy} = \sqrt{x}\sqrt{y}$$

in his head in just 30 seconds. The boy's name was George Bidder and he was a calculating prodigy, a child who possesses extraordinary powers of mental calculation. Bidder learned arithmetic by playing with buttons and marbles and became so fast at it that he soon became famous. Throughout his life, he was able to solve problems in his head in seconds that it took expert mathematicians using pencil and paper a long time to work out.*

Bidder's speed in solving the square-root problem is especially astonishing when you see the procedure written down. Reveal the solution shown on the left side of the transparency. *Although the procedure used is one that we have not learned, how could we check the answer to determine that it is correct?* After a student has explained that the answer could be multiplied by itself to see if the result is the original number, reveal this check shown on the right side of the transparency.

Show Transparency 12-3. *Here is the table of squares and approximate square roots given on page 563 of your book. Although this table falls far short of the twelve-digit number whose square root Bidder figured out, it can be used to find the approximate square roots of many numbers larger than those listed.*

---

*Information on Bidder and other calculating prodigies can be found in *Mathematical Recreations and Essays* by W.W. Rouse Ball and H.S.M. Coxeter (University of Toronto Press, 1974) and in *Mathematical Carnival* by Martin Gardner (Knopf, 1975).

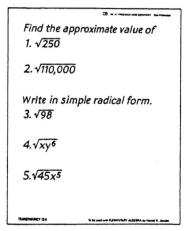

Find the approximate value of

1. $\sqrt{250}$

2. $\sqrt{110{,}000}$

Write in simple radical form.

3. $\sqrt{98}$

4. $\sqrt{xy^6}$

5. $\sqrt{45x^5}$

TRANSPARENCY 12-4   **TRANSPARENCY 12-5**

Remembering that we do not consider negative numbers to have square roots, this pattern is true for all numbers that are not negative. For example, if x = 4 *and* y = 9, *it is easy to verify that* $\sqrt{4 \cdot 9} = \sqrt{4} \cdot \sqrt{9}$. *If* x = 5 *and* y = 10, *it is not as easy to verify that* $\sqrt{5 \cdot 10} = \sqrt{5} \cdot \sqrt{10}$ *because the roots are not integers. Nevertheless, the equation is true. Students with calculators might compare* $\sqrt{50}$ *with* $\sqrt{5} \cdot \sqrt{10}$.

The examples on Transparency 12-4 can be used to develop the rest of the lesson in accord with the text.

Lesson 3
Square Roots of Quotients

Show Transparency 12-5. *Several years ago, a class of students at the Pacific Science Center in Seattle was given the following assignment: design and construct a package for an egg that could be dropped from a height of 200 feet without the egg breaking. After all of the packages had been turned in, they were taken up in a helicopter and dropped on the ground. About half of the packages landed with their eggs unbroken. Among the successful designs were an egg tucked inside a hot-dog bun in a shoe box filled with popcorn and an egg simply wrapped in 25 yards of aluminum foil.*

The time that it takes an object falling near the surface of the earth to hit the ground is given by the formula $t = \sqrt{\dfrac{h}{16}}$, *in which* h *represents the height in feet from which it is dropped and* t *represents the time in seconds. Write this formula at the top of the transparency and ask your students to use it to estimate the time that it took the packages to hit the ground.* $\left(t = \sqrt{\dfrac{200}{16}} = \sqrt{12.5} \right.$. *From the table on page 563, we see that the time is a number between 3.464 and 3.606 seconds.* $\Big)$

We can find out what the number is by observing that the problem is to find the square root of a quotient. From yesterday's lesson, we learned that the square root of a product can be written as the product of the square roots of its factors: $\sqrt{xy} = \sqrt{x} \cdot \sqrt{y}$. *In a similar fashion, we can write* $\sqrt{\dfrac{x}{y}} = \dfrac{\sqrt{x}}{\sqrt{y}}$. *This follows from the power of a quotient law. Letting* $a = \dfrac{1}{2}$ *in* $\left(\dfrac{x}{y}\right)^a = \dfrac{x^a}{y^a}$, *we get* $\left(\dfrac{x}{y}\right)^{1/2} = \dfrac{x^{1/2}}{y^{1/2}}$, *which is equivalent to saying that* $\sqrt{\dfrac{x}{y}} = \dfrac{\sqrt{x}}{\sqrt{y}}$.

Applying this principle to the expression for the time in which the packages fell, we can write $\sqrt{\dfrac{200}{16}} = \dfrac{\sqrt{200}}{\sqrt{16}} = \dfrac{10\sqrt{2}}{4} = 2.5\sqrt{2} \approx 2.5(1.414) = 3.535$. *The packages took a little more than 3.5 seconds to hit the ground.*

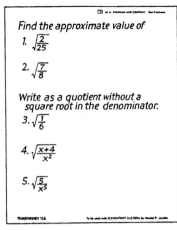

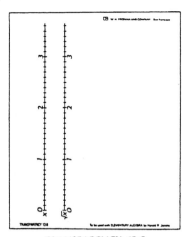
Additional examples of how to apply the "square root of a quotient" law are given on Transparency 12-6.

Transparency 12-7 is for use in discussing the Set IV exercise of Lesson 2.

Lesson 4
Adding and Subtracting Square Roots

A square-root race

Ask your students to turn a sheet of graph paper sideways and draw a pair of number lines spaced one inch apart as shown on Transparency 12-8. The lines should be labeled x and $\sqrt{x}$ respectively. (You may prefer to hand out duplicated copies of the transparency instead.)

Draw arrows from the numbers 0, 1, 2, and 3 on the x-line to their square roots, 0, 1, 1.41, and 1.73, on the $\sqrt{x}$-line, as shown in the figure below. Refer to the table on page 563.

Now we will locate the square roots of numbers between those shown. Although a calculator could be used to find them, working out the answers without a calculator will be a good review of yesterday's lesson.

To get everyone started, show how to find $\sqrt{1\frac{1}{2}}$.

$$\sqrt{1\frac{1}{2}} = \sqrt{\frac{3}{2}} = \sqrt{\frac{6}{4}} = \frac{\sqrt{6}}{2} \approx \frac{2.449}{2} \approx 1.22$$

Now give your students time to find $\sqrt{2\frac{1}{2}}$, $\sqrt{\frac{1}{2}}$, $\sqrt{\frac{1}{4}}$, $\sqrt{\frac{3}{4}}$, and $\sqrt{\frac{1}{10}}$. The results are shown in the figure above.

It is fun to think of the two lines as tracks in a race between a number and its square root. The race begins at 0, given that we do not consider negative numbers to have square roots. At the start, x and $\sqrt{x}$ are both at 0. From the arrows, we see that $\sqrt{x}$ immediately pulls ahead but x catches up at 1 and soon leaves $\sqrt{x}$ far behind.

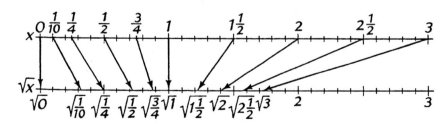

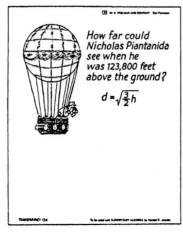

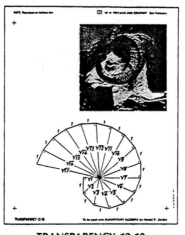

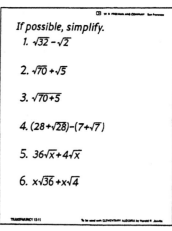

TRANSPARENCY 12-9　　**TRANSPARENCY 12-10**　　**TRANSPARENCY 12-11**

To conclude this exercise, you might ask the following questions. *What numbers are equal to their square roots? (0 and 1.) What numbers are larger than their square roots? ($x > \sqrt{x}$ if $x > 1$.) What numbers are smaller than their square roots? ($x < \sqrt{x}$ if $0 < x < 1$.)*

Transparency 12-9 is for use in discussing the Set IV exercise of Lesson 3.

Transparency 12-10 and its overlay might be used to introduce the new lesson in accord with the text. If so, show by tracing end to end the lengths representing the square roots of two numbers and comparing the result to the length of the square root of the sum of the numbers that $\sqrt{x} + \sqrt{y} \neq \sqrt{x + y}$. Examples for use in developing the rest of the lesson are given on Transparency 12-11.

Lesson 5
Multiplying Square Roots

A puzzle about a leaking can*

Show Transparency 12-12. *Suppose that a large can has three identical holes punctured in it in a vertical line along one side. The holes are spaced one-quarter,*

one-half, and three-quarters of the way down from the top of the can. If the can is kept full of water by running water in at the top, the water will spurt out of each hole in a different path. What do you suppose the paths would look like?

After giving everyone time to think about this, you might call on volunteers to draw sketches of their guesses on the chalkboard. The two most popular guesses are shown below.

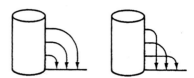

Add the overlay to the transparency and have your students make a sketch of the can and coordinate axes as shown. *The formula below the figure can be used to find the distance that the water spurts out from the side of the can, x, from the height of the hole, y. Can you think of any way to simplify it? (We can write $\sqrt{4y}$ in simple radical form to get $x = 2\sqrt{y} \cdot \sqrt{10 - y}$.) Before using the formula, we can use another fact that we know to write it in a more convenient form. Because $\sqrt{ab} = \sqrt{a} \cdot \sqrt{b}$, we know that $\sqrt{a} \cdot \sqrt{b} = \sqrt{ab}$. So $\sqrt{y} \cdot \sqrt{10 - y} = \sqrt{y(10 - y)}$ and the formula becomes $x = 2\sqrt{y(10 - y)}$. Use it to find x if $y = 2\frac{1}{2}$.*
$\left(x = 2\sqrt{2\frac{1}{2}\left(10 - 2\frac{1}{2}\right)} = 2\sqrt{\frac{5}{2} \cdot \frac{15}{2}} =\right.$

---

*Based on a question in *Millergrams II* by Julius Sumner Miller (Doubleday, 1970), p. 44.

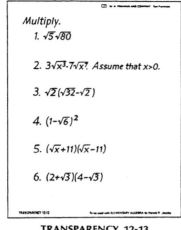

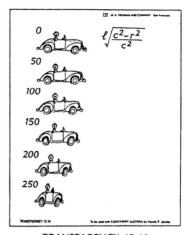

The three transparencies show:

Transparency 12-12: A puzzle about a leaking can

Transparency 12-13: Multiply.
1. $\sqrt{5}\sqrt{80}$
2. $3\sqrt{x^3}\cdot 7\sqrt{x^7}$. Assume that $x>0$.
3. $\sqrt{2}(\sqrt{32}-\sqrt{2})$
4. $(1-\sqrt{6})^2$
5. $(\sqrt{x}+11)(\sqrt{x}-11)$
6. $(2+\sqrt{3})(4-\sqrt{3})$

Transparency 12-14: $\ell\sqrt{\dfrac{c^2-r^2}{c^2}}$

$2\sqrt{\dfrac{75}{4}} = \sqrt{75} \approx 8.7.\Big)$ After checking this, give

your students time to find x if $y = 5$ and if $y = 7\frac{1}{2}$.

$\Big($If $y = 5$, $x = 10$; if $y = 7\frac{1}{2}$, $x \approx 8.7.\Big)$

Sketch the paths from the three holes, as shown below.

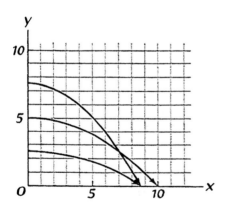

Note that the formula tells us only the points on the x-axis that the paths of water hit. It does not give us any information about any of the other points on each path.

Transparency 12-13 contains examples for possible use in presenting the new lesson.

Lesson 6
Dividing Square Roots

Show Transparency 12-14. *According to Einstein's theory of relativity, the fastest possible speed in the universe is the speed of light. If a moving object is made to go faster and faster—so fast that its speed approaches that of light—several strange things happen. Its length in the direction of motion appears to shrink, its mass increases, and time slows down for a clock moving with the object.*

An expression for the length of an object when it is in motion is $\ell\sqrt{\dfrac{c^2-r^2}{c^2}}$. *In this expression,* ℓ *represents its length when stationary,* c *represents the speed of light, and* r *represents the speed at which the object is moving.*

Light travels at a speed of 300 million meters per second, or, because a million meters is a megameter, 300 megameters per second. The drawings show a car at rest and what would happen if it could be driven 50, 100, 150, 200, and 250 megameters per second.

Suppose, for example, that the car is 5 meters long when it is standing still. What would its length become if it were going 150 megameters per second?

$\Big(\ell\sqrt{\dfrac{c^2-r^2}{c^2}} = 5\sqrt{\dfrac{300^2 - 150^2}{300^2}} = 5\sqrt{\dfrac{67,500}{90,000}} =$

$5\sqrt{\dfrac{3}{4}} = \dfrac{5}{2}\sqrt{3} \approx 2.5(1.732) = 4.33.$ *The   car would become about 4.33 meters long.* $)$

What would the length of the car become if it could go 299 megameters per second? $\left(\ell \sqrt{\dfrac{c^2 - r^2}{c^2}} = \right.$

$5 \sqrt{\dfrac{300^2 - 299^2}{300^2}} = 5 \sqrt{\dfrac{599}{300^2}}$. *Rounding 599 to*

600 and using the fact that $\sqrt{\dfrac{a}{b}} = \dfrac{\sqrt{a}}{\sqrt{b}}$, $5 \sqrt{\dfrac{599}{300^2}} \approx$

$5 \dfrac{\sqrt{600}}{300} = \dfrac{50\sqrt{6}}{300} = \dfrac{\sqrt{6}}{6} \approx \dfrac{2.449}{6} \approx 0.41$. *The car's length would shrink to about 0.41 meter or 41 centimeters!* $)$

What does the expression imply about what the length of the car would become if it could travel at the speed of light? (It would shrink to nothing.)

Transparency 12-15 is for use in discussing the Set IV exercise of Lesson 5.

Transparency 12-16 and its overlay might be used to introduce the new lesson in accord with the text. Example exercises are given on Transparency 12-17.

Lesson 7
Radical Equations

Show Transparency 12-18. *The method discovered by this scientist to slow down the speed of sound apparently works very well because it took his announcement so long to cross the room that the other man got tired of waiting to hear it. Although such a discovery might seem impossible, there is a way in which the speed of sound can be changed. It depends on the temperature.*

Show Transparency 12-19. *This formula shows how the speed of sound changes with the temperature. In it,* t *represents the temperature in degrees Celsius and* v *represents the speed of sound in meters per second. What is the speed of sound when the temper-*ature *is 0 degrees Celsius? (331 meters per second.)*

The colder the temperature, the slower sound travels. At what temperature does the speed of sound become half of what it is at $0°C$*? To find out, we can write the equation* $\dfrac{1}{2}(331) = 331 \sqrt{\dfrac{t}{273} + 1}$. *This is called a* radical equation *because the variable appears under a radical sign. We will learn how to solve radical equations in today's lesson.*

First, we might divide both sides of the equation by 331 to get $\dfrac{1}{2} = \sqrt{\dfrac{t}{273} + 1}$. *Notice that the square root is by itself on one side of the equation. We can now eliminate it by squaring both sides of the equation, getting* $\dfrac{1}{4} = \dfrac{t}{273} + 1$. *Subtracting 1 from each side and multiplying both sides of the resulting equation by 273, we get* $-\dfrac{3}{4} = \dfrac{t}{273}$, $t = 273\left(-\dfrac{3}{4}\right) \approx$ *-205. The temperature would have to be lowered to 205 degrees below zero on the Celsius scale.*

At what temperature would sound come to a complete stop? Having worked out the solution to the preceding question with you, your students should now try to solve this one on their own. Applying the same procedure to the equation $0 = 331 \sqrt{\dfrac{t}{273} + 1}$, *we get* $t = -273$. *The temperature would be -273 degrees Celsius (which is "absolute zero," the lowest possible temperature.)*

Further examples that might be used to present the new lesson are given on Transparency 12-20.

Transparency 12-21 is for use in discussing the Set IV exercise of Lesson 6. The "golden ratio" is the only positive number that becomes its own reciprocal by subtracting 1. More information on it is included in Martin Gardner's "Mathematical Games" column in the August 1959 issue of *Scientific American*, reprinted in *The Second Scientific American Book of Mathematical Puzzles and Diversions* (Simon and Schuster, 1961), and in *The Golden Mean*, by Charles F. Linn (Doubleday, 1974).

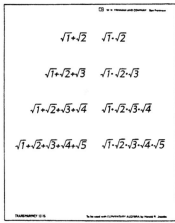

TRANSPARENCY 12-15

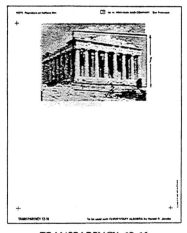

TRANSPARENCY 12-16

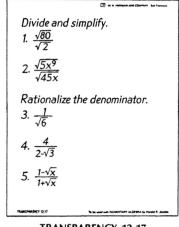

Divide and simplify.

1. $\dfrac{\sqrt{80}}{\sqrt{2}}$

2. $\dfrac{\sqrt{5x^9}}{\sqrt{45x}}$

Rationalize the denominator.

3. $\dfrac{1}{\sqrt{6}}$

4. $\dfrac{4}{2-\sqrt{3}}$

5. $\dfrac{1-\sqrt{x}}{1+\sqrt{x}}$

TRANSPARENCY 12-17

TRANSPARENCY 12-18

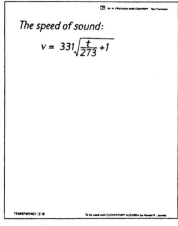

The speed of sound:

$$v = 331\sqrt{\dfrac{t}{273}+1}$$

TRANSPARENCY 12-19

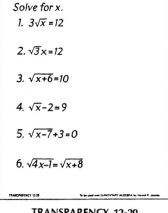

Solve for x.

1. $3\sqrt{x}=12$

2. $\sqrt{3}x=12$

3. $\sqrt{x+6}=10$

4. $\sqrt{x}-2=9$

5. $\sqrt{x-7}+3=0$

6. $\sqrt{4x-1}=\sqrt{x+8}$

TRANSPARENCY 12-20

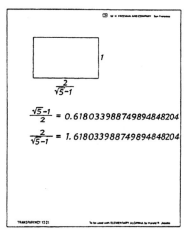

$\dfrac{\sqrt{5}-1}{2} = 0.618033988749894848204$

$\dfrac{2}{\sqrt{5}-1} = 1.618033988749894848204$

TRANSPARENCY 12-21

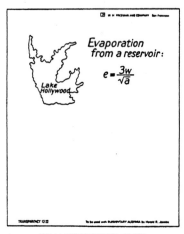

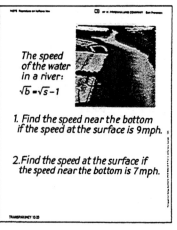

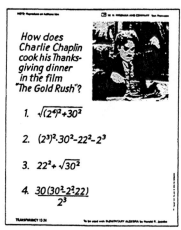

TRANSPARENCY 12-22   TRANSPARENCY 12-23   TRANSPARENCY 12-24

Review of Chapter 12

First day

Show either Transparency 12-22, which shows Lake Hollywood, a reservoir in the hills above Hollywood, California, or a map showing a reservoir in your own area. *The rate at which water evaporates from a reservoir depends on a variety of things, including the size of the reservoir and the temperature and motion of the air above it. A typical formula is* $e = \dfrac{3w}{\sqrt{a}}$*, in which e represents the evaporation in millimeters per day, w represents the average speed of the wind in meters per second, and a represents the surface area of the water in the reservoir in square meters.*

Find the rate of evaporation given by this formula if the average wind speed is 8 meters per second and the area of the water is 640,000 square meters.

$\left(e = \dfrac{3(8)}{\sqrt{640,000}} = \dfrac{24}{800} = 0.03. \text{ The rate of evap-}\right.$

oration would be 0.03 millimeters per day. $\Big)$

Find the rate of evaporation if the average wind speed is doubled but the area of the water remains the same. (It would be 0.06 millimeter per day.) Find the rate of evaporation if the average wind speed remains at 8 meters per second but the area of the wa-

ter is doubled. $\left(e = \dfrac{3(8)}{\sqrt{2(640,000)}} = \dfrac{24}{800\sqrt{2}} = \right.$

$\dfrac{24\sqrt{2}}{1600} \approx 0.02.$ *It would be about 0.02 millimeter*

per day. $\Big)$

Loss of water from a reservoir by evaporation is sometimes decreased by putting over its surface a protective layer of some chemical substance that is one molecule thick.

Transparency 12-23 is for use in discussing the Set IV exercise of Lesson 7.

Second day: Another calculator riddle*

Show Transparency 12-24, but keep the shoe in the photograph covered until the problem has been solved.

There is a famous scene in a Charles Chaplin film, The Gold Rush (1925), in which Chaplin, stranded in a mountain cabin without any food, cooks a strange Thanksgiving dinner. To find out how he made his dinner, solve the four problems, turning your calculator upside down to read the answers to the problems as words.

---

*The riddle is by Wallace Judd and appears in his book *Games Calculators Play* (Warner Books, 1975).

Answers to Set I

1. a) 6 and -6
   b) 1 and -1
   c) None
   d) $\frac{1}{2}$ and $-\frac{1}{2}$

2. a) 42
   b) 42
   c) 20
   d) 192
   e) 8
   f) 20

3. a) =
   b) =
   c) <
   d) =
   e) >
   f) =

4. a) $5\sqrt{2}$
   b) $6\sqrt{3}$
   c) $3\sqrt{89}$
   d) $10\sqrt{11}$

5. a) $2\sqrt{x}$
   b) x^2
   c) $x^3\sqrt{6}$
   d) $5x^2\sqrt{x}$

6. a) $\frac{\sqrt{5}}{4}$
   b) $\frac{\sqrt{6}}{10}$
   c) $\frac{\sqrt{11x}}{11}$
   d) $\frac{\sqrt{x}}{x^2}$

7. a) 6
   b) 4.3

c) 9.9
d) 9

8. a) $10\sqrt{3}$
   b) $3\sqrt{7}$
   c) $7\sqrt{2}$
   d) Not possible
   e) $\sqrt{5x}$
   f) Not possible

9. a) $3+6\sqrt{3}$
   b) $60+4\sqrt{6}$
   c) $6\sqrt{5}$

10. a) 50
    b) $27+10\sqrt{2}$
    c) $9-4\sqrt{5}$
    d) 27

TRANSPARENCY 12-25

11. a) 6
    b) 24
    c) $35+5\sqrt{7}$
    d) x^3
    e) $x+5\sqrt{x}$
    f) $14+10\sqrt{3}$
    g) $1-x$
    h) $x+3\sqrt{x}-10$

12. a) 3
    b) $10\sqrt{6}$
    c) $12\sqrt{6}+12$
    d) $\frac{\sqrt{10}-2}{3}$

13. a) 1.58
    b) 0.63
    c) 12.96

14. a) $4-\sqrt{2}$
    b) $\sqrt{5}$
    c) 63
    d) 434
    e) 196
    f) 4
    g) No solution
    h) No solution

TRANSPARENCY 12-26

The solutions are: (1) 34; (2) 57108; (3) 514; and (4) 3045. Read upside down, they form the sentence HE BOILS HIS SHOE.

The answers to the Set I exercises of the review lesson are given on Transparencies 12-25 and 12-26.

Chapter 13
QUADRATIC EQUATIONS

The solution of quadratic equations is often the last topic studied in an elementary algebra course. This seems unfortunate to me because students who are introduced to the quadratic formula in the last lessons are unlikely to be able to remember it or to have confidence in using it when it is needed in other courses.

The first three lessons set the stage for the rest of the chapter by discussing polynomial equations and functions in general. The first lesson explains how polynomial equations are classified and the second lesson reviews the graphing of linear and quadratic functions and extends it to polynomial functions of higher degree. The content of these two lessons is then applied in Lesson 3 to showing how the solutions of a polynomial equation can be read from the graph of the corresponding polynomial function.

The next four lessons deal with the solution of quadratic equations, including solving by factoring, by taking square roots, by completing the square, and by using the quadratic formula. The area diagrams introduced earlier in the course for illustrating polynomial multiplication and division are used to good advantage in making the derivation of the quadratic formula meaningful. An additional lesson

on the discriminant and the possible relationships of the graph of a second-degree function to the x-axis is intended to enhance the student's understanding of the nature of the solutions of a quadratic equation.

The final lesson of the chapter presents historical background about the solution of polynomial equations of higher degree and enables the student to apply techniques for solving quadratic equations to solving some simple cubic and quartic equations.

Lesson 1
Polynomial Equations

Show Transparency 13-1. *This is a page from an algebra book written in India in the twelfth century.* The man who wrote it, Bhaskara, was a famous mathematician of the time.* Add the overlay. *Translated into English, these lines say "unknown squared 18 and unknown 0 and units 0" and "unknown*

---

*Bhaskara's *Bija Ganita.* See D. E. Smith's *History of Mathematics* (Dover, 1958, volume 1, pp. 275-282, and volume 2, pp. 425-426.

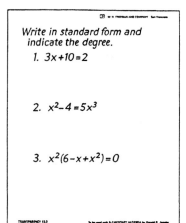

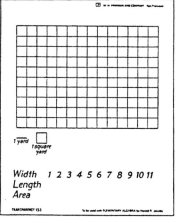

TRANSPARENCY 13-2   TRANSPARENCY 13-3

squared 16 and unknown 9 and units 18." Translated into twentieth-century symbols, they become the equation $18x^2 + 0x + 0 = 16x^2 + 9x + 18$, *or, more briefly, the equation* $18x^2 = 16x^2 + 9x + 18$. *This equation is called a* polynomial equation *because each side of it is a polynomial.* Show how to write it in standard form, explaining why it is called a second-degree equation.

From this chapter, we will learn a variety of ways for solving equations of this type. We will return to this equation in future lessons to see how it can be solved by some of them.

Additional examples of how to write polynomial equations in standard form are given on Transparency 13-2.

Lesson 2
Polynomial Functions

An experiment:
How big is the sheep pen?

Give each student a 12-inch pipe-cleaner and a duplicated copy of Transparency 13-3.* *Suppose*

---

*This exercise could be done without these materials but manipulating them makes it much easier to understand.

that a farmer has 24 yards of fencing with which to make a rectangular pen to hold sheep. How big will the pen be? To find out, we will use the pipe-cleaner to represent the fencing and the grid to represent some land.

First, have each student fold the pipe-cleaner in half and place the result on the grid to confirm that it represents 24 yards of fencing. (Illustrate this on the overhead projector.) Now fold the pipe-cleaner into a 2 × 10 rectangle and point out that, if the pen is 2 yards wide and 10 yards long, it contains 20 square yards of land. Write these numbers in the appropriate column of the table at the bottom of the transparency.

What would happen to the area of the pen if its width were changed to some other number? Would it also change, or would it remain the same? Form enough rectangles of other dimensions so that you can fill in the rest of the table.

Width	1	2	3	4	5	6	7	8	9	10	11
Length	11	10	9	8	7	6	5	4	3	2	1
Area	11	20	27	32	35	36	35	32	27	20	11

When your students have finished doing this, they will see that the area of the pen does depend on the width and that it seems to be largest when

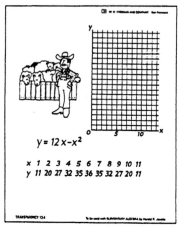

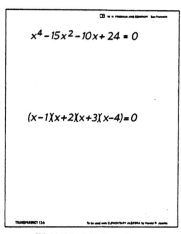

$$x^4 - 15x^2 - 10x + 24 = 0$$

$$(x-1)(x+2)(x+3)(x-4)=0$$

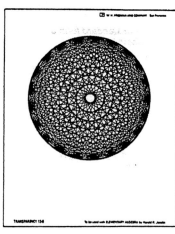

| TRANSPARENCY 13-4 | TRANSPARENCY 13-5 | TRANSPARENCY 13-6 |

the width is 6. *The area is a function of the width. Can you write a formula for this function, letting* x *represent the width and* y *represent the area?* [y = x(12 − x).] *Using the distributive rule, we can rewrite the right side of this formula to get* y = 12x − x². *Because the equation that results says that "y is equal to a second-degree polynomial in* x," it is called a second-degree polynomial function.

What does the graph of this sort of function look like? Show Transparency 13-4. *To find out, draw a pair of coordinate axes as shown and plot the points corresponding to the table.* The y-axis has been left unnumbered so that your students can help decide on a scale that will allow room for all of the points in the table.

After this has been done, have your students draw a smooth curve through the points and note that the graph gives a more complete picture of the function than the table did. *We can, for example, use the graph to estimate the areas of some pens whose widths are not integers.* Illustrate this with a couple of examples such as x = 1.5 and x = 8.5.

The symmetry of the graph about the line x = 6 *confirms the idea that the pen for which the width is 6 has the greatest area.*

The Set IV exercise of Lesson 1 is intended to permit the student to discover the zero-product property and its application to solving polynomial equations. Transparency 13-5 is for use in discussing this exercise. Contrast the simplicity of checking the solution x = −3 in the equation in factored form with the computation required to check it in the original equation.

Lesson 3
Solving Polynomial Equations by Graphing

Show Transparency 13-6. *This figure consists of a circle, a set of points spaced at equal intervals around it, and all of the lines connecting these points. Such a design is sometimes made by drawing a circle on a block of wood, pounding in nails at the positions of the points, and winding string or thread in every possible way around the nails to produce the lines. How many windings would be required to produce the complete pattern?* It depends on the number of points.

Show the upper figure on Transparency 13-7, keeping the lower figure covered. *Here is a circle with just five points spaced around it. How many lines can be drawn from each point?* (Four.) Add Overlay A to show the four lines from point A. *Because there are five points and four lines from each*

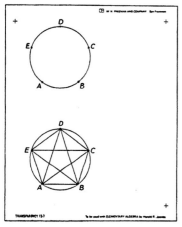

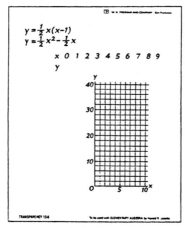

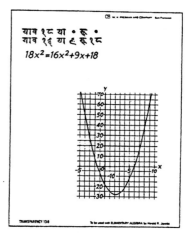

TRANSPARENCY 13-7   **TRANSPARENCY 13-8**   **TRANSPARENCY 13-9**

point, it would seem that there should be $5 \cdot 4 = 20$ lines altogether. This is not correct, however. Reveal the lower figure on the transparency. There are only ten. To see why we got twice this number, look at the four lines from point B. Add Overlay B, leaving Overlay A in place. What do you notice? (One of the lines from B is the same as one of the lines from A.) This reveals that our method counted one of the lines twice. It counted every line twice, in fact, which is why we got twice the correct answer.

Show Transparency 13-6 again. The figure we started with contains 21 points. How many lines are there from each of its points? (20.) How many lines are there altogether? $\left[\frac{1}{2}21(20) = 210.\right]$

Suppose that we represent the number of points spaced around a circle by x and the total number of lines by y. On the basis of the examples we have just considered, can you write a formula for y as a function of x? $\left[y = \frac{1}{2}x(x - 1).\right]$

Show that this is a polynomial function by using the distributive rule to rewrite the right side. $\left(y = \frac{1}{2}x^2 - \frac{1}{2}x.\right)$ What do you think the graph of this function would look like? Would it be straight or curved?

Show Transparency 13-8, having your students draw and label axes as shown, complete the table, and graph the function from the points contained in it. (Note that the factored form of the formula is easier to evaluate than the polynomial form.) After this has been done, point out that the curve dips below the x-axis between 0 and 1.

For what values of x is $\frac{1}{2}x^2 - \frac{1}{2}x = 0$? In other words, for what numbers of points on a circle would there be no lines joining them? The answer to this question can be seen by observing where the graph of $y = \frac{1}{2}x^2 - \frac{1}{2}x$ intersects the x-axis. It intersects the x-axis at 0 and 1. No lines can be drawn if there are no points on the circle and no lines can be drawn if there is only 1 point on the circle. One way, then, to find the solutions to a polynomial equation is to look at the graph of the corresponding polynomial function.

Here is another example. Show Transparency 13-9, keeping the graph covered. We will apply the method to finding the solutions of the equation from the twelfth-century algebra book that we saw in the first lesson: $18x^2 = 16x^2 + 9x + 18$. Show the students how to do this, showing the graph rather than spending time graphing the corresponding polynomial function. The solutions are –1.5 and 6.

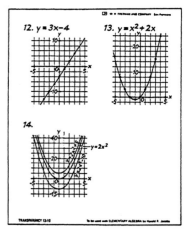

Transparencies 13-10, 13-11, 13-12, and 13-13 are for use in discussing the Set III and Set IV exercises of Lesson 2.

Lesson 4
Solving Quadratic Equations by Factoring

Show Transparency 13-14. *In the course of a day, the temperature usually goes up and then down again. Suppose that the temperature during a certain summer afternoon is given by the quadratic function,* $y = 82 + 6x - x^2$, *in which* y *is the temperature in degrees Fahrenheit and* x *is the time in the afternoon.*

Use this formula to find the Fahrenheit temperature at one o'clock. (87°.) *What is the temperature at seven o'clock?* (75°.) *We have found the temperatures at some given times. Now let's turn the problem around and try to find the time for a given temperature. For example, is there any time when the temperature is 90°? To find out, we have to solve the equation* $90 = 82 + 6x - x^2$ *for* x. *From yesterday's lesson, we learned one way to do this. Write the equation in standard form, graph the corresponding quadratic function, and look at the places in which it intersects the x-axis. Today we will learn how to solve this equation without drawing a graph. First, write the equation in standard form.* ($x^2 - 6x + 8 = 0$.) *Next, factor the polynomial on the left side.*

[$(x - 2)(x - 4) = 0$.] *We can now apply a useful fact: in order for a product to be equal to zero, at least one of its factors must be equal to zero. So either* $x - 2 = 0$ *or* $x - 4 = 0$. *The first of these equations is true if* $x = 2$ *and the second is true if* $x = 4$. *It looks as if the temperature hits 90° at two different times in the afternoon: two o'clock and four o'clock. Use the original equation to confirm that if* $x = 2$, $y = 90$ *and if* $x = 4$, $y = 90$.

As a second example of how to solve a quadratic equation by factoring, you might show Transparency 13-15 and solve the twelfth-century equation by this method.

Transparencies 13-16, 13-17, and 13-18 are for use in discussing the Set III and Set IV exercises of Lesson 3.

Lesson 5
Solving Quadratic Equations by Taking Square Roots

Show Transparency 13-19. *This photograph, taken on March 16, 1926, shows Robert Goddard, the father of modern rocketry, standing beside the first liquid-fuel rocket. The rocket was launched by means of a blowtorch and reached an altitude of 41 feet.*

Within ten years, Goddard succeeded in building a rocket that traveled faster than the speed of sound. In

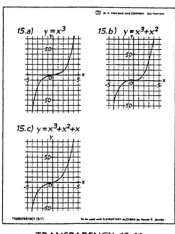

15.a) $y = x^3$ 15.b) $y = x^3 + x^2$

15.c) $y = x^3 + x^2 + x$

TRANSPARENCY 13-11

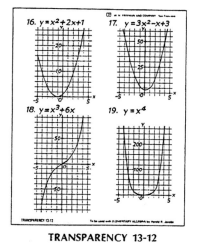

16. $y = x^2 + 2x + 1$ 17. $y = 3x^2 - x + 3$

18. $y = x^3 + 6x$ 19. $y = x^4$

TRANSPARENCY 13-12

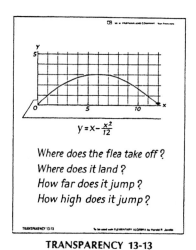

$y = x - \dfrac{x^2}{12}$

Where does the flea take off?
Where does it land?
How far does it jump?
How high does it jump?

TRANSPARENCY 13-13

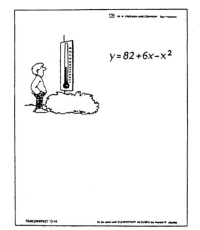

$y = 82 + 6x - x^2$

TRANSPARENCY 13-14

याव १८ या ० रू ०
याव १६ या ૯ रू १८
$18x^2 = 16x^2 + 9x + 18$

TRANSPARENCY 13-15

10.a) $y = 3x + 12$ 10.b) $y = 2x^2 - 7x$

10.c) $y = x^3 + 8$ 10.d) $y = x^4 + x^2 - 90$

TRANSPARENCY 13-16

12.a) $y = x^2 - 4x - 2$ 12.b) $y = x^2 + x - 5$

12.c) $y = x^2 + 2x + 8$ 12.d) $y = 2x^2 - 12x + 17$

TRANSPARENCY 13-17

$y = x^2 - 2x + 5$

$x^2 - 2x + 5 = 0$ has no solutions.
$x^2 - 2x + 5 = 4$ has one solution.
$x^2 - 2x + 5 = 8$ has two solutions.

TRANSPARENCY 13-18

TRANSPARENCY 13-19

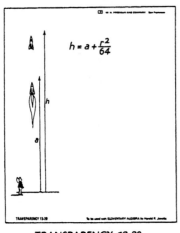

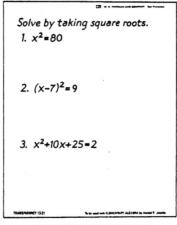

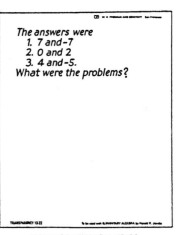

TRANSPARENCY 13-20   TRANSPARENCY 13-21   TRANSPARENCY 13-22

1960, the United States government paid his estate one million dollars in appreciation for the use of his discoveries.

Show Transparency 13-20, *The altitude that a rocket can reach is given by the formula* $h = a + \dfrac{r^2}{64}$, *in which* h *represents the altitude in feet,* a *represents the rocket's height at "burn-out" (the moment at which its fuel is used up), and* r *represents its speed at burn-out in feet per second.*

How high would a rocket go if its height and speed at burn-out are 1,250 feet and 240 feet per second respectively? $\left(h = 1,250 + \dfrac{240^2}{64} = 2,150.\ It\ would\ reach\ an\ altitude\ of\ 2,150\ feet. \right)$

How fast would a rocket have to be traveling at burn-out in order to reach an altitude of 10,000 feet if its height at burn-out is 7,500 feet? To answer this question, we have to solve the equation $10,000 = 7,500 + \dfrac{r^2}{64}$. *What type of equation is this? (A second-degree, or quadratic, equation.) We could solve this equation by either graphing or factoring, but there is another method that is more convenient. It consists of taking square roots. Guide your students through the solution of the problem using this method. Additional examples for illustrating the solution of quadratic equations by tak-*

ing square roots are given on Transparency 13-21.

Transparency 13-22 is for use in discussing the Set IV exercise of Lesson 4.

Lesson 6
Completing the Square

Before class, you might put the National Geographic Society's "Portrait U.S.A.," the first color photomosaic of the 48 contiguous United States, on the front wall. It was included in the July 1976 issue of the *National Geographic* and can be purchased from the National Geographic Society, Washington, D.C. 20036.

This picture is a composite of photographs taken by the Landsat satellites in orbit at an altitude of 570 miles above the earth. According to Tony Rossi, head of General Electric's Photographic Engineering Laboratory, it would not have been practical to produce such a picture without the use of satellites. Aerial photography from an altitude of 12 miles would have required more than 28,000 photographs, whereas "Portrait U.S.A." was put together from only 569 satellite images.

The amount of land that a satellite can photograph depends on its altitude. Show Transparency 13-23. A formula by which it can be determined is

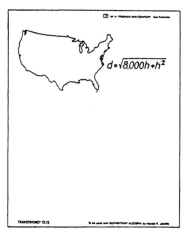

TRANSPARENCY 13-23

$d = \sqrt{8,000h + h^2}$, *in which* d *represents the distance in miles that can be seen in each direction and* h *represents the altitude in miles of the satellite.* If you have a calculator, use it to find* d *if* h = 570, *Landsat's altitude.* [$\sqrt{8,000(570) + (570)^2} \approx$ 2,210.] *Because the United States extends no more than about 1,500 miles in each direction from its geographic center, Lebanon, Kansas, the entire country can be seen from Landsat when it is over this spot.*

Suppose that Landsat were directly over Los Angeles. How high would it have to be in order to see all the way across the country to New York City, a distance of about 3,000 miles? To answer this question, we need to find h *if* d = 3,000. *Substituting 3,000 for* d *in the formula, we write* 3,000 = $\sqrt{8,000h + h^2}$. *What type of equation is this? (A radical equation.) Squaring both sides produces the quadratic equation* 9,000,000 = 8,000h + h^2, *or* h^2 + 8,000h = 9,000,000. *We can solve this equation by the square root method that we learned yesterday by first performing a little trick called "completing the square."*

First, we picture the left side of the equation, h^2 + 8,000h, *as a figure consisting of a square of*

area h^2 *and two rectangles each having an area of* 4,000h.

h^2	4,000h
4,000h	

9,000,000

	h	4,000
h	h^2	4,000h
4,000	4,000h	16,000,000

9,000,000 + 16,000,000 = 25,000,000

By adding a square having an area of 16,000,000 to the lower right corner, we can make the figure into a large square.

What we have done, called "completing the square," is equivalent to adding 16,000,000 to both sides of the equation h^2 + 8,000h = 9,000,000 *to get* h^2 + 8,000h + 16,000,000 = 9,000,000 + 16,000,000. *Rewriting the result and solving it by the square-root method, we get*

$$(h + 4,000)^2 = 25,000,000$$
$$h + 4,000 = \pm 5,000$$
$$h = \pm 5,000 - 4,000$$
$$= 1,000 \text{ or } -9,000$$

---

*The "distance to the horizon" formula discussed on pages 570–571 of the text is an approximation of this formula for use with small altitudes.

Solve by completing the square.

1. $x^2 + 8x = 33$

2. $x^2 - 20x + 36 = 0$

3. $x^2 - 6x = 1$

Find the width of the frame, given that it has the same area as the picture.

The second of these solutions does not seem to have any meaning with regard to the original problem. Checking the first solution in the original equation, we get

$$3,000 = \sqrt{8,000(1,000) + (1,000)^2}\,?$$
$$3,000 = \sqrt{8,000,000 + 1,000,000}\,?$$
$$3,000 = \sqrt{9,000,000}\,?$$
$$3,000 = 3,000$$

The satellite would have to be about 1,000 miles above Los Angeles.

Additional examples for teaching your students how to complete the square are given on Transparency 13-24. (Because the real purpose of learning how to solve a quadratic equation by completing the square is to enable the student to understand the derivation of the quadratic formula in the next lesson, the exercises in this lesson are deliberately restricted to solving equations to which the method can be easily applied. Equations of the form $x^2 + bx + c = 0$, in which b is an odd number, for example, are avoided as they would lead to having to work with fractions. The manipulative skills of some students could undoubtedly be improved by this sort of practice, but it seems to me that the added difficulties are too much for many algebra students to be able to handle at this point in the course.)

Transparency 13-25 is for use in discussing the Set IV exercise of Lesson 5.

Lesson 7
The Quadratic Formula

Because the derivation of the quadratic formula, together with examples of how to use it, takes a considerable amount of time, I suggest that you begin by showing Transparency 13-26 and then develop the lesson in accord with the text. I recommend having your students write the various equations for this lesson as you present them on the screen.

You might start by deriving a "linear formula" from the general linear equation written in standard form:

$$ax + b = 0$$
$$ax = -b$$
$$x = \frac{-b}{a}$$

If you do, point out that the equation $ax + b = 0$ represents all linear equations "rolled into one" and that, in solving it for x, we are in effect solving all linear equations at the same time. Be sure to show how to use the formula that results to find solutions to some equations such as $2x + 6 = 0$ and $5(x - 1) = 30$.

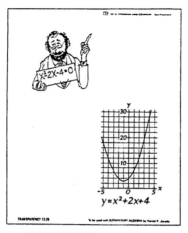

Now point out that the equation $ax^2 + bx + c = 0$ represents all quadratic equations "rolled into one" and that, in solving it for x, we will be solving all quadratic equations at the same time. After deriving the formula as in the text, show how to use it with some examples. Several are given on Transparency 13-27 for this purpose.

Transparency 13-28 is for use in discussing the Set IV exercise of Lesson 6. After showing what happens when attempting to solve the equation by completing the square, you might ask your students to try to solve it with the quadratic formula.

Lesson 8
The Discriminant

An iceberg in the middle of the quad

An annual tradition at our school is the "senior prank." During one weekend in the spring, some of the seniors set up some sort of prank for the rest of the school to discover on Monday morning. Some of the pranks have been relatively harmless, such as piling a huge stack of benches in the middle of the quad, but others have unfortunately turned out to be destructive, such as filling the second-story arcade between the main buildings with water. Almost invariably, there is a lot of cleaning up to be done afterward.

An ideal senior prank would be one that would attract a lot of attention, yet would cause no damage or require any cleanup later. Here is one that fits these specifications.

Show Transparency 13-29. *One idea would be to build a giant iceberg in the middle of the quad. It would certainly astonish everyone and, by melting, clean up after itself.*

How long would such an iceberg last? It certainly wouldn't be worth setting up if it were all melted by Monday morning. The rate at which the iceberg would melt depends on the time that had passed since it was put together. Suppose that the formula for the melting rate as a function of time is $r = 0.1t^2 - 8t + 160$, *in which* r *represents the rate in pounds per hour and* t *represents the time in hours. Use this formula to find the rate at which the iceberg would be melting at the beginning. (160 pounds per hour.) At what rate would it be melting 5 hours later? (122.5 pounds per hour.) What would the rate be after 10 hours had passed? (90 pounds per hour.) The rate is slowing down. What would it be at the very end? (0 pounds per hour.) We can use this fact to find out how long the iceberg would last. Letting* $r = 0$ *in the "rate of melting" equation, we write* $0 = 0.1t^2 - 8t + 160$. *Use the quadratic formula to solve this equation.*

$$t = \frac{-(-8) \pm \sqrt{(-8)^2 - 4(0.1)(160)}}{2(0.1)}$$

$$= \frac{8 \pm \sqrt{64 - 64}}{0.2} = \frac{8}{0.2} = 40$$

How long would the iceberg last? (40 hours.)

This follows from the fact that the expression under the radical sign in the quadratic formula for the solutions turned out to be zero. Show Transparency 13-30. We can also see this from the graph of the quadratic function. It intersects the x-axis in exactly one point.

The rest of the new lesson on the discriminant can now be developed in accord with the text. Transparencies 13-31 and 13-32 can be used for this purpose.

Transparency 13-33 is for use in discussing the Set IV exercise of Lesson 7. Actually stretching a thread across the figure is preferable to drawing lines on it.

Lesson 9
Solving Higher-Degree Equations

To begin this lesson, you might show a copy of Cardan's *The Great Art* to your class. A translation by T. Richard Witmer with a foreword by Oystein Ore has been published by the M.I.T. Press (1968).

The title page of the first edition of the book, written in Latin and published in 1545, is reproduced on Transparency 13-34. *The heading says: "Jerome Cardan, outstanding mathematician, philosopher, and physician, The Great Art, or the Rules of Algebra." The opening sentences below the picture of Cardan are also of interest: "In this book, learned reader, you have the rules of algebra. It contains so many new discoveries and demonstrations by the author . . . that its forerunners . . . are washed out."*

Cardan's book soon became famous as the cause of one of the most bitter feuds in the history of science.

Here is how it came about.* People have known how to solve certain quadratic equations as far back as the Babylonian civilization of about 1800 B.C. There is evidence that the Greeks were solving simple cubic equations about 400 B.C. The famous Persian poet and mathematician Omar Khayyam made some progress in the solution of cubic equations about 1100 A.D., but it was not until the sixteenth century that general methods for solving such equations were discovered.

Cardan, whose book explained these methods, had learned one of them from a man named Niccolo Tartaglia. Tartaglia later claimed that Cardan had sworn to him that he would keep the method secret. Tartaglia had planned to write a book explaining it, hoping that the book would make him famous. Later, Cardan found out that another man, Scipione del Ferro, had discovered the method before Tartaglia had. Feeling that he was then free to give away the secret, Cardan included it in his book. Although he began by giving both Ferro and Tartaglia credit for the discovery, Tartaglia was furious. Cardan had beaten him to the punch and Tartaglia resented it the rest of his life.

Although we will not study any of the methods for solving cubic equations included in Cardan's book, we will consider some examples of how equations of degree higher than 2 can be found. Examples for doing this, different from those in the text, are given on Transparencies 13-35 and 13-36.

Transparencies 13-37 and 13-38 are for use in discussing exercise 12 of Set III and the Set IV exercise of Lesson 8.

---

*The historical information is excerpted from the article "Solution of Polynomial Equations of Third and Higher Degrees" by Rodney Hood in the N.C.T.M. yearbook *Historical Topics for the Mathematics Classroom* and from Dr. Ore's foreword to Witmer's translation of *The Great Art*.

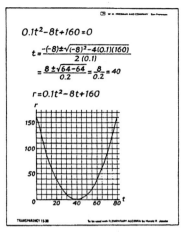

$0.1t^2 - 8t + 160 = 0$

$t = \dfrac{-(-8) \pm \sqrt{(-8)^2 - 4(0.1)(160)}}{2(0.1)}$

$= \dfrac{8 \pm \sqrt{64-64}}{0.2} = \dfrac{8}{0.2} = 40$

$r = 0.1t^2 - 8t + 160$

TRANSPARENCY 13-30

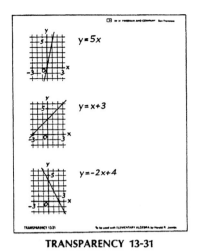

$y = 5x$

$y = x+3$

$y = -2x+4$

TRANSPARENCY 13-31

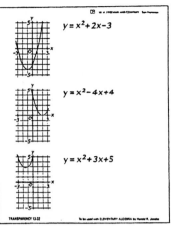

$y = x^2 + 2x - 3$

$y = x^2 - 4x + 4$

$y = x^2 + 3x + 5$

TRANSPARENCY 13-32

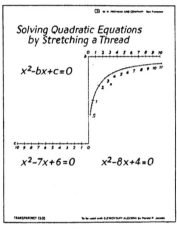

Solving Quadratic Equations
by Stretching a Thread

$x^2 - bx + c = 0$

$x^2 - 7x + 6 = 0$ $x^2 - 8x + 4 = 0$

TRANSPARENCY 13-33

HIERONYMI CAR
DANI, PRÆSTANTISSIMI MATHE
MATICI, PHILOSOPHI, AC MEDICI.
ARTIS MAGNÆ,
SIVE DE REGVLIS ALGEBRAICIS,
Lib. unus. Qui & totius operis de Arithmetica, quod
OPVS PERFECTVM
inscripsit, est in ordine Decimus.

TRANSPARENCY 13-34

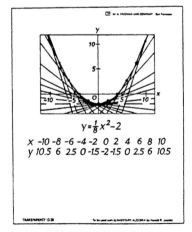

How many solutions can the
equation $x^4 - 33x^2 - 28x + 60 = 0$
have?

Use this graph of the function
$y = x^4 - 33x^2 - 28x + 60$ to estimate
their values.

TRANSPARENCY 13-35

Solve for x.

1. $2x^3 + 2x^2 - 40x = 0$

2. $x^5 - 6x^4 + 7x^3 = 0$

TRANSPARENCY 13-36

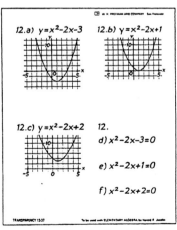

12.a) $y = x^2 - 2x - 3$ 12.b) $y = x^2 - 2x + 1$

12.c) $y = x^2 - 2x + 2$ 12.

d) $x^2 - 2x - 3 = 0$

e) $x^2 - 2x + 1 = 0$

f) $x^2 - 2x + 2 = 0$

TRANSPARENCY 13-37

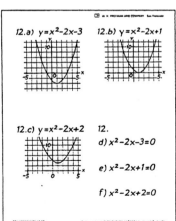

$y = \tfrac{1}{8}x^2 - 2$

x	-10	-8	-6	-4	-2	0	2	4	6	8	10
y	10.5	6	2.5	0	-1.5	-2	-1.5	0	2.5	6	10.5

TRANSPARENCY 13-38

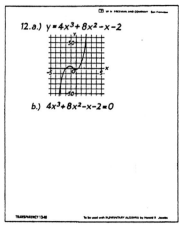

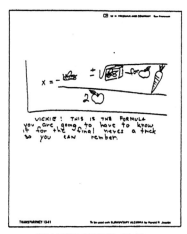

TRANSPARENCY 13-39   TRANSPARENCY 13-40   TRANSPARENCY 13-41

Review of Chapter 13

First day

Show Transparency 13-39. *When someone walks to the end of a diving board, the board bends because of the weight. Suppose that a rather heavy person is standing at the end of a diving board that is 4 meters long. How far each point of the board sags below its normal position depends on its distance from the fixed end of the board: the greater the distance, the greater the sag.*

Add the overlay. *A typical formula for this function is* $y = 0.1x^3 - 1.2x^2$, *in which x represents the distance in meters from the end of the board and y represents the change in position in centimeters of a point on the board at that location. Complete the table for this function.*

x	0	1	2	3	4
y	0	-1.1	-4	-8.1	-12.8

and graph it on a pair of axes like those shown. (The bending is exaggerated because of the different units used for x and y.) After this has been done, remove the overlay.

Now solve the cubic equation $0.1x^3 - 1.2x^2 = 0$ *for* x. *[Factoring, we get* $0.1x^2(x - 12) = 0$. *The*

solutions of the equation are 0 and 12.] Do both make sense with regard to the diving board? (The first one does because the board does not sag at the fixed end. The second does not because it corresponds to a point 12 meters from the end of the board and the board is only 4 meters long.)

Transparency 13-40 is for use in checking exercise 12 of Lesson 9.

Second day

Several years ago, I found the note reproduced on Transparency 13-41 on a piece of paper on the floor of my classroom. It was evidently written for an algebra student by one of her friends. Your students might enjoy seeing it and explaining what it means. (Note that b^2 has been drawn as if it were b^3.) My initial reaction was one of dismay that students would have to resort to such devices but, after thinking about it, I realized that Vickie's friend had come up with a very vivid device for making the formula easier to remember.

Answers to the Set I exercises of the review lesson are given on Transparencies 13-42 and 13-43.

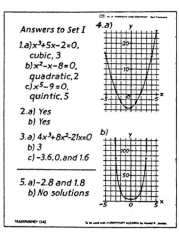

TRANSPARENCY 13-42

Answers to Set I

1.a) $x^3+5x-2=0$, cubic, 3
b) $x^2-x-8=0$, quadratic, 2
c) $x^5-9=0$, quintic, 5

2.a) Yes
b) Yes

3.a) $4x^3+8x^2-21x=0$
b) 3
c) $-3.6, 0,$ and 1.6

4.a)

b)

5.a) -2.8 and 1.8
b) No solutions

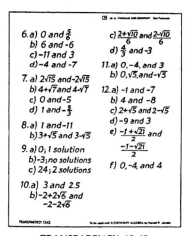

TRANSPARENCY 13-43

6.a) 0 and $\frac{5}{6}$
b) 6 and -6
c) -11 and 3
d) -4 and -7

7.a) $2\sqrt{15}$ and $-2\sqrt{15}$
b) $4+\sqrt{7}$ and $4-\sqrt{7}$
c) 0 and -5
d) 1 and $-\frac{5}{3}$

8.a) 1 and -11
b) $3+\sqrt{5}$ and $3-\sqrt{5}$

9.a) 0; 1 solution
b) -3; no solutions
c) 24; 2 solutions

10.a) 3 and 2.5
b) $-2+2\sqrt{6}$ and $-2-2\sqrt{6}$

c) $\frac{2+\sqrt{10}}{6}$ and $\frac{2-\sqrt{10}}{6}$
d) $\frac{4}{3}$ and -3

11.a) $0, -4,$ and 3
b) $0, \sqrt{5},$ and $-\sqrt{5}$

12.a) -1 and -7
b) 4 and -8
c) $2+\sqrt{5}$ and $2-\sqrt{5}$
d) -9 and 3
e) $\frac{-1+\sqrt{21}}{2}$ and $\frac{-1-\sqrt{21}}{2}$
f) $0, -4,$ and 4

Chapter 14

THE REAL NUMBERS

In his article titled "Number" in the September 1964 issue of *Scientific American*, Philip J. Davis states:

> The number systems employed in mathematics can be divided into five principal stages, going from the simplest to the most complicated. They are: (1) the system consisting of the positive integers only; (2) the next higher stage, comprising the positive and negative integers and zero; (3) the rational numbers, which include fractions as well as integers; (4) the real numbers, which include the irrational numbers, such as π; (5) the complex numbers, which introduce the "imaginary" number $\sqrt{-1}$.

At this point in their study of algebra, the students have progressed through the first four of these stages, the work at the beginning being limited to the counting numbers and zero, with the negative integers introduced in Chapter 3, the rational numbers in Chapter 4, and irrational numbers in the form of square roots in Chapter 12. They are now ready to put their knowledge of these sets of numbers in perspective.

The chapter begins with a lesson that reviews the properties of rational numbers and explains that the rational numbers are equivalent to the numbers that have repeating decimal forms. The next three lessons deal with the properties of the irrational num-

bers, with special attention being given to square roots and higher ones and the number pi. The chapter closes with a lesson on the real numbers that reviews the relationships of the counting numbers, integers, and rational numbers to the real number system.

Lesson 1
Rational Numbers

A trick with the number 7

In preparation for this trick, write a student's name that contains six different letters below the digits 1 4 2 8 5 7 on a piece of paper, but do not show it to your students.

Example: *142857*
           ALISON

Write four other letters below the digits 0 3 6 9.

Example: *0369*
           EHBT

Tell your students to choose an integer between 1 and 6 and divide it by 7 to 12 decimal places. Now call off the ten digits in order from 0 to 9, telling the letter that goes with each digit. Doing this for the example, we have

0123456789
EAIHLOBNST

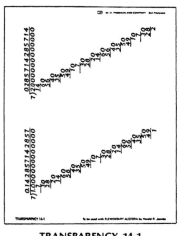

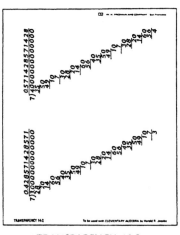

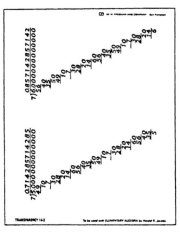

TRANSPARENCY 14-1   TRANSPARENCY 14-2   TRANSPARENCY 14-3

Tell your students to write the letter corresponding to each digit above the appropriate digit or digits in their answers. If they do all of this correctly, they should be able to find the name that you have chosen somewhere in the result.

Example:

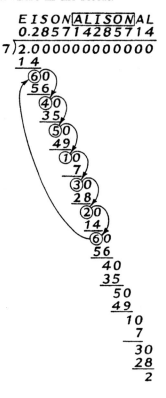

To show how the trick works, ask a student to tell the number into which he or she divided 7. Show the half of the transparency (14-1, 14-2, or 14-3) that illustrates the division, writing the appropriate letters above the digits of the answer.

The trick has as its basis the fact that, when 7 is divided into a number that is not a multiple of 7, there are only six possible remainders: 1, 2, 3, 4, 5, or 6. Circle these on the transparency as indicated in the example above. The result is a cycle that repeats endlessly as the division is continued further. Because the remainders repeat endlessly, so do the digits in the answer. When a number between 1 and 6 is divided by 7, the six digits 1 4 2 8 5 7 are always in the answer in that order. By assigning the six letters of someone's name to them, we can be certain that the name will appear at the end, regardless of the number originally chosen.

After using the example to explain the meaning of the word *period*, point out to your students that the case we have just considered is not just an accident. Every *rational number begins repeating digits when changed to decimal form. (If it repeats zeros, we sometimes say that it "ends," or "terminates.") Conversely, every number in decimal form that has an endlessly repeating pattern can be expressed as the quotient of two integers and is therefore rational.*

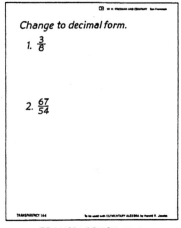

Change to decimal form.

1. $\frac{3}{8}$

2. $\frac{67}{54}$

TRANSPARENCY 14-4

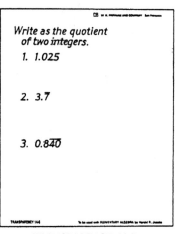

Write as the quotient of two integers.

1. *1.025*

2. *3.7*

3. *0.8$\overline{40}$*

TRANSPARENCY 14-5

The first person to find what number times itself equals seven may do whatever he or she wants for the rest of the period.

TRANSPARENCY 14-6

Additional examples for use in developing these ideas are given on Transparencies 14-4 and 14-5.

Lesson 2
Irrational Numbers

Show the cartoon at the top of Transparency 14-6, keeping the statement at the bottom covered. *Substitute teachers often have a difficult time of it, especially if the regular teacher has not left any assignment for the students to do. An article in* Saturday Review *several years ago titled "Survival Guide for Substitutes" (April 1973, pp. 51–52) offered suggestions for substitute teachers of various subjects.*

Uncover the statement at the bottom of the transparency. *Here is one of the suggestions for something to do in a math class. The problem is simple in one sense and yet impossible in another. There are two simple answers to it. Can you tell what they are? (The positive and negative square roots of 7:* $\sqrt{7}$ *and* $-\sqrt{7}$.) *The impossible aspect of the problem is that the exact decimal forms of these numbers cannot be written out.* Show the left column of Transparency 14-7. *The positive square root of 7 is clearly between 2 and 3, but it is not 2.1 or 2.2 or 2.3 or any of the other numbers shown here.* Show the middle column. *Neither is it 2.61 or 2.62 or 2.63. . . .* Show the right column. *Regardless of how many digits of its decimal form we may figure out, we will never be finished because the decimal form never comes to an end nor*

does it eventually begin a repeating pattern of digits. This means that the square roots of 7 are not rational numbers. They are irrational.* *The rest of the new lesson might now be presented in accord with the text.*

Transparency 14-8 is for use in discussing the Set IV exercise of Lesson 1.

Lesson 3
More on Irrational Numbers

Show Transparency 14-9. *This cat, one of the world's heaviest, tops the scales at 42 pounds. Its heart beats more slowly than that of any ordinary cat because of its great weight. The rate of a cat's heartbeat at rest is given by the formula* $n = \frac{294}{\sqrt[4]{w}}$, *in which* n *represents the number of heartbeats per*

---

*Note to the teacher: To see why $\sqrt{7}$ is not rational, notice that, if $\sqrt{7} = \frac{m}{n}$, then $7n^2 = m^2$. Therefore $\sqrt{7}$ could be rational only if $7n^2 = m^2$ has a solution in which m and n are nonzero integers. But there can be no integer solutions to $7n^2 = m^2$ because, if there were, the number m^2 (which is the same as $7n^2$) would have an even number of sevens as prime factors (because it is a perfect square) and at the same time an odd number of sevens as prime factors (because it is seven times a perfect square). This would contradict the uniqueness of the prime factorization of positive integers. So there are no nonzero integer solutions to $7n^2 = m^2$ and none for $7 = \frac{m}{n}$. Thus 7 is not rational.

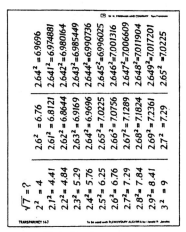

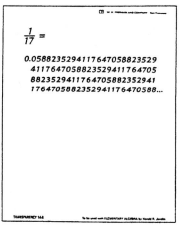

TRANSPARENCY 14-7   **TRANSPARENCY 14-8**   **TRANSPARENCY 14-9**

minute and w represents the cat's weight in pounds. What do you think the symbol $\sqrt[4]{}$ means? (The fourth root of w, or the number whose fourth power is w.) Suppose that a cat weighs 16 pounds. Can you figure out the rate of its heartbeat from this formula?

$$\left(\text{n} = \frac{294}{\sqrt[4]{16}} = \frac{294}{2} = 147.\right.\ \text{It is 147 beats per}$$

minute.$)$

To find the rate of the 42-pound cat's heartbeat, we need to know the decimal value of $\sqrt[4]{42}$. Make a table of some fourth powers to find the integers between which it lies.

x	x⁴
1	1
2	16
3	81

Because $(\sqrt[4]{x})^2 = \sqrt{x}$, $\sqrt[4]{x} = \sqrt{\sqrt{x}}$. Use either the table on page 563 or your calculator to estimate $\sqrt{\sqrt{42}}$ to the nearest tenth. (2.5.) What is the approximate heartbeat rate of the cat in the photograph? (About 118 beats per minute.)

Explain that $\sqrt[4]{42}$, like $\sqrt{42}$, is irrational. The table on Transparency 14-10 might be used to develop the rest of the lesson in accord with the text.

Transparency 14-11 is for use in discussing the Set IV exercise of Lesson 2.

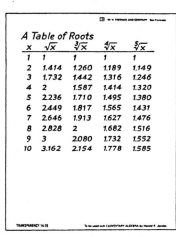

A Table of Roots

x	$\sqrt{x}$	$\sqrt[3]{x}$	$\sqrt[4]{x}$	$\sqrt[5]{x}$
1	1	1	1	1
2	1.414	1.260	1.189	1.149
3	1.732	1.442	1.316	1.246
4	2	1.587	1.414	1.320
5	2.236	1.710	1.495	1.380
6	2.449	1.817	1.565	1.431
7	2.646	1.913	1.627	1.476
8	2.828	2	1.682	1.516
9	3	2.080	1.732	1.552
10	3.162	2.154	1.778	1.585

TRANSPARENCY 14-10

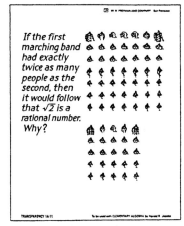

If the first marching band had exactly twice as many people as the second, then it would follow that $\sqrt{2}$ is a rational number. Why?

TRANSPARENCY 14-11

Lesson 4
Pi

Every calculator has ten keys for the numbers zero through nine. Many calculators have an eleventh key for another number. Do you know what it is? (Pi.) Show the symbol for pi at the top of Transparency 14-12, keeping the rest of the transparency covered.

People have been aware of the number pi since ancient times. It was not until the eighteenth century, however, that the number became identified with the name and symbol by which we now know it, the sixteenth letter of the Greek alphabet,

Do you know what the number pi means? Let someone explain before revealing the figures on the transparency and discussing the terms circumference, diameter, radius, and area.

Two numbers that are often used as approximations of pi are $\frac{22}{7}$ and 3.14. These numbers can only be approximations of pi because both are rational and pi is irrational. Show Transparency 14-13. *The first 5,000 places of its decimal form are shown on this transparency. To date, pi has been computed to more than one million decimal places.*

Two famous equations in which pi appears are the formulas for the circumference and area of a circle: $c = 2\pi r$ and $a = \pi r^2$. The exercises on Transparency 14-14 can be used to demonstrate how these formulas are used.

Transparencies 14-15 and 14-16 are for use in discussing exercises 14 and 15 and the Set IV exercise of Lesson 3.

Lesson 5
The Real Numbers

Show Transparency 14-17. *Although it is harder to push a heavy person in a swing than it is to push a light one, the time that it takes the person to swing back and forth once depends only on the length of the swing.* Add the overlay. *The formula for the time as a function of the length is $t = 2\pi\sqrt{\frac{l}{32}}$, in which l represents the length in feet and t represents the time in seconds. Notice that the formula contains the number π.*

Before using the formula, it is helpful to rewrite it so that a fraction does not appear under the radical sign. Doing this, we get $t = \frac{\pi\sqrt{2l}}{4}$. Use the formula to find the swing time, to the nearest tenth of a second, of a swing that is 8 feet long. $\left[t = \frac{\pi\sqrt{2(8)}}{4} = \pi \approx 3.1.$ *The time is about 3.1 seconds.* $\right]$

What would happen to the swing time if the length were doubled to 16 feet? $\left[t = \frac{\pi\sqrt{2(16)}}{4} = \pi\sqrt{2} \approx 4.4.$ *The time increases to about 4.4 seconds.* $\right]$

Solve the formula for l in terms of t. $\left(l = \frac{8t^2}{\pi^2}. \right)$ *Use the result to figure out how long a swing would have to be in order to have a swing time of one minute.* $\left[l = \frac{8(60)^2}{\pi^2} \approx 2,918.$ *The swing would have to be more than half a mile long.* $\right]$

What kind of numbers are the exact answers to the problems we have solved? (Irrational.) What kind of numbers are the approximate answers? (Rational.)

Transparencies 14-18 and 14-19 are for use in discussing exercises 15 and 16 and the Set IV exercise of Lesson 4.

To introduce the new lesson, you might show Transparency 14-20 with the remark that the word "irrational" does not have the same meaning in mathematics that it has in everyday speech. *To say that a person is irrational means that he is incoherent or unbalanced. To say that a number is irrational means that it cannot be written as the quotient of two integers. The irrational numbers, together with the*

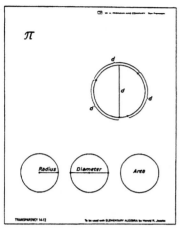

TRANSPARENCY 14-12

TRANSPARENCY 14-13

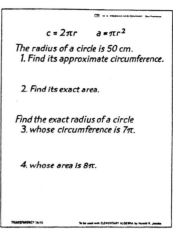

TRANSPARENCY 14-14

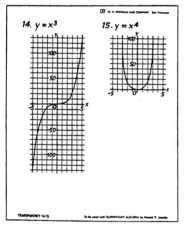

TRANSPARENCY 14-15

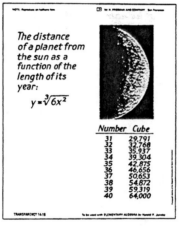

TRANSPARENCY 14-16

TRANSPARENCY 14-17

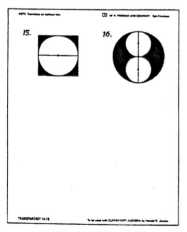

TRANSPARENCY 14-18

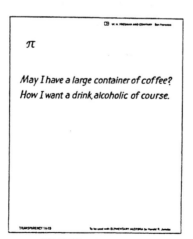

TRANSPARENCY 14-19

TRANSPARENCY 14-20

TRANSPARENCY 14-21

rational numbers, *make up the set of numbers called* the *real numbers.*

Show Transparency 14-21, asking your students to copy the equations and then, working with you, tell the solution of each and what sort of number it is. You might write the results as shown below.

Equation	Solution	Type of number
$x - 3 = 2$	$2 + 3 = 5$	Counting number
$x + 3 = 2$	$2 - 3 = -1$	Integer
$3x = 2$	$\frac{2}{3} = 0.66666...$	Rational number
$x^3 = 2$	$\sqrt[3]{2} = 1.25992...$	Irrational number

Have your students draw a diagram comparable to the one at the top of page 694 of the text illustrating the relationships of these types of numbers.

The solutions to the four equations can be used to explain the meanings of *closure* and *order*. For example, the counting numbers are not closed with respect to subtraction because the solution to the second equation, $2 - 3 = -1$, is not a counting number. The solutions to the third and fourth equations reveal that the integers are neither closed with respect to division nor taking roots. The order of the four solutions can be illustrated on a number line.

Review of Chapter 14

First day: A cubic formula

Perhaps the hardest thing to learn in our study of algebra has been the quadratic formula. It is worth knowing, however, because every quadratic equation can be solved by it. Now that we are familiar with cube roots, we will look at an example of a formula for solving cubic equations. Show Transparency 14-22. Don't worry. You aren't expected to memorize it! It is called a cubic formula rather than the cubic formula because it applies only to cubic equations in standard form for which a = 1 *and* b = 0.* *See if you can use it to solve the equation* $x^3 + 9x - 26 = 0$. *When you are finished, check your solution. (Although we have learned that cubic equations can have as many as three solutions, this equation has only one.)*

SOLUTION

$$x = \sqrt[3]{\sqrt{\left(\frac{9}{3}\right)^3 + \left(\frac{-26}{2}\right)^2} - \frac{-26}{2}} -$$

$$\sqrt[3]{\sqrt{\left(\frac{9}{3}\right)^3 + \left(\frac{-26}{2}\right)^2} + \frac{-26}{2}}$$

$$= \sqrt[3]{\sqrt{27 + 169} + 13} - \sqrt[3]{\sqrt{27 + 169} - 13}$$

$$= \sqrt[3]{14 + 13} - \sqrt[3]{14 - 13}$$

$$= 3 - 1 = 2$$

Transparency 14-23 is for use in discussing the Set IV exercise of Lesson 5.

*How this formula is derived is discussed in *An Introduction to the History of Mathematics*, third edition, by Howard Eves (Holt, 1969), pp. 217-218.

A cubic formula:
$$ax^3+bx^2+cx+d=0$$
$$x=\sqrt[3]{\sqrt{\left(\frac{c}{3}\right)^3+\left(\frac{d}{2}\right)^2}-\frac{d}{2}}-\sqrt[3]{\sqrt{\left(\frac{c}{3}\right)^3+\left(\frac{d}{2}\right)^2}+\frac{d}{2}}$$

TRANSPARENCY 14-22

Arrange in order from smallest to largest.

1. 1^{2^3} 1^{3^2} 2^{1^3} 2^{3^1} 3^{1^2} 3^{2^1}

2. 2^{3^4} 2^{4^3} 3^{2^4} 3^{4^2} 4^{2^3} 4^{3^2}

TRANSPARENCY 14-23

Decimal Reciprocals

TRANSPARENCY 14-24

Second day

Give each student a duplicated copy of Transparency 14-24. *We have learned that, whenever a rational number is written in decimal form, it always falls into a repeating pattern of digits. Here is a table listing the decimal forms of the reciprocals of the integers from 1 through 30. It is easy to see from it that the decimal reciprocals of 1 and 2 repeat zeros. Circle them and the other numbers whose reciprocals repeat zeros. What do they have in common?* (They have no factors other than 1, 2, and 5.) *What does the reciprocal of 3 repeat?* (3.) *Draw a bar over the first 3 after the decimal point to indicate that it is the period. What does the reciprocal of 7 repeat?* (142857.) *Draw a bar over the first group of these digits. Do the same for all of the other reciprocals that do not repeat zeros.*

When this has been done, point out that the period of the reciprocal of 7 contains the maximum number of digits possible, six, because, when 7 is divided into 1, only six remainders other than 0 are possible. *Which reciprocals of numbers larger than 7 have the maximum number of digits possible?* (17, 19, 23, and 29.) *This can easily be seen by placing a ruler or the edge of a sheet of paper diagonally on the table (as shown in the figure at the right).*

Decimal Reciprocals

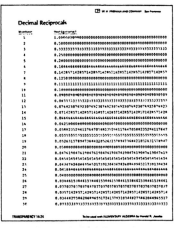

Number	Reciprocal
①	1.000000000000000000000000000000000000000000000000
②	0.500000000000000000000000000000000000000000000000
3	0.333333333333333333333333333333333333333333333333
④	0.250000000000000000000000000000000000000000000000
⑤	0.200000000000000000000000000000000000000000000000
6	0.166666666666666666666666666666666666666666666666
7	0.142857142857142857142857142857142857142857142857
⑧	0.125000000000000000000000000000000000000000000000
9	0.111111111111111111111111111111111111111111111111
⑩	0.100000000000000000000000000000000000000000000000
11	0.090909090909090909090909090909090909090909090909
12	0.083333333333333333333333333333333333333333333333
13	0.076923076923076923076923076923076923076923076923
14	0.071428571428571428571428571428571428571428571428
15	0.066666666666666666666666666666666666666666666666
⑯	0.062500000000000000000000000000000000000000000000
17	0.058823529411764705882352941176470588235294117647
18	0.055555555555555555555555555555555555555555555555
19	0.052631578947368421052631578947368421052631578947
⑳	0.050000000000000000000000000000000000000000000000
21	0.047619047619047619047619047619047619047619047619
22	0.045454545454545454545454545454545454545454545454
23	0.043478260869565217391304347826086956521739130434
24	0.041666666666666666666666666666666666666666666666
㉕	0.040000000000000000000000000000000000000000000000
26	0.038461538461538461538461538461538461538461538461
27	0.037037037037037037037037037037037037037037037037
28	0.035714285714285714285714285714285714285714285714
29	0.034482758620689655172413793103448275862068965517
30	0.033333333333333333333333333333333333333333333333

W. H. FREEMAN AND COMPANY San Francisco

Answers to Set I

1. a) False
   b) False
   c) True
   d) True
   e) True
2. a) 3.2
   b) 0.225
   c) $0.\overline{72}$
   d) $0.208\overline{3}$
3. a) $\frac{103}{10}$
   b) $\frac{31}{3}$
   c) $\frac{17}{20}$
   d) $\frac{85}{99}$
4. a) Rational
   b) None because it is not an integer.
5. a) −5
   b) Not possible
   c) 15
   d) 10
6. $0.345, 0.\overline{345}, 0.3\overline{45}, 0.34\overline{5}$
7. a) 5
   b) None
   c) −10
   d) 1 and −1
8. a) >
   b) <
   c) <
   d) >
9. a) $\sqrt[n]{0.9}$ gets larger.
   b) 1
   c) $\sqrt[n]{1.1}$ gets smaller.
   d) 1

TRANSPARENCY 14-25 To be used with ELEMENTARY ALGEBRA by Harold R. Jacobs

TRANSPARENCY 14-25

W. H. FREEMAN AND COMPANY San Francisco

10. a) $\frac{157}{50}$
    b) 3.14
    c) Smaller
11. a) 1.6π
    b) 12π
    c) $\frac{15}{\pi}$
    d) $\sqrt{17}$
12. a) 32
    b) $8x^2$
    c) $32 - 8\pi$
    d) $8x^2 - 2\pi x^2$
13. a) 25,120 miles
    b) 28 sq. meters
14. a) $\sqrt[3]{27}$ and 2^7
    b) −27
    c) 2.7, $2.\overline{7}$, and $\frac{2}{7}$
    d) $\sqrt{27}$ and 27π
15. a) Yes
    b) No
    Example: $-1 - -2 = 1$
    c) No
    Example: $(-1)(-2) = 2$
    d) No
    Example: $(-1)^2 = 1$
    e) Yes
16. a) $-\frac{19}{4}$; rational
    b) $5\sqrt{3}$; irrational, real
    c) $\frac{\sqrt{7}}{6}$ and $-\frac{\sqrt{7}}{6}$; irrational, real
    d) No solution

TRANSPARENCY 14-26 To be used with ELEMENTARY ALGEBRA by Harold R. Jacobs

TRANSPARENCY 14-26

Would we notice patterns comparable to these if we had a table of the decimal forms of the square roots *of the integers from 1 through 30? Why not?*

Answers to the Set I exercises of the review lesson are given on Transparencies 14-25 and 14-26.

TRANSPARENCY 15-1

	Height of model	Actual height
King Kong	2 feet	50 feet
Building	?	450 feet

TRANSPARENCY 15-2

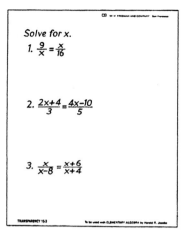

Solve for x.

1. $\frac{9}{x} = \frac{x}{16}$

2. $\frac{2x+4}{3} = \frac{4x-10}{5}$

3. $\frac{x}{x-8} = \frac{x+6}{x+4}$

TRANSPARENCY 15-3

Chapter 15

Fractional Equations

This chapter gives students an opportunity to strengthen many skills learned earlier in the course, including working with rational expressions, solving linear and quadratic equations, and working with formulas.

The chapter begins with a lesson on ratio and proportion from which the student learns how to solve proportions by multiplying means and extremes. This is followed by a couple of lessons on solving fractional equations that are not proportions. The final lesson deals with the use and manipulation of formulas that contain fractions.

Lesson 1
Ratio and Proportion

Show Transparency 15-1. *One of the most popular films of all time is the original version of* King Kong, *released by RKO-Radio Pictures in 1933. If you have seen either it or the remake, you know that many special effects were used. In one of them,* King Kong, *whose size seems tremendous, is shown climbing up the Empire State Building. For this and many other scenes, small models of the ape were used.* Show Transparency 15-2. *This photograph shows one of these models being used in the production of the film.*

If King Kong was supposed to create the illusion of being 50 feet tall but was represented by a model only 2 feet tall, how tall should a model of the upper part of the Empire State Building be if it is actually 450 feet tall? Use this problem to develop the ideas of the lesson in accord with the text. A proof that the product of the means of a proportion is equal to the product of the extremes should be presented after observing that it is true in the King Kong proportion. Additional examples for use in showing how to solve proportions are given on Transparency 15-3.

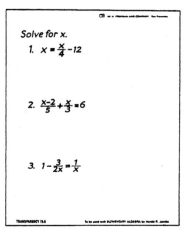

| TRANSPARENCY 15-4 | TRANSPARENCY 15-5 | TRANSPARENCY 15-6 |

Transparency 15-5 content:

Solve for x.

1. $x = \frac{x}{4} - 12$

2. $\frac{x-2}{5} + \frac{x}{3} = 6$

3. $1 - \frac{3}{2x} = \frac{1}{x}$

Transparency 15-6 content:

What is the natural scale of this map?

Lesson 2
Equations Containing Fractions

Show Transparency 15-4. *According to one modern dictionary* (The American Heritage Dictionary of the English Language), *"to keep an ear to the ground" means "to give attention to or watch the trends of public opinion."** The phrase was once taken literally because the American Indians used to put their ears to the ground to listen for the sound of an approaching horseback rider. They did this because they knew that sound travels more quickly through the ground than it travels through the air.*

It is possible to figure out how far away the horseback rider is by timing the interval between the moment the sound is first heard in the ground and the moment that it is first heard in the air. To learn how, we will estimate the distance for a time interval of 8 seconds. Sound travels through the air at a speed of about 1,100 feet per second. How much time would it take sound traveling at this rate to travel a distance of x feet? $\left(\frac{x}{1,100}\ seconds.\right)$ *It travels through the ground about ten times as fast as it does through the air. How much time would it take sound traveling at*

this faster rate to travel x feet? $\left(\frac{x}{11,000}\ seconds.\right)$ *The difference between the two times is 8 seconds. Use this information to write an equation.*

$$\left(\frac{x}{1,100} - \frac{x}{11,000} = 8\right)$$

Although this equation contains fractions, it is not a proportion; so we cannot eliminate the fractions by multiplying means and extremes. We can eliminate them by another method, however. The method is to multiply both sides of the equation by the least common denominator of the fractions. Show your students how to do this for the horseback-rider equation. *Approximately how far away is the horseback rider? (About 9,800 feet, or a little less than 2 miles.)*

Additional examples for illustrating the solution of equations containing fractions are given on Transparency 15-5.

Transparency 15-6 is for use in discussing the Set IV exercise of Lesson 1.

Lesson 3
More on Fractional Equations

Peter Newell created two clever books for children whose pictures are meant to be looked at both

---

*This exercise is adapted from a problem by Morris Kline in *Mathematics and the Physical World* (Crowell, 1959), pp. 60–62.

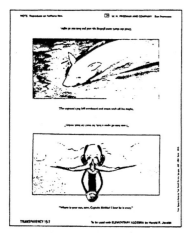

TRANSPARENCY 15-7

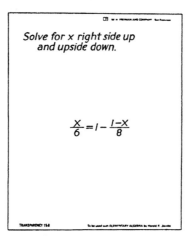

Solve for x right side up and upside down.

$$\frac{x}{6} = 1 - \frac{1-x}{8}$$

TRANSPARENCY 15-8

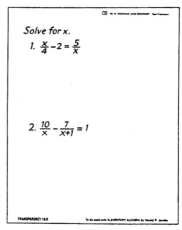

Solve for x.

1. $\frac{x}{4} - 2 = \frac{5}{x}$

2. $\frac{10}{x} - \frac{7}{x+1} = 1$

TRANSPARENCY 15-9

If a watermelon weighs $\frac{9}{10}$ of its weight and $\frac{9}{10}$ of a pound, how many pounds does it weigh?

TRANSPARENCY 15-10

rightside up and upside down. Called *Topsys and Turvys,* the books were originally published in 1893 and 1894. A collection of seventy-four of the pictures is still available in a Dover edition.

Two topsy-turvys are reproduced on Transparency 15-7. After showing them to your class (keep the second one covered while showing the first), show the equation on Transparency 15-8 and explain that, like the topsy-turvy drawings, it is actually two different equations, one of which is seen by turning the other upside down. Ask your students to copy the rightside up equation in the center of a sheet of paper and solve it, writing the work below it. After they have done this, they should turn their papers upside down and solve the second equation. When you discuss the solutions, point out that the first equation leads to a linear equation and hence has one solution, whereas the second leads to a quadratic equation and turns out to have two solutions.

Additional examples of fractional equations that lead to quadratic equations are given on Transparency 15-9.

Transparency 15-10 is for use in discussing the Set IV exercise of Lesson 2.

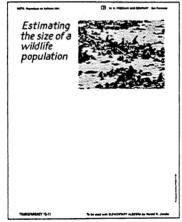

TRANSPARENCY 15-11

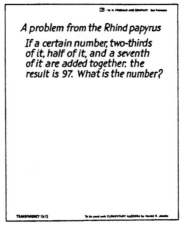

TRANSPARENCY 15-12

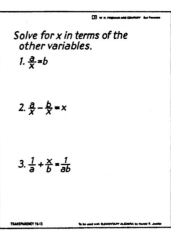

TRANSPARENCY 15-13

Lesson 4
Solving Formulas

Estimating the size of a wildlife population

Show Transparency 15-11. *Today we are going to consider an interesting way in which algebra can be used to estimate the size of a wildlife population.* Our example has as its basis an experiment conducted in Alaska several years ago to estimate the number of fur seal pups on St. Paul Island.*

The experiment employed something called the capture-recapture method. The idea is to catch some animals from the population to be counted, tag them, and then release them. After some time has passed, more animals are caught and the number of tagged ones in the new catch determined. The total number of animals in the population is estimated from this information.

Suppose that we let a *represent the number of animals originally caught and tagged,* b *represent the total number of animals in the second catch,* c *represent the number of tagged animals in the second catch, and* x *represent the total number of animals in*

the population. *It seems reasonable to assume that the ratio of the tagged animals in the second catch to the total number of animals in that catch is equal to the ratio of the total number of tagged animals to the total number of animals in the population. Write a proportion based upon this assumption.* $\left(\dfrac{c}{b} = \dfrac{a}{x}.\right)$

In the experiment to determine the fur seal pup population on St. Paul Island, 4,965 seals were caught and tagged by shaving off some of the black fur on the tops of their heads. Later 900 seals were caught and it was found that 218 of them were tagged. Use these numbers to estimate the total number of fur seal pups on the island to the nearest thousand. $\left(\dfrac{218}{900} = \dfrac{4,965}{x}; x \approx 20,000.\right)$

A formula for the size of a wildlife population as determined by the capture-recapture method can be derived from the proportion $\dfrac{c}{b} = \dfrac{a}{x}$ *by solving it for* x. *What is the formula?* $\left(x = \dfrac{ab}{c}.\right)$ *Does it agree with the method you used to find the number of fur seal pups?* $\left[Yes; \dfrac{(4,965)(900)}{218} \approx 20,000. \right]$

Transparency 15-12 is for use in discussing the Set IV exercise of Lesson 3.

Examples for use in presenting the new lesson are given on Transparency 15-13.

---

*The information in this lesson is from the article "Estimating the Size of Wildlife Populations" by Samprit Chatterjee in *Statistics by Example: Exploring Data*, edited by Frederick Mosteller, William H. Kruskal, Richard F. Link, Richard S. Pieters, and Gerald R. Rising (Addison-Wesley, 1973), pp. 99–104.

TRANSPARENCY 15-14   TRANSPARENCY 15-15

Lesson 5
More on Solving Formulas

In 1842, an Austrian physicist named Doppler used mathematics to make a surprising prediction about the behavior of sound. He predicted that, if a source of sound were to move past someone, the sound would seem to drop in pitch. The prediction was confirmed by an experiment with fifteen trumpet players on a moving train and is now known as the Doppler effect.

A recording demonstrating the Doppler effect is available on Folkways Records FX-6136. The record, titled *The Science of Sound*, was produced by the Bell Telephone Laboratories and includes the sounds of a train whistle, a train crossing bell, and racing cars.

Show Transparency 15-14. *Here are Doppler's equations for the apparent frequency of a moving sound source. In them,* f′ *represents the apparent frequency,* f *represents the actual frequency,* r *represents the speed of the sound source, and* s *represents the speed of sound. The equations are easier to use if the right sides are rewritten as simple fractions. Do this.* $\left(\text{Toward you: } f' = \dfrac{sf}{s-r}; \text{ away from you: } f' = \dfrac{sf}{s+r}.\right)$

Suppose that a car drives past you with its horn producing a sound with a frequency of 500 cycles per second, that the car is traveling 30 miles per hour (which is 44 feet per second), and that the speed of sound is 1,100 feet per second. Find the apparent frequency of the car horn as the car moves toward you and as it moves away. [*Toward you:*

$$f' = \frac{(1,100)(500)}{1,100 - 44} = \frac{550,000}{1,056} \approx 521;$$

approximately 521 cycles per second. Away from you:

$$f' = \frac{(1,100)(500)}{1,100 + 44} = \frac{550,000}{1,144} \approx 481;$$

approximately 481 cycles per second.]

How fast would the car have to be going for the apparent frequency of its horn to be twice its actual frequency as it came toward you?

$$\left[1,000 = \frac{(1,100)(500)}{1,100 - x}; x = 550.\right.$$

The car would have to go 550 feet per second, which is 375 miles per hour.]

If we had to solve several problems of this sort, it would be convenient to solve the general formula for r *in terms of the other variables first. Do this with your students.* [*Solving the first formula for* r *gives*

$$r = \frac{sf' - sf}{f'} \text{ or } r = \frac{s(f' - f)}{f'}.\right]$$

Transparency 15-15 is for use in discussing the Set IV exercise of Lesson 4.

Additional examples for use in the new lesson are given on Transparency 15-16.

$$\frac{424 + 224}{18} = 36 \qquad \frac{424 - 224}{18} = 11.\overline{1}$$

Review of Chapter 15

First day

Show Transparency 15-17. *This problem is from a book by an Indian mathematician named Mahavira who lived in the ninth century. Can you translate it into an equation? Let* x *represent the number of camels in the herd.* $\left(\frac{1}{4}x + 2\sqrt{x} + 3 \cdot 5 = x.\right)$ *The equation contains both a fraction and a square root. Clearing of fractions first, we get* $x + 8\sqrt{x} + 60 = 4x$. *Isolating the term containing the square root before squaring, we get* $8\sqrt{x} = 3x - 60$. *Squaring both sides, we get* $64x = 9x^2 - 360x + 3600$. *Writing the result in standard form yields* $9x^2 - 424x + 3600 = 0$. *Applying the quadratic formula and using a calculator, we get*

$$x = \frac{424 \pm \sqrt{(424)^2 - 4(9)(3600)}}{2(9)}$$

$$= \frac{424 \pm \sqrt{179,776 - 129,600}}{18}$$

$$= \frac{424 \pm \sqrt{50,176}}{18} = \frac{424 \pm 224}{18}$$

The herd contained 36 camels, a solution that is much easier to check than it was to obtain!

Transparency 15-18 is for use in discussing the Set IV exercise of Lesson 5.

Second day

Show Transparency 15-19. *The time that it takes each planet in our solar system to travel in its orbit once around the sun varies from only 88 days for Mercury to 248 years for Pluto. This diagram, drawn to correct scale, shows the orbits of the five planets closest to the sun. Suppose that the five planets are in a line, all on the same side of the sun as shown. Where would they be six months later? Because six months is half of an earth year, we know that the earth would have traveled half-way around its orbit to the other side of the sun. Add the overlay. This shows where the other planets would have gone in this time. Mercury, the planet closest to the sun, would have circled it twice, ending up in the spot shown. Venus, meanwhile, would have gone somewhat more than three-quarters of the way around the sun, Mars slightly more than one-quarter of the way, and Jupiter not very far at all.*

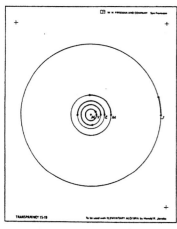

TRANSPARENCY 15-19

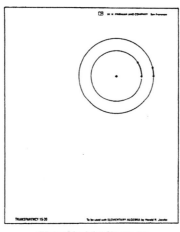

TRANSPARENCY 15-20

Answers to Set I

1. a) $\frac{2}{13}$ c) All numbers except 1. b) $\frac{1-a}{a}$

b) $\frac{2}{13} = \frac{4}{26}$ d) 4 c) $\frac{a}{2}$

c) 13 and 4 e) 30 d) $\frac{a}{a-b}$

d) 2 and 26 f) -16 e) $\frac{ab}{1-ab}$

e) They are equal. g) -1

h) $2\sqrt{6}$ and $-2\sqrt{6}$ 7. a) 37°C

2. a) $\frac{1}{3}$ i) $\frac{5+\sqrt{5}}{10}$ and b) $F = \frac{9C+160}{5}$

b) $\frac{2}{3}$ $\frac{5-\sqrt{5}}{10}$ c) 320°F

c) Not possible

d) $\frac{4}{5x}$ j) $\frac{2}{3}$ and 1 8. a) 190

3. a) x-16 5. a) 180π cu. cm. b) $x = \frac{1+\sqrt{1+8n}}{2}$

b) 4+3x b) $h = \frac{3v}{\pi r^2}$ c) 80

c) 7x-3

d) $x^2+5x+20$ c) 12 cm.

4. a) 2.5 6. a) $\sqrt{a^2-1}$

b) 7.5 and $-\sqrt{a^2-1}$

TRANSPARENCY 15-21

How long would it take the five planets to line up with each other again? This is a hard question to answer and so we will consider a simpler one.

Show Transparency 15-20. *This diagram shows the orbits of just two planets around the sun. Suppose that the planet closer to the sun takes a days to travel once around it and that the planet farther from the sun takes b days to travel once around it. One trip around an orbit corresponds to 360°. Through how many degrees of their orbits would the planets travel in one day?* $\left(\frac{360}{a} \text{ and } \frac{360}{b}.\right)$ *The difference between these numbers,* $\frac{360}{a} - \frac{360}{b}$, *is the number of degrees that the inner planet gains on the outer one in the course of one day. If the planets are originally in line on one side of the sun and it takes x days for them to line up in the same way again, then* $x\left(\frac{360}{a} - \frac{360}{b}\right) = 360$. *Solve this equation for x, simplifying the resulting equation as much as possible.* $\left(x = \frac{ab}{b-a}.\right)$

It takes the earth about 365 days to travel once in its orbit around the sun and Venus about 225 days. Use the equation to find how many days pass between two times that the earth and Venus are in line on one side of the sun. $\left(x = \frac{225 \cdot 365}{365 - 225} = \frac{82,125}{140} \approx 587.\right.$

Approximately 587 days. $\Big)$

Answers to the Set I exercises of the review lesson are given on Transparency 15-21.

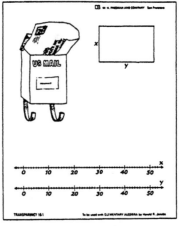

TRANSPARENCY 16-1

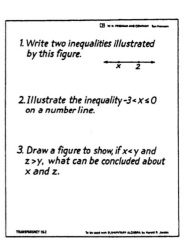

TRANSPARENCY 16-2

Chapter 16

Inequalities

This chapter is a brief introduction to algebraic inequalities. Although the inequality symbols, $>$ and $<$, first appeared in Chapter 3 and the students have worked with them throughout the course, the use of these symbols until now has been restricted primarily to expressing relationships of the numbers of arithmetic.

From the first lesson, the student learns how to represent linear inequalities by shading a number line. The rules for solving linear inequalities are developed in the next lesson and applied both in it and in the following lessons to solving simple inequalities in one or more variables. A final lesson shows how equations and inequalities containing absolute values can be solved both with and without the use of a number line.

Lesson 1
Inequalities

An amusing way to introduce this lesson would be to prepare some "postcards" of the following sizes: $1\frac{1}{4}$ by 2 inches, 3 by $4\frac{1}{4}$ inches, $3\frac{1}{2}$ by $5\frac{1}{2}$ inches (the

standard dimensions), 14 by 22 inches, and 28 by 44 inches (the dimensions of a sheet of poster board).

Hold up the postcard of standard size. *Most postcards come in this size. Suppose, for a gag, that you decided to mail a small postcard to someone* (hold up the $1\frac{1}{4}$-by-2-inch card) *or, perhaps, a large one* (hold up the 14-by-22-inch card). *The post office would accept one of these cards but not the other. The small one is too small; according to postal regulations, a postcard has to be at least $4\frac{1}{4}$ inches long and 3 inches wide.* (Hold up the 3-by-$4\frac{1}{4}$-inch card.) *There is also a limit to how large a postcard may be.* (Hold up the 28-by-44-inch card.) *Although the mailman who would have to deliver it wouldn't be too happy, it is legal to mail a card this big. If the card has the standard shape, it may be as long as 44 inches and as wide as 28 inches. Such a postcard would, of course, require a lot more postage than one of standard size.*

Transparency 16-1 can now be used to show how the limits on postcard dimensions can be written as inequalities and represented on a number line. Additional examples for use in developing the rest of the lesson are given on Transparency 16-2.

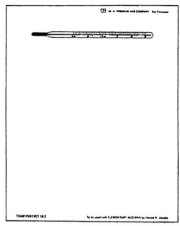

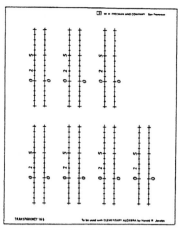

TRANSPARENCY 16-3   TRANSPARENCY 16-4   TRANSPARENCY 16-5

*"All animals are equal
But some animals are
    more equal than others."*

*The typical life-span of a horse is
    more than twice that of a dog.
Dogs live longer than cows or pigs.
Cows outlive pigs by at least eight
    years.
The typical life-span of a pig is ten years.*

*What can you conclude about
    the life-span of a horse?*

Lesson 2
Solving Linear Inequalities

Show Transparency 16-3. *This is a picture of the standard thermometer used for measuring a person's temperature. Normal temperature is marked with an arrow on the scale. Write an equation for it, letting* t *represent temperature. (*t = 98.6.*) Write an inequality indicating the range of temperatures that the thermometer can measure. (*$96 \leq t \leq 106$.*) Write an inequality representing the temperatures that would break the thermometer by overexpansion (*$t > 106$.*)*

According to the Guinness Book of World Records, *marathon runners in hot weather sometimes reach temperatures of 105.8°. Cases have been recorded in which people having temperatures as low as 60° or as high as 112° have survived.*

Transparency 16-4 is for use in discussing the Set IV exercise of Lesson 1.

To introduce the new lesson, you might show the Peanuts strip of September 23, 1968, to your class. The first panel shows Lucy sitting at her stand with the sign, "Psychiatric help 5 cents." Snoopy comes along with a hammer and some wood and sets up a stand beside her with the sign "Friendly advice 2 cents." The strip is reproduced in the Peanuts book titled *My Anxieties Have Anxieties* (Peanuts Parade Paperback No. 18).

After showing the cartoon, you might hand out duplicated copies of Transparency 16-5. Starting with the inequality, $2 < 5$, to show that Snoopy charges less for his advice than Lucy does for hers, some examples such as those illustrated below and on the next page can be used to help your students discover the rules of operation for inequalities.

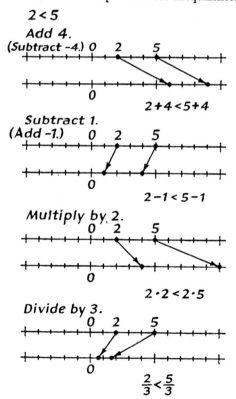

$2 < 5$

Add 4.
(Subtract −4.)

$2 + 4 < 5 + 4$

Subtract 1.
(Add −1.)

$2 − 1 < 5 − 1$

Multiply by 2.

$2 \cdot 2 < 2 \cdot 5$

Divide by 3.

$\dfrac{2}{3} < \dfrac{5}{3}$

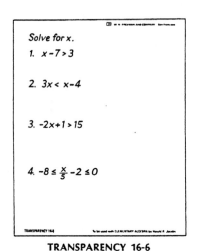

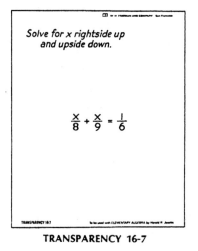

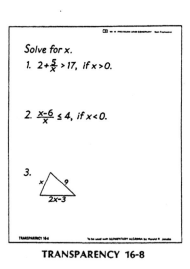

TRANSPARENCY 16-6 | **TRANSPARENCY 16-7** | **TRANSPARENCY 16-8**

Multiply by 0.

$$0 \cdot 2 = 0 \cdot 5$$

Multiply by -1.

$$-1 \cdot 2 > -1 \cdot 5$$

Divide by -2.

$$\frac{2}{-2} > \frac{5}{-2}$$

Examples for use in explaining how to use these rules to solve inequalities are given on Transparency 16-6.

Lesson 3
More on Solving Inequalities

Show Transparency 16-7. *Here is another problem that is actually two-in-one. Copy it in the center of a sheet of paper and solve it in each direction. (The*

solution to the rightside-up version is $x > \dfrac{12}{17}$. *In the upside-down version, the fact that the variable appears in the denominators of the fractions makes the solution more complicated and two cases must be considered.*

$$\frac{9}{1} < \frac{6}{x} + \frac{8}{x}$$

$$9 < \frac{14}{x}$$

Case 1. $x > 0$.

$$9x < 14$$

$$x < \frac{14}{9}$$

Case 2. $x < 0$. *In this case, no solution is possible because* $\dfrac{14}{x} < 0$ *if* $x < 0$. *Thus* $9 < \dfrac{14}{x}$ *is impossible.*

The solution to the upside-down version, then, is
$$0 < x < \frac{14}{9}.$$

Additional examples for developing the new lesson are given on Transparency 16-8.

Transparency 16-9 is for use in discussing the Set IV exercise of Lesson 2.

TRANSPARENCY 16-9

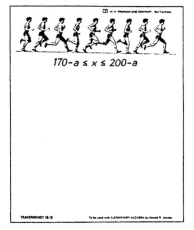

$$170-a \leq x \leq 200-a$$

TRANSPARENCY 16-10

Lesson 4
Absolute Value and Inequalities

Show Transparency 16-10. *Running is one of the best ways to keep physically fit. Ideally, the runner's heart rate should be raised to about 75 percent of its maximum capacity and kept there for about 15 minutes. To accomplish this, the runner should run at a speed such that his pulse is raised to the level indicated by the inequality 170 − a ≤ x ≤ 200 − a, in which a represents his age and x represents the rate in heartbeats per minute.**

Copy this inequality, substituting your own age for a and simplifying. (For a 16-year-old, the pulse rate inequality is 154 ≤ x ≤ 184. Thus the speed at which a 16-year-old runs should be such that his heart rate is raised to between 154 and 184 beats per minute.) The rest of the new lesson can now be developed along the following lines.

Illustrate the inequality that you have written on a number line.

154   184

---

*Easy Running by Bronnie Storch Kupris (Dell, 1978), p. 22.

Another way to represent this inequality is by means of absolute value. What number is midway between the minimum and maximum pulse rate for your age? (For a 16-year-old, it is 169.) Mark this point on the number line.

154   169   184
15   15

What is its distance from each endpoint of the pulse-rate interval? (15.) Suppose that you run at a speed such that your pulse rate is raised to x beats per minute and that x is some number in this interval. This means that the distance of x from 169 is less than or equal to what number? (15.) Because the distance between x and 169 can be represented by |x − 169|, we can write the inequality, |x − 169| ≤ 15.

Verify that this inequality is equivalent to the original one by checking both endpoint values in it. (|154 − 169| = |−15| = 15, |184 − 169| = |15| = 15.)

Additional examples for use in illustrating the solution of equations and inequalities containing

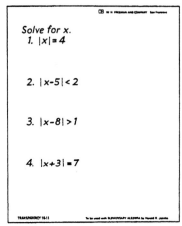

Solve for x.
1. |x| = 4

2. |x-5| < 2

3. |x-8| > 1

4. |x+3| = 7

TRANSPARENCY 16-11

A given play in poker is favorable if $\frac{wp}{c} > 1$.

TRANSPARENCY 16-12

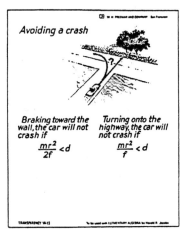

Avoiding a crash

Braking toward the wall, the car will not crash if $\frac{mr^2}{2f} < d$

Turning onto the highway, the car will not crash if $\frac{mr^2}{f} < d$

TRANSPARENCY 16-13

absolute values are given on Transparency 16-11. Be sure to show how to solve each example both with and without a number line.

Transparency 16-12 is for use in discussing the Set IV exercise of Lesson 3.

Review of Chapter 16

Avoiding a crash*

Show the diagram at the top of Transparency 16-13. *Imagine that you are driving along a road that ends in a T-shaped intersection with a highway. You are driving much faster than you should be and you see that you are in danger of hitting a brick wall along the far side of the intersection. There are no cars on the highway in either direction.*

What should you do to avoid hitting the wall? Would it be better to steer straight at it, putting on your brakes as hard as you can, or, without touching the brakes, to turn onto the highway?

Show the inequalities at the bottom of the transparency. *The two choices are represented by these inequalities. In them,* m *represents the mass of the car,* r *represents its speed,* f *represents the frictional force and* d *represents the distance of the car from the wall. Solve each of these inequalities for* f. *(Noting that* f > 0, *we get for braking toward the wall,* $f > \frac{mr^2}{2d}$, *and for turning onto the highway,* $f > \frac{mr^2}{d}$.) *In which inequality is the frictional force greater?* $\left[\textit{The second, because } \frac{mr^2}{d} = 2\left(\frac{mr^2}{2d}\right). \right]$ *To try to avoid crashing into the wall, what should the driver do?* (Brake toward the wall.)

---

*This exercise is from *Science Brain-Twisters, Paradoxes, and Fallacies* by Christopher P. Jargocki (Scribners, 1976), pp. 27 and 104–105.

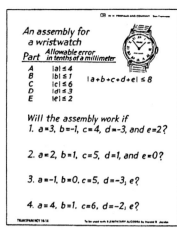

An assembly for a wristwatch

Part — Allowable error in tenths of a millimeter

A	$|a| \le 4$
B	$|b| \le 1$
C	$|c| \le 6$
D	$|d| \le 3$
E	$|e| \le 2$

$|a+b+c+d+e| \le 8$

Will the assembly work if

1. $a=3$, $b=-1$, $c=4$, $d=-3$, and $e=2$?

2. $a=2$, $b=1$, $c=5$, $d=1$, and $e=0$?

3. $a=-1$, $b=0$, $c=5$, $d=-3$, e?

4. $a=4$, $b=1$, $c=6$, $d=-2$, e?

TRANSPARENCY 16-14

Answers to Set I

1. a) >
   b) <
   c) <
   d) >
   e) =
   f) >
   g) =
   h) >

   c) None
   d) >

   5. a) No
   b) Yes
   c) Yes
   d) Yes
   e) No
   f) Yes
   g) Yes
   h) Yes
   i) No

2. a) $x \ge -3$
   b) $2 < x < 7$
   c) $x < 9$
   d) $-4 \le x \le 1$

3. a) >
   b) None
   c) $\ge$
   d) None

4. a) <
   b) >

   6. a) $x > -1$
   b) $x > \frac{3}{4}$
   c) $x > -8$
   d) $x \ge 8$
   e) $2 < x < 6$
   f) $x > -2$
   g) $x \le 24$

   h) $x > 3$
   i) $x < -5$
   j) $x > -3$
   k) $x < -3$
   l) $x \le 22$

   7. a) $x < \frac{b+1}{a}$
   b) $x < \frac{ab}{c}$
   c) $x \le \frac{c}{a-b}$
   d) $a \le x < \frac{x}{2a+b}$

   8. a) 5 and -5
   b) 11 and -7
   c) 4 and -16
   d) $x+4$ and $x-4$

   9. a)

   b)

   c)

   d)

   e)

   f)

   10. a) 5 and -5
   b) $-15 < x < 15$
   c) 3 and -3
   d) 3 and -17
   e) $x \ge 8$ or $x \le -8$
   f) $x \ge 8$ or $x \le -4$
   g) No solution
   h) All nos. except -1.

TRANSPARENCY 16-15

Transparency 16-14 is for use in discussing the Set IV exercise of Lesson 4.

The answers to the Set I exercises of the review lesson are given on Transparency 16-15.

TRANSPARENCY 17-1

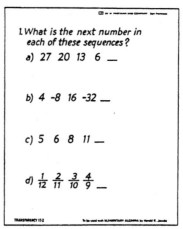

1. What is the next number in each of these sequences?

a) 27 20 13 6 __

b) 4 -8 16 -32 __

c) 5 6 8 11 __

d) $\frac{1}{12}$ $\frac{2}{11}$ $\frac{3}{10}$ $\frac{4}{9}$ __

TRANSPARENCY 17-2

2. Find the first three terms of the sequence whose nth term is given by the formula

a) $t_n = 2 + 4n$

b) $t_n = 2 \cdot 4^n$

3. Find a formula for the nth term of each of these sequences.

c) 3 11 19 27 35...

d) 1 $\frac{1}{4}$ $\frac{1}{9}$ $\frac{1}{16}$ $\frac{1}{25}$ ...

TRANSPARENCY 17-3

Chapter 17
Number Sequences

A brief survey of the interesting properties and applications of number sequences brings the course to an end. The primary emphasis is on arithmetic and geometric sequences. Their relationship to linear and exponential functions affords a nice review of the latter, whereas the work with infinite geometric sequences prepares the student for those lessons in geometry in which the formulas for the circumference and area of a circle are obtained by treating the circle as the limiting case of a regular polygon. Although the formulas introduced in this chapter are worth knowing and your students should understand how they are derived, it is probably best not to expect that they be memorized.

Lesson 1
Number Sequences

Show Transparency 17-1. *The longest taxi ride on record was taken by two women from Hoboken, New Jersey, Mrs. Ann Drache and Mrs. Nesta Sgro, in 1976. The ride covered a total distance of 6,752 miles and lasted four weeks.*

Rather than paying the usual meter rate, the women agreed with the driver at the beginning of the trip on a fare of $2,500. Ordinarily, taxi fares begin with a set amount, which then increases at a set rate. For example, a typical rate (the one in Los Angeles as this is being written) is 80 cents at the beginning and 20 cents for each quarter of a mile traveled. This means that the meter would read 0.80 when you got in the cab and then change to 1.00, 1.20, 1.40, 1.60, and so on, as the trip progressed. Write these numbers as a sequence below the photograph and use them to introduce the lesson as in the text. Example exercises are given on Transparencies 17-2 and 17-3.

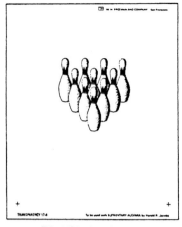

TRANSPARENCY 17-4

TRANSPARENCY 17-5

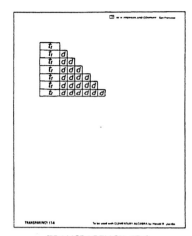

TRANSPARENCY 17-6

Find the following sums.

1. 35+39+43+47+51+55+59+63+67

2. 1+3+5+7+···+99

Find the given terms.

3. 13 24 35 ...; the 20ᵗʰ term

4. 75 70 65 ...; the 51ˢᵗ term

TRANSPARENCY 17-7

Lesson 2
Arithmetic Sequences

A bowler's nightmare

Show Transparency 17-4. *If you enjoy bowling, you know that the pins are set up in four rows with one pin in the first row, two pins in the second, three pins in the third, and four in the last. To make a strike, you have to knock over all ten pins at the same time.* Add the overlay. *Suppose that you are dreaming about bowling and in your dream there are ten rows of pins instead of four. If each successive row still has one more pin than the row preceding it, how many pins would you have to knock over to make a strike?*

After your students have had time to work out the answer (55) the long way, point out that the numbers of pins in the rows form an arithmetic sequence

$$1 \quad 2 \quad 3 \quad 4 \quad 5 \quad 6 \quad 7 \quad 8 \quad 9 \quad 10$$

and that the answer is the sum of these numbers. Use Transparency 17-5 with its overlay to show how this can be done by imagining the array upside down beside itself, as in the example of counting the number of people in the people pyramid on page 784 of the text.

To develop the general formula for the sum of the terms of an arithmetic sequence, reproduce the pattern on the last page of the book of transparency masters, cut it out (as one piece), and use it with Transparency 17-6 as done on pages 784–785 of the text. The figure on Transparency 17-6 can also be used to develop the formula for the nth term of an arithmetic sequence.

Some example exercises are given on Transparency 17-7.

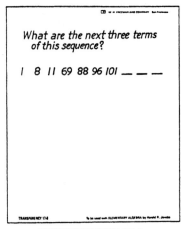

What are the next three terms of this sequence?

1 8 11 69 88 96 101 __ __ __

TRANSPARENCY 17-8

Transparency 17-8 is for use in discussing the Set IV exercise of Lesson 1.

Lesson 3
Geometric Sequences

One of the old Fibber McGee and Molly radio programs included a humorous segment about a geometric sequence. It is available on a long-playing record titled "Son of Jest Like Old Times" and the first two minutes of it would make an amusing introduction to this lesson. The record can be purchased for $6.98, postpaid, from the Radiola Company, Dept. C, Sandy Hook, Conn. 06482.

The dialogue in the segment I suggest playing goes like this:

FIBBER: Hey Molly! Molly, where are you? Hey Molly, where are you? Hey Molly!

MOLLY: Here I am dearie. You can stop shouting. Somebody chasing you or something? Are the police. . . ?

FIBBER: No, no, no, no. Look, we're going to be rich. I've got the idea of a lifetime.

MOLLY: All right, all right. Take off your coat and hat and tell me about it.

FIBBER: Well, my gosh, kiddo, aren't you excited? You don't want to be rich? Don't you want to know my plans?

MOLLY: Of course I do, McGee; I always like to know how you're going to make us rich.

FIBBER: Well.

MOLLY: You've had thousands of ideas like that and I've been interested in every one of them.

FIBBER: Well, this one is it, Snookie. This is fool proof. This is air tight and it's legal. See this can of soup?

MOLLY: Yes, what kind is it?

FIBBER: Mock turtle. I got it down. . . .

MOLLY: Why didn't you get tomato? I don't care for mock turtle soup.

FIBBER: The kind of soup it is ain't important. I bought this can of soup because. . . .

MOLLY: Well, as long as you're buying soup, you might at least have got the same kind we like, you know. My goodness, you don't need to. . . .

FIBBER: But I tell you that's *why* I bought it; because neither of us like it.

MOLLY: Let me feel your forehead, dearie. You're running a little fever.

FIBBER: No, no, no, I'm not. I'm cuke as a coolcumber, I mean as . . . Listen, this is a new soup, see, MacBender's mock turtle soup, and here's what they advertise. Now get this. They say "if you don't like this soup, return it to your dealer and get double your money back."

MOLLY: Yes, but you won't know if you like it till you eat it and if you eat it how can you return it?

FIBBER: We can taste it, can't we? And we won't like it, will we, so we return it, don't we, so we get twice our money back, aren't we?

MOLLY: Yes, but dearie, you can't. . . .

FIBBER: Look, I paid thirty-two cents for this soup. Double my money is sixty-four. I take the

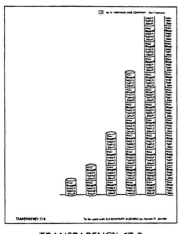

TRANSPARENCY 17-9

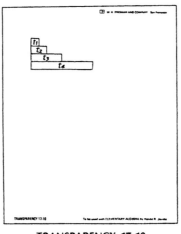

TRANSPARENCY 17-10

3 12 48 192 ...
1. Write a formula for the nth term.

2. Write an expression for the 7th term.

3. Write a formula for the sum of the first n terms.

4. Write an expression for the sum of the first 7 terms.

TRANSPARENCY 17-11

sixty-four and buy two cans of soup. I take them back and get a dollar twenty-eight. Then I buy four cans of soup. I return them. Two dollars fifty-six. You catch on?

MOLLY: Why, I never heard of such a thing.

FIBBER: Ha, ha, ha, ha. I've got it all figured out, baby. I've figured out that, by doubling my money every time, at the end of two weeks I'll have three million four hundred thousand bucks!

MOLLY: But darling, the company that makes the soup. Why, you'll put them out of business!

FIBBER: That's their tough luck. They should've thought of that.*

After playing the recording, you might show Transparency 17-9 and ask your students to write the first ten terms of the sequence of numbers of soup cans in McGee's scheme. (1 2 4 8 16 32 64 128 256 512.) *We see from this sequence that, if the clerk at the grocery store cooperated with the scheme, McGee would be taking back 512 cans of soup on his tenth trip. How many cans would he return on the nth trip? In other words, what is a formula for the nth term of the sequence?* ($t_n = 1 \cdot 2^{n-1}$.)

To make the amount of money he has in mind, Fibber McGee would have to make twenty-five trips to the grocery store. How many cans would he return on the twenty-fifth trip? (2^{24}, *which, as students with calculators can discover, is 16,777,216.*)

*© Radiola Company, 1970. Reproduced by permission.

Some of your students may notice that the number in the script, $3,400,000, does not make sense in this sequence: twenty-four trips would net McGee $2,684,354.56 and twenty-five trips would net him $5,368,709.12. It is probably best not to call attention to this, however, unless someone brings it up.

How many cans in all would McGee have bought and returned in twenty-five trips? (It is the sum: $1 + 2^1 + 2^2 + 2^3 + \ldots + 2^{24}$.) *Is there a shortcut for finding this sum? For example, could we find it by using the method that we use to find the sum of the terms of an arithmetic sequence?* Show Transparency 17-6 together with its duplicate pattern to remind everyone of this method. Then show Transparency 17-10, together with the duplicate pattern from the last page of the book of transparency masters to show that the same method will not work. Use the second pattern with the figure on the transparency to help your students discover the formula as done on pages 793–794 of the text. After deriving the general formula for the sum of the terms of a geometric sequence, apply it to finding the sum of the numbers of cans in the twenty-five trips: $1 + 2^1 + 2^2 + 2^3 + \ldots + 2^{24}$

$$= \frac{1(2^{25} - 1)}{2 - 1} = 2^{25} - 1 = 33,554,431.$$

Additional examples are given on Transparency 17-11.

Graphs for checking exercises 14 and 15 of Lesson 2 are given on Transparency 17-12. Transparency 17-13 is for use in discussing the Set IV exercise.

Lesson 4
Infinite Geometric Sequences

The problem of the miniature horses presented in the text is based upon an article titled "A step backward in horse history" in the March 1974 issue of the *Smithsonian* magazine and was suggested by Michael Nevard, who teaches at Chatsworth High School, Chatsworth, California. The article includes several interesting pictures in addition to the one on page 802 of the text.

Transparencies 17-14 and 17-15 are for use in introducing and developing this lesson. To use Transparency 17-16, help your students construct the following table and ask them to guess the numbers at the bottom before developing the formula for the "sum" of the terms of an infinite geometric sequence having ratio r such that $|r| < 1$.

n	t_n	S_n
1	24	24
2	6	30
3	1.5	31.5
4	0.375	31.875
5	0.09375	31.96875
∞	?	?

You might cut out some strips like those on the transparency and stack them up to illustrate the sum.

An example of how the sum formula can be used to express a repeating decimal ($0.\overline{63}$ is a good one) as the quotient of two integers should also be considered.

Graphs for checking exercises 13 and 14 of Lesson 3 are given on Transparency 17-17.

Review of Chapter 17

Show Transparency 17-18. *Maurits Escher created this woodcut in 1939, the same year that he created the picture on the cover of our textbook. It was one of his first attempts to represent infinity. As Escher described it, "the figures that were used to construct this picture are subjected to a constant radial reduction in size, working from the edges toward the center, the point at which the limit is reached of the infinitely many and the infinitely small."*[*]

Can infinitely many figures really be imagined in a picture that is not infinite? We can use our knowledge of infinite geometric sequences to answer this question. As we look from the edges to the center, each successive reptile shown in the darkest color is six-tenths as long as the one before it. Show Transparency 17-19. If the first reptile is 9 centimeters long (the length in Escher's original woodcut), what would the sum of the lengths of an infinite number of them be?
$$(S = \frac{9}{1 - 0.6} = 22.5. \text{ It would be 22.5 centimeters.})$$

After checking this answer, show Transparency

---

[*] Quoted in the last chapter, "An Artist's Approach to Infinity," of *The Magic Mirror of M. C. Escher* by Bruno Ernst (Random House, 1976).

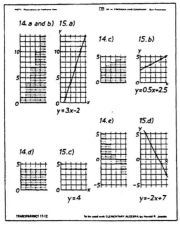

TRANSPARENCY 17-12

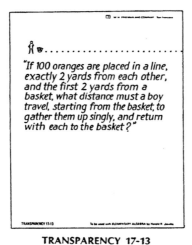

"If 100 oranges are placed in a line, exactly 2 yards from each other, and the first 2 yards from a basket, what distance must a boy travel, starting from the basket, to gather them up singly, and return with each to the basket?"

TRANSPARENCY 17-13

50 45 40.5 36.45 ...

TRANSPARENCY 17-14

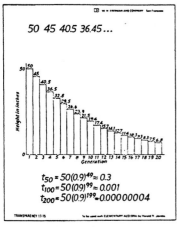

50 45 40.5 36.45 ...

$t_{50} = 50(0.9)^{49} \approx 0.3$
$t_{100} = 50(0.9)^{99} \approx 0.001$
$t_{200} = 50(0.9)^{199} \approx 0.00000004$

TRANSPARENCY 17-15

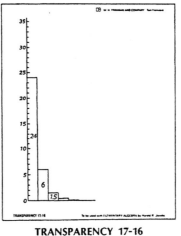

TRANSPARENCY 17-16

TRANSPARENCY 17-17

TRANSPARENCY 17-18

TRANSPARENCY 17-19

Chapter 17, Review   175

TRANSPARENCY 17-21

n	t_n	S_n
1	64	64
2	32	96
3	16	112
4	8	120
5	4	124
6	2	126
7	1	127
8	0.5	127.5
9	0.25	127.75
10	0.125	127.875
⋮	⋮	⋮

Answers to Set I

1.a) Geometric, ratio 4
 b) Arithmetic, difference 2.5
 c) Neither
 d) Arithmetic, difference $-\frac{1}{2}$
 e) Geometric, ratio 10
 f) Neither

2.a) 6
 b) $\frac{5}{36}$
 c) 40.5
 d) 21
 e) 1
 f) $\frac{1}{3}$

3.a) 48
 b) 98
 c) 5

d) 3

4.a) 2,5,8,11
 b) 1,16,81,256
 c) 4,16,64,256
 d) 15,20,25,30
 e) 10,20,40,80
 f) 2,2$\frac{1}{2}$,3$\frac{1}{3}$, 4$\frac{1}{4}$
 g) a and d
 h) c and e

5.a) $t_n = n+4$, 104
 b) $t_n = -2n$, -30
 c) $t_n = 5^n$, 78,125
 d) $t_n = \frac{n}{6}$, $\frac{4}{6}$
 e) $t_n = n^3$, 1,000

f) $t_n = 4 \cdot 7^{n-1}$, 67,228
 g) $t_n = \frac{1}{n^2}$, $\frac{1}{196}$
 h) $t_n = \sqrt{n-1}$, $\frac{1}{7}$

6.a) 208
 b) -63
 c) 2,583
 d) 1,740

7.a) 65,535
 b) 156,248
 c) -2,184

8.a) 0.01
 b) -0.7
 c) 1.5

9.a) 18
 b) 625

TRANSPARENCY 17-22

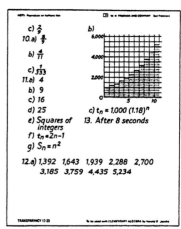

c) $\frac{2}{9}$
10.a) $\frac{4}{7}$
 b) $\frac{4}{11}$
 c) $\frac{1}{333}$
11.a) 4
 b) 9
 c) 16
 d) 25
 e) Squares of integers
 f) $t_n = 2n-1$
 g) $S_n = n^2$

b)

c) $t_n = 1,000 (1.18)^n$
13. After 8 seconds

12.a) 1,392 1,643 1,939 2,288 2,700
  3,185 3,759 4,435 5,234

TRANSPARENCY 17-23

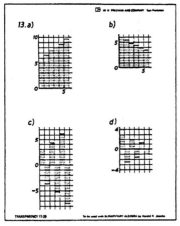

13.a)   b)

c)   d)

TRANSPARENCY 17-20

17-18 again. *Escher's original woodcut measures slightly more than 32 centimeters from each corner to the center; so there is room, in effect, for infinitely many of these reptiles without even putting them end to end!*

Graphs for checking exercise 13 of Lesson 4 are given on Transparency 17-20. Transparency 17-21 is for use in discussing the Set IV exercise.

Answers to the Set I exercises of the review lesson are given on Transparencies 17-22 and 17-23.

Final Review

Ten sample questions representative of one type of question used in the mathematics sections of the SAT are reproduced on Transparency F-1. These questions provide an interesting way in which to begin the final review. According to the booklet *Taking the SAT*, published by the Educational Testing Service on behalf of the College Board, "some questions in the mathematical sections of the SAT require you to apply numerical, graphic, spatial, symbolic, and logical techniques to situations familiar to you; these may be similar to exercises in

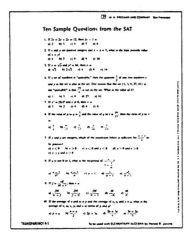

TRANSPARENCY F-1

Final Review : Test I

1. 0.003, 0.01, 0.2
2. x+−y
3. 6x
4. ≥
5. 6x⁵ → $6x^5$
6. True
7. 5√2 → $5\sqrt{2}$
8. a > c
9. 2x+8
10. 3 and 13
11. $\frac{21}{5}$
12. 10x
13. $4x^2$
14. (2x+1)(2x−1)
15. $3x^2$
16. False
17. $(x-3y)^2$
18. $\frac{2}{3}$
19. $4x^2-20x+25$
20. $\frac{2x}{5}$
21. True
22. $\frac{7}{5}$
23. $5\sqrt{2}$
24. $\frac{x+2}{x-2}$
25. 121
26. y
27. $\frac{x^2}{8}$
28. $x+\sqrt{y}$
29. 73
30. True
31. 4.4
32. $\frac{\sqrt{2x}}{4}$
33. $-4x^2+x+8$
34. x < −5
35. False
36. xy(x+y)(x−y)
37. x^2+4x+2
38. 100.48
39. $\frac{3}{11}$
40. −2 < x < 3
41. $\frac{1}{5}$
42. (x+7)(x−6)
43. 1
44. $\sqrt{5}$

TRANSPARENCY F-2

45. $\frac{x}{x+y} - \frac{y}{x+y}$
46. $0.\overline{45}$
47. y = |x|
48. 5.1×10^{-3}
49. 0 and 3
50. 1.4×10^4
51. x^3+3x^2+2x
52. 14x
53. $x^2-5x-14$
54. 4.8
55. 14
56. $1, \frac{1}{4}, \frac{1}{5}$
57. (7,4)
58. Irrational
59. x = ab
60. $\frac{4}{3x}$
61. y
62. $\frac{x}{2}$
63. It remains the same.
64. 9 and −9
65. $2x^2(2x-3)$
66. 820
67. True
68. $-2+\sqrt{5}$ and $-2-\sqrt{5}$
69. $x < -\frac{6}{5}$
70. $y = \frac{100}{x}$
71. Inversely
72. Y
73. $\frac{1.5x}{1.2(x+20)}$
74. 1.5x = 1.2(x+20)
75. 120 meters
76. $\frac{t+2h+2w \leq 100}{100}$
77. t ≤ 30
78. 10
79. 320
80. 1 hour

TRANSPARENCY F-3

your textbook. Other questions may require you to do some original thinking. The mathematical preparation expected is a year of algebra and some geometry."

The final review in the textbook consists of two exercise sets, each containing eighty items, which can be taken as practice final exams. Answers to Test I are given on Transparencies F-2 and F-3.

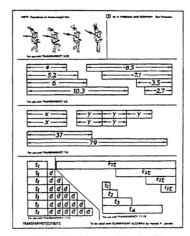

TRANSPARENCY CUTOUTS

Chapter 17, Final Review   177

ANSWERS
to Exercises in Sets I, III, and IV

ANSWERS
to Exercises in Sets I, III, and IV

Introduction

1. a)

Think of a number:	1	2	3	4	5
Double it:	2	4	6	8	10
Add six:	8	10	12	14	16
Divide by two:	4	5	6	7	8
Subtract the number that you first thought of:	3	3	3	3	3

   b) No.
   c) Think of a number: □

   Double it: □□

   Add six: □□ ∘∘∘ / ∘∘∘

   Divide by two: □ ∘∘∘
   Subtract the number that
      you first thought of: ∘∘∘
   d) Yes.
2. Step 1. Think of a number.
   Step 2. Multiply it by four.
   Step 3. Add eight.
   Step 4. Divide by four.
   Step 5. Add three.
   Step 6. Subtract the number that
      you first thought of.
3. a) 2.
   b) The result at the end is 4 instead of 2.
   c) The result at the end is 1.
   d) The result at the end now depends on the
      number first thought of.
4. (A possible answer.)
   Think of a number.
   Triple it.
   Add twelve.
   Divide by three.
   Subtract the number that
      you first thought of.
   The result is four.

Chapter 1, Lesson 1

Set I
1. 1,776.   **2.** 1,107.   **3.** 1.984.   **4.** 20.202.
5. 1,370.   **6.** 1,370.   **7.** 4.664.   **8.** 10.631.
9. 2.8.   **10.** 1.605.

Set III
17. a) $3 + 11$ or 14.   b) $3 + x$.   c) $y + 11$.
d) $y + x$.   e) $7 + 2$ or 9.   f) $7 + x$.   g) $9 + 1 + 4$
or 14.   h) $x + 1 + 4$ or $x + 5$.   i) $9 + y + 4$ or
$y + 13$.   j) $x + y + 4$.   **18.** a) $8 + 3$.   b) 11.
c) $x + 6$.   d) 7.   e) 11.   **19.** a) 17.   b) $x + 14$.
c) $x + y + 16$.   d) $x + x + x$.   e) $x + y + x + y$.

20. a) ∘ ∘ □ and □ ∘ ∘

   b) ∘ ∘ ∘ ∘ ∘ ∘ ∘ ∘ □ ∘ and □ ∘ ∘ ∘ ∘ ∘ ∘ ∘ ∘ ∘

   c) □ □ ∘ ∘ ∘ and □ ∘ ∘ ∘ □

21. a) $2 + 1 + y$ or $3 + y$.   b) $0 + 1 + y$ or $1 + y$.
c) $x + 1 + 6$ or $x + 7$.   d) $x + 1 + 9$ or $x + 10$.
e) $3 + 1 + 7$ or 11.   **22.** a) 19 miles.   b) $x + 1$
miles.   c) $y + 3$ miles.   d) $y + z$ miles.

Set IV
It would happen with any four numbers because:

a	b	$a+b$
x	y	$x+y$
$a+x$	$b+y$	$a+b+x+y$ $= a+x+b+y$

Chapter 1, Lesson 2

Set I

1. 20,222. **2.** 589. **3.** 877. **4.** 3.321.
5. 4.221. **6.** 0. **7.** 0.1. **8.** 0.01.
9. 1,793.88. **10.** 179.388.

Set III

18. a) 9 − 3 or 6. b) $x − 5$. c) $5 − x$. d) 20 − 8
or 12. e) $x − 2$. f) $y − x$. g) $7 − x$. h) $x − 7$.
19. a) 5. b) $y − 7$. c) 3. d) $10 − y$. e) $y − 15$.
f) $z − y − 21$. **20.** a) 7. b) 8. c) 18. d) The
value of $x − 2$ gets larger. e) 7. f) 6. g) 0.
h) The value of $8 − x$ gets smaller. **21.** a) 11.
b) 14. c) 11. d) 14. e) Each expression is
$x + 7 − y$. **22.** a) $12 − x$ feet. b) $x + y$ feet.
c) $x − y$ feet. **23.** a) $2 − x$ dollars. b) $2 + y$
dollars. c) $z − 2$ dollars. **24.** a) $100 − x$.
b) $100 + y$. c) $100 + y − z$.

Set IV

The clerk is giving a customer the change for an
$8.47 purchase: $20 − $8.47 = $11.53. The problem
is being solved by addition.

Chapter 1, Lesson 3

Set I

1. 36,000. **2.** 714,285. **3.** 77,777. **4.** 1.
5. 12.345. **6.** 12.345. **7.** 1.001. **8.** 10.01.
9. 100.1. **10.** 10,000.

Set III

18. a) b) ▢▢▢ c)

19. a) 21. b) 10. c) $7x$. d) $7 + x$. e) xy.
f) $x + y$. g) yy. h) $5x$. i) $x − 5$. j) $10 + x$.
k) $24x$. l) $y + 17$. m) $60y$. n) $x + y + 2$.
o) $2xy$. **20.** a) $3 \cdot 10$. b) $10 \cdot 3$. c) $3x$. d) $15 \cdot 4$.
e) $4x$. f) xy. g) $19 + 19$. h) $x + x + x + x + x$.
i) $3 + 3 + \cdots + 3$ (y of them).
j) $x + x + \cdots + x$ (x of them).
21. a) 72. b) $16x$. c) $7xy$. d) xxx. **22.** a) 108
cents. b) $9x$ cents. c) $108y$ cents. **23.** a) $20x$.
b) xx. **24.** a) 1,000. b) $1,000x$. c) 3,000.
d) $1,000y$. e) 100,000. f) $100,000y$. g) $1,000,000y$.

Set IV

1.

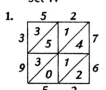

Yes: 3952 is correct.

2.

Yes. If we "carry" from each
slanting column to the next
going clockwise, we get 2407.

Chapter 1, Lesson 4

Set I

1. 50. **2.** 0.02. **3.** 1234. **4.** 3003.
5. 1.6. **6.** 16. **7.** 0.625. **8.** 0.0625.

Set III

17. a) 4. b) 6. c) $\frac{x}{5}$. d) $\frac{5}{x}$. e) $\frac{3}{x}$. f) $3x$. g) $\frac{x}{12}$.

h) $x − 12$. i) $\frac{y}{x}$. j) yx. **18.** a) 21. b) 42.

c) 77. d) The value of $7x$ gets larger. e) 0. f) 4.

g) 17. h) The value of $\frac{x}{3}$ gets larger. i) 2. j) 1.8.

k) 0.4. l) The value of $\frac{18}{x}$ gets smaller. **19.** a) 7.

b) 8. c) $\frac{15}{x}$. d) $\frac{x}{4}$. e) $\frac{x}{y}$.

20. a)

$$\begin{array}{rl} 966 & \\ -840 & \text{40 twenty-ones} \\ \hline 126 & \\ -126 & \underline{\text{6 more twenty-ones}} \\ \hline 0 & \text{46 twenty-ones subtracted} \end{array}$$

b)

$$\begin{array}{rl} 125 & \\ 7\overline{)875} & \quad 875 \\ -700 & -700 \quad \text{100 sevens} \\ \hline 175 & \quad 175 \\ -140 & -140 \quad \text{20 more sevens} \\ \hline 35 & \quad 35 \\ -35 & \underline{-35 \quad \text{5 more sevens}} \\ \hline 0 & \quad 0 \quad \text{125 sevens subtracted} \end{array}$$

21. a) 20 minutes.   b) $\frac{2000}{x}$ minutes.   **22.** a) xy.

b) $\frac{80}{x}$.   c) $\frac{x}{8}$.   **23.** a) 150 square inches.   b) $6x$ square inches.   c) 16 square inches.   d) $\frac{y}{6}$ square inches.   **24.** a) 1.08 dollars.   b) 0.09 dollars.   c) $\frac{x}{10}$ dollars.   d) $\frac{x}{120}$ dollars.

Set IV

Because $\frac{3}{4}$ of the traveler's bread came from the first pilgrim, he should receive $\frac{3}{4}$ of the coins, or 9 coins.

Chapter 1, Lesson 5

Set I

1. 25.   **2.** 32.   **3.** 1,000.   **4.** 10,000,000.
5. 1.   **6.** 1.   **7.** 1.69.   **8.** 9.61.   **9.** 0.064.
10. 0.004096.

Set III

18. a) "x cubed" and "the third power of x."   b) An exponent.   **19.** a) 6^2.   b) 4^2.   c) y^2.   d) 5^3.
e) y^3.   **20.** a) 2^3.   b) 5^{10}.   c) x^2.   d) x^4.   e) 9^y.
f) y^z.   **21.** a) 6^5.   b) 11^3.   c) y^7.   d) 3^{10}.   e) 3^z.
f) x^y.   g) $1 \cdot 1 \cdot 1 \cdot 1$.   h) $x \cdot x \cdot x \cdot x \cdot x$.
i) $5 \cdot 5 \cdot \ldots \cdot 5$ (x of them).
j) $x \cdot x \cdot \ldots \cdot x$ (y of them).
22. a) 7^3.   b) 81^2.   c) 9^4.   d) 3^8.   e) 10^3.   f) 10^7.
g) Every power of 1 is 1.   **23.** a) The last digit of 6^{100} is 6.
b) $5^2 = 25$
$ 5^3 = 125$
$ 5^4 = 625$
$ 5^5 = 3{,}125$
$ 5^6 = 15{,}625$
c) The last two digits of 5^{100} are 25.   d) The last digit of 9^{100} is 1.

Set IV

2^{19} minutes, which is more than 52 times as many minutes as are in a week. In fact, in the fourteenth week he would spend more minutes studying than there are minutes in a week.

Chapter 1, Lesson 6

Set I

1. a) ☐   b) ☐ o o o o
c) ☐☐☐ o o o o o o o o o o o o
d) ☐☐☐ o o o   e) ☐ o   f) o

2. a) $3x$.   b) $3 + x$.   c) $3 - x$.   d) $\frac{3}{x}$.   e) x^3.
f) 3^x.   **3.** a) $2a$.   b) $b + b + b + b + b$.   c) c^3.
d) $d \cdot d \cdot d \cdot d$.   e) xe.   f) f^y.

Set III

11. a) 1.   b) 0.   c) 1.   d) 1.   e) $\frac{0}{0}$ is meaningless.
f) 0.   g) x.   h) 0.   **12.** a) The product $2 \cdot 5$ would have two rows.   b) The product $1 \cdot 5$ would have one row.   c) No; $0 \cdot 5$ would have no rows.   **13.** a) 0 because zero times zero is zero.   b) 0.   c) 0.
14. a) x.   b) $y - x$.   c) y.   d) 0.   e) $y + x$.   f) 0.
g) y.   h) x.
15. a) $\begin{array}{r} 45 \\ -5 \\ \hline 40 \end{array}$

b) The answer is 4 tens and 0 ones.   **16.** a) $x + 1$.
b) $x - 1$.   c) 1.   **17.** a) If $\frac{7}{0}$ were equal to 0, then
$0 \cdot 0$ would be equal to 7.   b) No. If $\frac{7}{0}$ were equal to
7, then $0 \cdot 7$ would be equal to 7.   c) $\frac{0}{7} = 0$ because
$7 \cdot 0 = 0$.   d) No; because we could also say $\frac{0}{0} = a$
for any number a.   **18.** a) No, because no matter how many times zero is subtracted from 12, the result is always 12.   b) The calculator endlessly subtracts zero from zero without getting an answer.*

Set IV

If x is not equal to 0, then $x^0 = 1$. If x is equal to 0, it would seem that $0^0 = 0$.

---

*This is the way that electrically driven calculators work. An electronic calculator will give an error message.

Chapter 1, Lesson 7

Set I

1. a) 5^3. b) Not possible. c) 2^6.   **2.** a) $x + 1$.
b) $17 - x$. c) $x + 26$.   **3.** a) $7x$. b) $\dfrac{1000}{x}$ days.
c) $15,000 + 10x$.

Set III

10. a) Figure 1. b) Figure 6. c) Figure 4.
d) Figure 3. e) Figure 5. f) Figure 6. g) Figure
2.   **11.** a) 17. b) 2. c) 117. d) 17. e) 34.
f) 66. g) 128. h) 48. i) 25. j) 25. k) 27. l) 27.
m) 20. n) 0.5. o) 6. p) 147. q) 147.
12. a) $7x + 1$. b) $x^3 - y^3$. c) $3x - 3y$. d) $12 + \dfrac{x}{6}$.
e) $5x^2$. f) $x + y^5$. g) $\dfrac{1}{xy}$. **13.** a) 4. b) 8.
c) 119. d) 333.   **14.** a) 44. b) 11. c) 49.
d) 1,100. e) 620. f) 5.   **15.** a) 51 square
centimeters. b) $y^2 - x^2$ square centimeters.

Set IV

1. 72.   **2.** 40.   **3.** The calculator would do the
operations in order from left to right.   **4.** Write
down some of the intermediate steps.

Chapter 1, Lesson 8

Set I

1. a) 0. b) 100. c) 0. d) Not possible. e) 0.01.
f) 100.   **2.** a) 7. b) 6. c) x. d) $\dfrac{20}{x}$. e) x. f) $\dfrac{y}{x}$.
3. a) 5. b) $\dfrac{x}{10}$.

Set III

9. a) Yes. b) Yes. c) No. d) No. e) Yes. f) No.
g) Yes. h) Yes.   **10.** a) 48. b) 144. c) 29.
d) 85. e) 23. f) 55. g) 1. h) 40. i) 3. j) 4.
k) 6. l) 14. m) 16. n) 16. o) 13. p) 160.
q) 961.   **11.** a) Figure 1. b) Figure 3. c) Figure
2. d) Figure 5. e) Figure 4. f) Figure 6.
g) Figure 2. h) Figure 6.   **12.** a) $(x + 11)y$.
b) $11y + x$. c) $\dfrac{x}{3} - 1$. d) $\dfrac{x - 1}{3}$. e) $(x + y)^2$.
f) $x^2 + y^2$. g) $(x - y)x$. h) $9 - (2x)^3$.

i) $(9 - 2x)^3$. j) $\dfrac{x + y}{5y}$.   **13.** a) 0. b) 20. c) 128.
d) 273. e) 0. f) 20. g) 128. h) 273.

Set IV

1. All multiplication signs except for the first one
which should be an addition sign:
$1 + 2 \cdot 3 \cdot 4 \cdot 5 \cdot 6 \cdot 7 \cdot 8 \cdot 9 \cdot 10$.   **2.** 3,628,801.
3. Use the same symbols of operation as before, but
put parentheses around the $1 + 2$:
$(1 + 2) \cdot 3 \cdot 4 \cdot 5 \cdot 6 \cdot 7 \cdot 8 \cdot 9 \cdot 10$.   **4.** 5,443,200.

Chapter 1, Lesson 9

Set I

1. a) $5a$. b) b^3. c) $a + b$. d) 0.   **2.** a) $20 + x$.
b) $5x$. c) $\dfrac{y}{20}$ minutes.   **3.** a) $1000 - x$ pounds.
b) $\dfrac{1000}{y}$ pounds. c) $1000 - 10z$ pounds.

Set III

11. a) $2(5 + 4) = 2(5) + 2(4)$.
b) $7(3 + 1) = 7(3) + 7(1)$.
c) $4(8 - 5) = 4(8) - 4(5)$.
d) $5(4 - 1) = 5(4) - 5(1)$.   **12.** a) $7x^2$. b) $3(5x)$
or $15x$. c) $2(x + 7)$. d) $10(x + y)$.
e) $x^3 + x^3 + x^3 + x^3$. f) $7x + 7x$.
g) $(x + 8) + (x + 8) + (x + 8)$.
13. a) $2(x + 6) = (x + 6) + (x + 6)$
$\qquad\qquad\quad = x + x + 6 + 6$
$\qquad\qquad\quad = 2x + 12$
b) $4(x + y) = (x + y) + (x + y) + (x + y) + (x + y)$
$\qquad\qquad\quad = x + x + x + x + y + y + y + y$
$\qquad\qquad\quad = 4x + 4y$
c) $3(x^2 + 2) = (x^2 + 2) + (x^2 + 2) + (x^2 + 2)$
$\qquad\qquad\quad = x^2 + x^2 + x^2 + 2 + 2 + 2$
$\qquad\qquad\quad = 3x^2 + 6$
14. a) $2x + 10$. b) $4y - 28$. c) $3x + x^2$.
d) $y^2 - y$. e) $10x + 80$. f) $6x + x^2$. g) $5y - 20$.
h) $xy - y^2$. i) $3x^2 + 27$. j) $x^3 - 2x^2$.
15.

a)
$$
\begin{array}{r}
84 \\
\times\ 21 \\
\hline
84 \\
1680 \\
\hline
1764
\end{array}
$$

b) $21 \cdot 84 = (20 + 1)84$
$\qquad\quad = 20 \cdot 84 + 1 \cdot 84$
$\qquad\quad = 1680 + 84$
$\qquad\quad = 1764$

c)
$$\begin{array}{r} 21 \\ \times\ 84 \\ \hline 84 \\ 1680 \\ \hline 1764 \end{array}$$

d) $84 \cdot 21 = (80 + 4)21$
$= 80 \cdot 21 + 4 \cdot 21$
$= 1680 + 84$
$= 1764$

16. a) $3(x + 7)$ and $3x + 21$. b) $y(2 + y)$ and $2y + y^2$. c) $4(x + y + 6)$ and $4x + 4y + 24$. d) $x(y^2 + y + 1)$ and $xy^2 + xy + x$.
17. a) $x + y$. b) $40(x + y)$ cents. c) $40x$ cents. d) $40y$ cents. e) $40x + 40y$ cents.

Set IV

1. a) 4. b) 36. c) 49. d) 100. e) 4. f) 36. g) 25. h) 82. **2.** $(x + y)^2$ and $x^2 + y^2$ are sometimes equal and sometimes not equal.

Chapter 1, Review

Set I

1. a) $4 \cdot 7$. b) 7^4. c) $x + x$. d) $y \cdot y \cdot y \cdot y \cdot y \cdot y$.
2. a) w^2. b) $3x$. c) $17 - y$. d) z^5.
3. a) Think of a number: ☐

Multiply by five: ☐☐☐☐☐

Add eight: ☐☐☐☐☐ °°°°/°°°°

Subtract three: ☐☐☐☐☐☐ ° ° ° ° °

Divide by five: ☐°

Subtract the number that
you first thought of: °

b) The number 1. c) Steps 3 and 4. d) Add 5.
4. a) Figure 3. b) Figure 1. c) Figure 2.
5. a) $4^2 =$ 16
$4^3 =$ 64
$4^4 =$ 256
$4^5 = 1,024$
$4^6 = 4,096$
b) The last digit of 4^{100} is 6. **6.** a) 64. b) 8^2.
7. a) Perimeter, 22; area, 28. b) Perimeter, $2x + 6$; area, $3x$. c) Perimeter, $4y$; area, y^2. **8.** a) 14.

b) 28. c) 65. d) 125. **9.** a) $x + 151$.
b) $160 - x - y$ or $160 - (x + y)$. **10.** a) If $\frac{2}{0} = a$, then $0 \cdot a = 2$. But $0 \cdot a = 0$.
b) No. **11.** a) $600x$. b) $\dfrac{10,000}{x}$.
12. a) $5x + 1$. b) $(x + 3)^2$. c) $x^6 - 7$.
13. a) $7a + 14$. b) $b - b^2$. c) $5c + 45$. **14.** a) $3x$ carbon atoms and $8x$ hydrogen atoms. b) $3x + 5x$. c) 11. d) $11x$.

Chapter 2, Lesson 1

Set I

1. a) 3. b) 81. c) 9. d) 27. **2.** a) 0. b) 1. c) 0. **3.** a) $100x$. b) $400 - x$.

Set III

10. a)
x	0	1	2	3	4
y	0	5	10	15	20

b)
x	3	4	5	6	7
y	0	1	2	3	4

c)
x	1	2	3	4	5
y	11	21	31	41	51

d)
x	1	2	3	4	5
y	1	8	27	64	125

e)
x	0	1	2	3	4
y	0	2	6	12	20

11. a)
x	0	1	2	3
y	2	6	34	110

b)
x	2	3	4	5
y	6	24	52	90

12. a) $y = x + 3$. b) $y = 5x$. c) $y = \dfrac{x}{2}$.
d) $y = x - 6$. e) $y = x^3$. f) $y = 5x - 1$.
g) $y = x^3 + 2$. h) $y = 10x + 4$. i) $y = 11x$.
j) $y = 10 - x$. **13.** a) The longer the senator spoke, the fewer the people who were listening to him. b) No. **14.** a) 7.5 dollars. b) $d = 7.5h$.
c) 375 dollars. **15.** a) 2. b) 14. c) $n = 2d$.

1. Taking first and second differences of the temperatures, we get

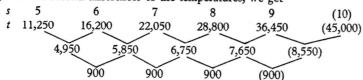

s	5		6		7		8		9		(10)
t	11,250		16,200		22,050		28,800		36,450		(45,000)

	4,950		5,850		6,750		7,650		(8,550)

	900		900		900		(900)

2. 45,000°C.

Chapter 2, Lesson 2

Set I

1. a) $5n$. b) n^5. c) $7 \cdot 7 \cdot 7 \cdot 7$.
d) $4 + 4 + 4 + 4 + 4 + 4 + 4$.   **2.** a) 1. b) x.
c) $\dfrac{x^2}{2}$.   **3.** a) $4x - 9$. b) $(x - 9)4$. c) $a^2 + b^2$.
d) $(a + b)^2$.

Set III

11. A, (1, 2); B, (5, 0); C, (8, 1); D, (10, 5);
E, (9, 8); F, (2, 9); G, (0, 5); O, (0, 0).
12.

a)

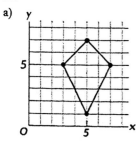

b)

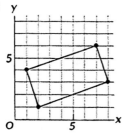

13. a) $y = x^2$. b) 4. c)

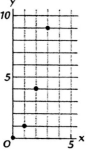

14. a)

x	1	2	3	4
y	0	1	2	3

b) $y = x - 1$.

15. a)

x	0	1	2	3	4
y	5	4	3	2	1

b) $y = 5 - x$.

16. a)

x	2	3	4	5	6
y	1	3	5	7	9

b)

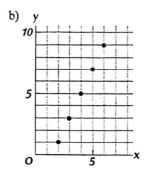

17. a)

x	0	2	4	6	8
y	5	6	7	8	9

b)

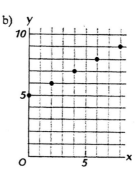

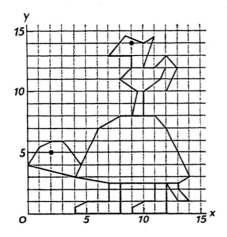

Chapter 2, Lesson 3

Set I

1. a) Yes. b) No. c) Yes. d) No.
2. a) $3x + 3y$. b) $5x - 35$. c) $2x - x^2$.
d) $y^3 + y^2$. **3.** a) $6x$. b) $24 - x$.

Set III

8. a)

x	1	2	3	4
y	3	4	5	6

b) $y = x + 2$. **9.** a) 3. b) 2 and 4. **10.** a) Yes.
b) Yes. c) No. d) Yes. e) Yes. f) No. g) Yes.
h) No. i) Yes. j) No.

11. a)

x	0	1	2	3	4
y	3	4	5	6	7

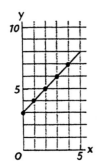

b)

x	0	1	2	3	4
y	0	3	6	9	12

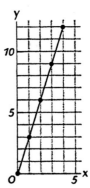

c)

x	1	2	3	4
y	1	4	7	10

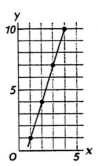

d)

x	0	0.5	1	1.5	2
y	0	0.125	1	3.375	8

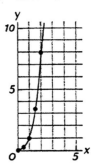

e)

x	0	1	2	3
y	1	3	5	7

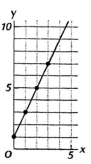

f)

x	0	1	2	3
y	1	2	5	10

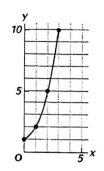

g)

x	0	2	4	6	8
y	0	0.5	1	1.5	2

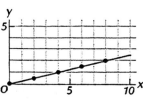

h)

x	1	2	3	4	5
y	4	2	1.3	1	0.8

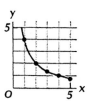

Set IV

1. There is more than one value of y corresponding to $x = 0.5$, $x = 2$, $x = 3$, $x = 3.5$, and $x = 4.5$.
2. Remove the vertical line segments and either the right or left endpoints of the horizontal line segments.

Chapter 2, Lesson 4

Set I

1. a) Figure 3. b) Figure 2. c) Figure 4. d) Figure 1. **2.** a) $y = 6 - x$. b) $y = 10x + 1$.
c) $y = x^3$. **3.** a) $120x$ minutes. b) $120 - y$.

Set III

11. a)

x	0	1	2	3	4
y	0	3	6	9	12

b) y is doubled. c) A direct variation. d) $y = 3x$.
e) 30.
12. a)

x	2	4	6	12	18
y	7	14	21	42	63

b) $\frac{7}{2}$. c) $y = \frac{7}{2}x$ or $y = 3.5x$. **13.** a) Yes; the speed of light is the constant of variation. b) No; after awhile the cat stops growing. c) No; too many cooks and there is no soup. d) Yes, if the bottles are uniform in size; the constant of variation is the refund per bottle. **14.** a) Yes. b) No. c) No.
d) Yes. e) No. **15.** a) $y = 2x$. b) $y = x$.
c) $y = 3x$. d) $y = 2.5x$. e) 2 for line 1; 1 for line 2; 3 for line 3; 2.5 for line 4. f) Line 3. g) The larger the constant of variation, the steeper the line.
16. a) 600 cubic centimeters. b) 250 degrees Kelvin.
c) The greater the constant of variation, the greater the size of the balloon. **17.** a) $y = 3.2x$. b) A straight line passing through the origin.

Set IV

1. There is a discount if you buy more than 11 cassettes. **2.** It would be cheaper to buy 12.

Chapter 2, Lesson 5

Set I

1. a) $x - y$. b) $2x + 3$. c) $y + 1$. **2.** a) $2x$.
b) $4x$. c) $8x$. **3.** a) $3^2 + 4^2 = 9 + 16 = 25$;
$5^2 = 25$; true. b) $3^3 + 4^3 + 5^3 = 27 + 64 + 125 = 216$; $6^3 = 216$; true. c) $3^4 + 4^4 + 5^4 + 6^4 = 81 + 256 + 625 + 1,296 = 2,258$; $7^4 = 2,401$; false.

Set III

10. a)

x	0	1	2	3
y	1	3	5	7

b)

x	0	1	2	3
y	4	6	8	10

c)

x	0	1	2	3
y	5	7	9	11

d)

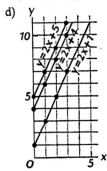

e) They are parallel. f) $y = 2x + 1$ at $(0, 1)$;
$y = 2x + 4$ at $(0, 4)$; $y = 2x + 5$ at $(0, 5)$. g) At $(0, 8)$.

11. a)

x	0	1	2	3
y	3	4	5	6

b)

x	0	1	2	3
y	3	5	7	9

c)

x	0	1	2	3
y	3	6	9	12

d)

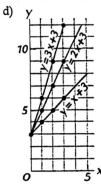

e) All pass through the same point. f) At $(0, 3)$.
g) At $(0, 3)$.

12. a)

x	0	1	2	3	4
y	2	2	2	2	2

b)

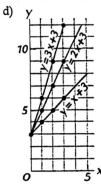

c) It is a horizontal line. d) $y = 2$. **13.** a) It is
the difference between successive values of y. b) It
is the value of y corresponding to $x = 0$.
c) $y = 6x + 2$. d) $y = 3x + 1$. e) $y = 5x + 9$.
f) $y = 9x + 5$. **14.** a) $y = 2x + 5$. b) Linear.
c) No.

15. a)

x	1	2	3	4	5
y	80	80.5	81	81.5	82

b) $y = 0.5x + 79.5$. c) No. d) \$104.50.

Set IV

1. $y = 6x + 4$. **2.** 4°C.

Chapter 2, Lesson 6

Set I

1. a) $x - 4$. b) $2(x + 5)$. c) $x^3 - 1$. **2.** a) The
axes. b) The origin. c) Coordinates.
3. a) Directly. b) A straight line passing through the
origin.

Set III

11. a)

x	1	2	3	4	5
y	10	5	$3\frac{1}{3}$	$2\frac{1}{2}$	2

b) If x is doubled, y is halved. c) An inverse
variation. d) $y = \dfrac{10}{x}$. e) 0.5.

12. a)

x	2	4	6	12	16
y	24	12	8	4	3

b) 48. c) $y = \dfrac{48}{x}$. **13.** a) No; if this were true, a
baby would have a very long attention span. b) Yes;
if someone swims twice as fast, the time required is
cut in half. c) Yes; if the price is doubled, you can
buy half as much. d) No; the more dogs, the faster
the sled. **14.** a) Yes. b) No. c) No. d) Yes.
e) No.

15. a)

x	1	2	3	4	5
y	5	2.5	1.7	1.25	1

b)

x	1	2	3	4	5
y	8	4	2.7	2	1.6

c)

x	1	2	3	4	5
y	12	6	4	3	2.4

d)

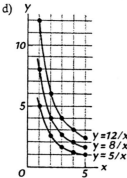

e) 5 for $y = \dfrac{5}{x}$; 8 for $y = \dfrac{8}{x}$; 12 for $y = \dfrac{12}{x}$.   f) The larger the constant of variation, the greater the distance of the curve from the origin.

16. a) $y = \dfrac{36}{x}$.   b) 36.   c) Inversely.

17. a) $n = \dfrac{318}{d}$.   b) 636.

Set IV

1. $f = \dfrac{600}{d}$.   **2.** 7.5 kilograms.

Chapter 2, Review

Set I

1.

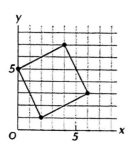

2. a)

x	1	2	3	4	5
y	6	5	4	3	2

b)

x	0	1	2	3	4
y	1	2	9	28	65

c)

x	4	5	6	7	8
y	0	5	12	21	32

3. a) True.   b) True.   c) False.   d) True.
e) False.   **4.** a) $y = x - 5$.   b) $y = 2x - 1$.
c) $y = \dfrac{x}{3}$.   d) $y = x^2 + 1$.

5. a)

0	2	4	6	8	10
0	1	2	3	4	5

b) y is tripled.   c) A direct variation.   d) $y = \dfrac{x}{2}$.
e) 15.

6. a)

x	1	2	3	4
y	0	1	2	3

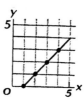

b)

x	1	2	3	4
y	5	7	9	11

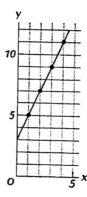

c)

x	1	2	3	4
y	8	4	2.7	2

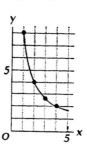

7. a)

x	1	2	3	4
y	3	6	9	12

b)
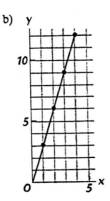

c) A direct variation. d) 7.5 seconds.

8. a)

x	4	8	12	16
y	4	5	6	7

b)
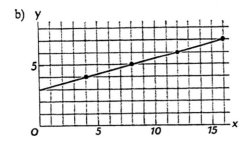

c) No. d) No; the graph does not pass through the origin.

9. a)

x	3	4	5	6
y	4	3	2.4	2

b)
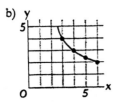

c) Inversely. d) It becomes very small.

Chapter 3, Lesson 1

Set I

1. a) 2,592. b) The four digits are the same.

2. a) 72. b) 120. c) 120. d) 0.   **3.** a) $\frac{x}{3}$.

b) 21y. c) Direct.

Set III

11. a) Mexico City is 2,300 meters above sea level.
b) Archimedes died 212 years before the birth of
Christ. c) It went down 48 points. d) Mr.
Micawber is 800 dollars in debt.
12. a) $+15° > -40°$. b) $-8 < 2$. c) $12 > -12$.
d) $-60 < -50$.   **13.** a) $2 < 8$. b) $11 > 5$.
c) $0 > -6$. d) $-4 < 4$. e) $7 > -9$. f) $-1 > -3$.
g) $-10 < 1$. h) $-12 < -8$.   **14.** a) –7. b) +6.
c) –7 –5 –3 –1 0 +2 +4 +6.   **15.** a) 4. b) 4.
c) 7. d) 7. e) 4. f) 12. g) 11. h) 11.   **16.** a) 4.
b) –4. c) 5. d) –5. e) 0. f) 0. g) 1. h) –1.
17. a) $x > -2$. b) $x^3 < 20$. c) $4x > 1$.
d) $x - 2 < 0$.

Set IV

1. No; they're going in opposite directions.   **2.** Yes;
the Rabbit is going faster than the Cougar.

Chapter 3, Lesson 2

Set I

1. a) By finding the sum of the four sides.
b) $2x + 6$. c) By finding the product of two adjacent
sides. d) 3x.   **2.** a) $1^3 + 2^3 = 1 + 8 = 9$;
$(1 + 2)^2 = 3^2 = 9$; true. b) $1^3 + 2^3 + 3^3 =$
$1 + 8 + 27 = 36$; $(1 + 2 + 3)^2 = 6^2 = 36$; true.
c) $1^3 + 2^3 + 3^3 + 4^3 = (1 + 2 + 3 + 4)^2$.
d) $1^3 + 2^3 + 3^3 + 4^3 = 100$;
$(1 + 2 + 3 + 4)^2 = 10^2 = 100$; yes.
3. a) Directly. b) 0.5 meters per minute.
c) $d = 0.5t$.

Set III

10. A, (2, 0); B, (0, –8); C, (–1, –9); D, (–3, –5);
E, (–4, 0); F, (–5, 7); O, (0, 0).   **11.** a) In the third
quadrant. b) In the fourth quadrant. c) On the
negative x-axis. d) On the negative y-axis.

12. a)

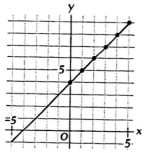

b) $y = x + 4$.   c) (See graph.)

d)

x	-5	-4	-3	-2	-1
y	-1	0	1	2	3

13. a)

y	0	1	2	3	4
y	0	2	4	6	8

b) $y = 2x$.

c)

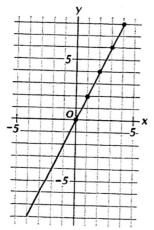

d)

x	-4	-3	-2	-1
y	-8	-6	-4	-2

14. a)

x	4	5	6	7	8
y	0	1	2	3	4

b and c)

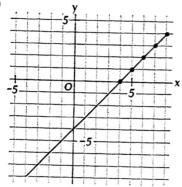

d)

x	-4	-3	-2	-1	0	1	2	3
y	-8	-7	-6	-5	-4	-3	-2	-1

15. a)

x	0	1	2	3
y	5	6	7	8

x	0	1	2	3
y	3	2	1	0

b)

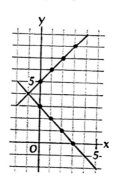

c) $(-1, 4)$.

Set IV

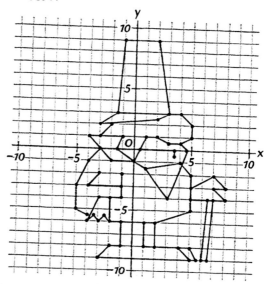

Chapter 3, Lesson 3

Set I

1. a) $0 > -10$.   b) $-25 < -2$.   c) $34 > -43$.
2. a) 80.   b) 100.   c) 9.   d) 16.   **3.** a) $y = 1,890x$.
b) The average number of calls per week.

Set III

9. a)

b)

c)

d)

e)

10. a) –5. b) 7. c) 0. d) 12. e) –a. f) b.
11. a) 0. b) 0. c) 10. d) –13. e) –9. f) –15.
g) –12. h) –18. i) –40. j) –4. k) 9. l) 17.
12. a) –7. b) 3. c) –3. d) –7. e) 7. f) –7. g) –6.
h) 6. i) 20. j) –16. k) –9. l) 5. 13. a) –12.
b) –2. c) –8. d) 0. e) –6. f) –16. g) –10.
h) –14. i) –20. j) –21. k) –3. l) 3.

Set IV

1. 0; each number occurs with its negative.
2. 1,005; 96 + 97 + 98 + 99 + 100 + 101 + 102 + 103 + 104 + 105.

Chapter 3, Lesson 4

Set I

1. a) -4. b) 0. c) -1. 2. a) The x-coordinate is negative and the y-coordinate is negative. b) The x-coordinate is positive and the y-coordinate is negative. c) The x-coordinate is positive and the y-coordinate is 0. d) The x-coordinate is 0 and the y-coordinate is negative.
3. a)

x	1	2	3	4	5
y	1	3	5	7	9

b) Linear. c) $y = 2x - 1$. d) 49.

Set III

10. a) $8 - 1 = 7$

b) $-6 - -4 = -2$

c) $2 - 3 = -1$

d) $-4 - -6 = 2$

e) $7 - -2 = 9$

f) $-1 - 5 = -6$

11. a) $8 + -1 = 7$. b) $-6 + 4 = -2$.
c) $2 + -3 = -1$. d) $-4 + 6 = 2$. e) $7 + 2 = 9$.
f) $-1 + -5 = -6$. 12. a) 11. b) –11. c) 29.
d) –29. e) –11. f) 3. g) 16. h) –10. i) –8. j) 8.
k) 0. l) –10. 13. a) 10. b) 10. c) 12. d) 12.
e) –9. f) –9. g) 11. h) 11. i) –10. j) –10. k) 0.
l) 0. m) $x - y - z$. n) $x - y + z$. 14. a) 2,654
meters. b) 6,280 meters. c) 311 meters.
15. a) –3, –2, +4. b) +4 + –3 + –2 + +4. c) +3.

Set IV

1.

4	9	2	–6	–1	–8
3	5	7	–7	–5	–3
8	1	6	–2	–9	–4

2. Yes. 3. Each number is the opposite of a symmetric number of the original.

Chapter 3, Lesson 5

Set I

1. a) –10. b) 0. c) –909. 2. a) $x(x + 1)$.
b) $\dfrac{1 + y}{4}$. c) $\dfrac{1}{y + 4}$. d) $5 - z^3$. e) $(5 - z)^3$.
3. a) $0.05x$. b) $20y$.

Set III

12. a) $4(5) = 5 + 5 + 5 + 5 = 20$.
b) $4(-5) = -5 + -5 + -5 + -5 = -20$.
c) $5(-4) = -4 + -4 + -4 + -4 + -4 = -20$.
13. a) –20. b) –42. c) –72. d) 6. e) 0. f) 84.
g) –75. h) 16. i) –99. j) 350. 14. a) 0. b) 0.
c) $-x$. d) y. e) $-xy$. f) $-xy$. g) xy.
15. a) –121. b) 121. c) 9. d) –9. e) –9. f) 7.
g) 0. h) –1. i) –2. j) –2. 16. a) 120. b) –120.
c) 120. d) –120. e) 1. f) –1. g) 1. h) 900.
i) –27,000. j) 64. k) 64. l) 0. 17. a) $>$. b) $<$.
c) $=$. d) $<$. e) $>$. 18. a) –14. b) 15. c) 49.
d) –10. e) –4. f) 21.
19. a)

x	–1	–2	–3	–4	–5
x^2	1	4	9	16	25

b) No.

c)
$$x \quad -1 \quad -2 \quad -3 \quad -4 \quad -5$$
$$x^3 \quad -1 \quad -8 \quad -27 \quad -64 \quad -125$$
d) The cube of a negative integer is negative.

Set IV

	4	1	-5
3	12	3	-15
-6	-24	-6	30
2	8	2	-10

The diagram is a multiplication table. Choosing three numbers in different rows and columns insures that each of the six original factors (shown at the top and on the left) is included exactly once:

$$(4)(1)(-5)(3)(-6)(2) = 720$$

Chapter 3, Lesson 6

Set I

1. a) 4. b) 5. c) 15. **2.** a) 4. b) 5. c) 15.
3. a) $x - y$. b) $y - x$.

Set III

10. a) –3. b) –3. c) –8. d) 8. e) –9. f) 0. g) 1.
h) –1. **11.** a) –13. b) 6. c) 4. d) –2. e) 9.
f) –16. g) –1. h) 7. **12.** a) $\frac{x}{y}$. b) 0. c) 0.
d) –x. e) –y. f) x. g) –1. h) 1. **13.** a) $<$.
b) =. c) =. d) =. e) $>$. f) =. **14.** a) –4.
b) 4. c) –25. d) 25. e) 22. f) –22. g) –7.
h) 7. **15.** a) –5. b) –2. c) –4. d) 1. e) 20.
f) 5.

Set IV

```
1.   -20          2.     20
   - -5  1           - -5  1
   -----             -----
    -15                25
   - -5  2           - -5  2
   -----             -----
    -10                30   .
   - -5  3                  .
   -----                    .
     -5
   - -5  4 times
   -----
      0
```

3. In the second problem the numbers get larger.

Chapter 3, Lesson 7

Set I

1. a)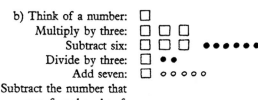

2. a) $y = -x^2$. b) $y = 3 - x$. c) $y = 4x + 10$.
3. a) –1,000,001. b) –999,999.

Set III

10. a)

Think of a number:	4	1	0	-5	-12
Multiply by three:	12	3	0	-15	-36
Subtract six:	6	-3	-6	-21	-42
Divide by three:	2	-1	-2	-7	-14
Add seven:	9	6	5	0	-7
Subtract the number that you first thought of:	5	5	5	5	5

b)
Think of a number: □
Multiply by three: □ □ □
Subtract six: □ □ □ ● ● ● ● ● ●
Divide by three: □ ● ●
Add seven: □ ○ ○ ○ ○ ○
Subtract the number that you first thought of: ○ ○ ○ ○ ○

11. a) 48. b) –36. c) –40. d) 250. e) 36.
f) –216. **12.** a) –3. b) –3. c) 17. d) 3. e) –8.
f) 10. g) 64. h) –64. i) –91. j) –91. k) 900.
l) –28. m) –64. n) 132. o) –21. p) –1.
13. a) –22. b) –22. c) 30. d) 30. e) –20.
f) –20. **14.** a) 80 meters. b) 0 meters. c) –55
meters. d) The ball is level with the roof of the second building. e) The ball is 55 meters below the roof of the second building. **15.** a) 212 degrees.
b) 23 degrees. c) –40 degrees.

Set IV

1. The balloons are lighter than air. **2.** –0.25
pound. **3.** No.

Chapter 3, Review

Set I

1. a) False; the opposite of –1 is 1. b) True.
c) True. d) False; the sum of –1 and –2 is –3, but
the difference of –1 and –2 is 1. e) True.
2. a) –12. b) –18. c) –45. d) –5. e) –18. f) –12.
g) 45. h) 5.   **3.** a) 3 + 7. b) –8 + –4.
c) $x + -y$.   **4.** a) $x > 0$. b) $x^3 = 9x$.
c) $x - 3 < 2$.
5. a)

Think of a number:	2	0	–6	–12
Add six:	8	6	0	–6
Multiply by two:	16	12	0	–12
Subtract eighteen:	–2	–6	–18	–30
Divide by two:	–1	–3	–9	–15
Add five:	4	2	–4	–10
Subtract the number that you first thought of:	2	2	2	2

b)

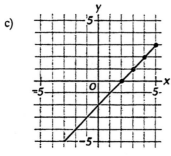

Think of a number:
Add six:
Multiply by two:
Subtract eighteen:
Divide by two:
Add five:
Subtract the number that
you first thought of:

6. a) 11. b) –1. c) –8. d) –12. e) 6. f) –2.
7. a) <. b) >. c) >. d) =. e) <.
8. a)

x	2	3	4	5
y	0	1	2	3

b) $y = x - 2$.

c)

d)

x	–2	–1	0	1
y	–4	–3	–2	–1

9. a) 2. b) 10. c) –5. d) 22.   **10.** a) –8. b) 5.
c) x. d) –1.

---

11. a) 60. b) –19. c) –71. d) –2,002. e) 17.
12. a) Ollie got 5 more wrong than right. b) 11.
c) 8.   **13.** a) 17. b) 17. c) 6. d) –6. e) –23.
14. a) 2(–4 + –6) = 2(–10) = –20;
     2(–4) + 2(–6) = –8 + –12 = –20.
  b) –5(8 + –1) = –5(7) = –35;
     –5(8) + –5(–1) = –40 + 5 = –35.
  c) –7(–3 + 9) = –7(6) = –42;
     –7(–3) + –7(9) = 21 – 63 = –42.

Chapter 4, Lesson 1

Set I

1. a) 1. b) 9. c) 46. d) 24.
2. a)

x	1	2	4	5
y	10	5	2.5	2

b) It is divided by 2. c) Inverse variation. d) $y = \dfrac{10}{x}$.

3. a) $x + y$. b) $x + 2y$ dollars. c) $x + 2y - 150$
dollars.

Set III

9. a) 2.75. b) 0.275. c) 0.0275. d) 0.008. e) 0.08.
f) 0.8. g) 4. h) 40. i) 0.4.   **10.** a) $\dfrac{6}{1}$. b) $\dfrac{1}{1}$.
c) $\dfrac{2}{10}$ or $\dfrac{1}{5}$. d) $\dfrac{9}{100}$. e) $\dfrac{45}{10}$ or $\dfrac{9}{2}$. f) $\dfrac{-31}{10}$. g) $\dfrac{15}{2}$.
h) $\dfrac{-26}{5}$.
11. a)

$$-8\ -7\ -6\ -5\ -4\ -3\ -2\ -1\ \ 0\ \ 1\ \ 2\ \ 3\ \ 4\ \ 5\ \ 6\ \ 7\ \ 8$$
$$-5.4 \qquad -2\tfrac{1}{2}\ \ -0.5\tfrac{1}{4}\ \ 1.6\ \ 3.1\ 4\tfrac{1}{3}$$

b) $4\dfrac{1}{3}$. c) –5.4.   **12.** a) 2.7 > 2.6.
b) –2.7 < –2.6. c) 4.5 > –5.4. d) –4.5 > –5.4.
e) 0.8 > 0.08. f) –0.8 < 0.08. g) 3.11 > 3.1.
h) –3.11 < –3.1.   **13.** a) 5.34. b) 4.86. c) 1.224.
d) 21.25. e) 1.089. f) 0.891. g) 0.09801. h) 10.

Set IV

1. $9\left(\dfrac{1}{11} = 0.090909\ldots\right)$. **2.** $8\left(\dfrac{1}{7} = 0.142857142857\ldots\right)$.

Chapter 4, Lesson 2

Set I

1. a) 0. b) 3. c) 0. d) –3. e) 0.   **2.** a) $5x - 5y$.

b) $-5x + -5y$. c) $xy - y$. d) $-y + xy$.   **3.** a) 49.
b) $45 - -4$. c) $45 + 4$.

Set III

11. a) 2. b) 9. c) 7.1. d) 0. e) 0.3. f) x.
g) $-x$.   **12.** a) 15. b) 15. c) 5. d) –5. e) 48.
f) 48. g) 5. h) 0.2.   **13.** a) $=$. b) $>$. c) $>$.
d) $=$. e) $<$. f) $>$. g) $<$. h) $>$.   **14.** a) False;
example, $|0| = 0$. b) False; example,
$|1| - |2| \neq |1 - 2|$. c) True. d) True.
15. a) –9.3. b) 3.3. c) –7. d) 7. e) 6.2. f) –4.8.
g) –5.98. h) –0.74.   **16.** a) –2.92. b) 2.81.
c) 0.885. d) 0.397.   **17.** a) 4.43. b) –1.25.
c) –1.77. d) 1.77. e) 10.97. f) –10.97.

Set IV

1. $|x| + |y|$.   **2.** $\left| |x| - |y| \right|$.

Chapter 4, Lesson 3

Set I

1. a) $\frac{-7}{1}$. b) $\frac{3}{10}$. c) $\frac{-16}{10}$ or $\frac{-8}{5}$. d) $\frac{9}{4}$.
2. a) $(-2)^4$. b) $(-3)^5$. c) $(-10)^3$. d) $-1,000,000$
cannot be expressed as a power of -10.
3. a)

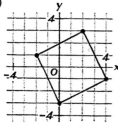

b) A square.

c) $(0, 2), (-1, -1), (2, -2)$.

Set III

11. a) $4.5 + 5.4 = 9.9$. b) $-7.3 + 7.3 = 0$.
c) $-2 + -8.9 = -10.9$. d) $1.06 + -6 = -4.94$.
12. a) 5.2. b) –8.2. c) –8.43. d) –1.9.   **13.** a) 14.
b) –14. c) –44.44. d) –44.44 e) 28.8. f) 0.288.
14. a) 8. b) 8. c) –0.9. d) –0.9. e) 4.5. f) –0.045.
15. a) –1.8. b) –1. c) 7. d) 3.5. e) 4.4. f) –0.1.
g) 9.3. h) –4.1.
16.

	a) 1.2	5.1	–3.5
	3.8	–0.1	8.5
	–7.6	0.2	–17
	–3.6	4.2	–13
	–1.8	2.1	–6.5
	–3	–3	–3

b) –3.   **17.** a) –3.73. b) –15.141. c) –80.19.
d) 12.5. e) –7.07. f) 111.111.

Set IV

1. 531,770 (OLLIES).   **2.** 35,009 (GOOSE).

Chapter 4, Lesson 4

Set I

1. a) $=$. b) $<$. c) $>$. d) $<$. e) $>$.   **2.** a) –5.
b) 0. c) 1.5.   **3.** a) $18 + x$ points. b) $6y + 2$
points. c) $6y + 2 > 18 + x$.

Set III

10. a) 4. b) 3. c) –7. d) –8. e) 0. f) –11.
11. a) 3. b) 2.7. c) 2.72. d) 2.718. e) 2.7183.
f) 2.71828.   **12.** a) 0.44. b) 0.56. c) –1.63.
d) –0.16. e) –0.02. f) –0.00.   **13.** a) –7.8.
b) 5.1. c) 7.2. d) –7.2. e) 12.3. f) 1.2. g) 0.1.
h) –0.7. i) –7.1. j) –0.1.   **14.** a) –11.0. b) –11.0.
c) –50.1. d) 30.3.   **15.** a) 7.67. b) –10.13.
c) –10.95. d) 1.51. e) –7.24.

Set IV

1. The museum keeps track of its daily admission.
2. 115,500 to 116,449 (if our convention for rounding
is used).

Chapter 4, Lesson 5

Set I

1. a) –0.85. b) 1.65. c) –0.5. d) –0.32.
2. a) 1.008, 1.08, 1.8. b) –1.8, –1.08, –1.008.
c) 0.5, 0.55, 0.555. d) –0.555, –0.55, –0.5.
3. a) $x + y + 9$. b) $5x + 10y + 57$.
c) $0.05x + 0.1y + 0.57$.

Set III

9. a)

x	0	1	4	5	8
y	4	3.5	2	1.5	0

b and c)

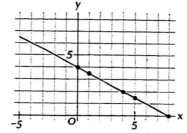

d) (6, 1), (3, 2.5), (–4, 6). e) (6, 1), (3, 2.5), (–4, 6).

10. a)

x	0	0.1	0.2	0.3	0.4	0.5	0.6	0.7	0.8	0.9	1.0
y	1	0.99	0.96	0.91	0.84	0.75	0.64	0.51	0.36	0.19	0

b and c)

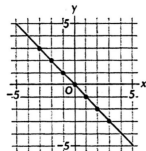

11. a) Linear. **b)** b is the y-coordinate of the point in which the line crosses the y-axis. **c)** $y = 2x + 2$.
d) $y = 2x - 5$.

12. a)

x	–3	–2	–1	0	1	2	3
y	3	2	1	0	–1	–2	–3

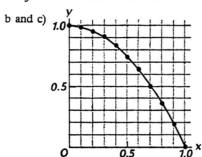

b)

x	–1	0	1	2	3	4	5
y	–10	–7	–4	–1	2	5	8

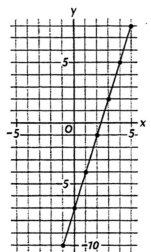

c)

x	–1	0	1	2
y	–8	–7	–6	1

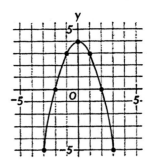

d)

x	–3	–2	–1	0	1	2	3
y	–5	0	3	4	3	0	–5

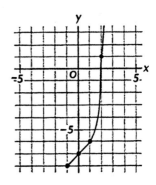

e)

x	–8	–7	–6	–5	–4	–3	–2	–1	0
y	–1	–0.88	–0.75	–0.63	–0.5	–0.38	–0.25	–0.13	0

x	1	2	3	4	5	6	7	8
y	0.13	0.25	0.38	0.5	0.63	0.75	0.88	1

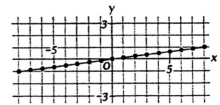

f)

x	-8	-7	-6	-5	-4	-3	-2	-1	0
y	-1	-1.14	-1.33	-1.6	-2	-2.67	-4	-8	-

1	2	3	4	5	6	7	8
8	4	2.67	2	1.6	1.33	1.14	1

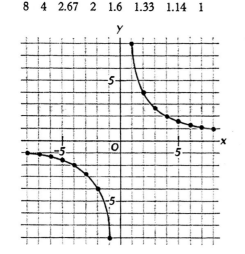

13. a) $y = -x$, $y = 3x - 7$, and $y = \frac{x}{8}$. b) $y = -x$ and $y = \frac{x}{8}$. c) $y = x^3 - 7$, $y = 4 - x^2$, and $y = \frac{8}{x}$. d) $y = \frac{8}{x}$.

Set IV

1.

x	0	0.5	1.0	1.5	2.0	2.5	3.0	3.5	4.0	4.5	5.0	5.5	6.0
y	0	2.75	5.0	6.75	8.0	8.75	9.0	8.75	8.0	6.75	5.0	2.75	0

2.

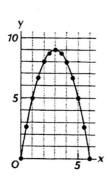

3. The graph looks like the first bounce.

Set I

1. a) $\frac{-3}{1}$. b) $\frac{1}{1}$. c) $\frac{9}{100}$. d) $\frac{-11}{2}$. **2.** a) 1.125. b) –0.08. c) 8.5. **3.** a) $>$. b) $<$. c) $>$. d) $>$. e) $<$. f) $=$. **4.** a) 7.7. b) –2.2. c) –9.2. d) 1. e) 20.91. f) –1.8. g) –3.6. h) –2.43. i) –3. j) –0.32. k) 0.0256. l) –12.5. **5.** a) –1.3. b) 0. c) 8.8. d) –3. e) 6.1. f) –0.1.

6.

a) 1.6	0.5	-2.4
-1.4	-2.5	-5.4
-0.7	-1.25	-2.7
5.8	5.25	3.8
11.6	10.5	7.6
10	10	10

b) 10. **7.** a) 4. b) –1.3. c) 0.58. d) –7.
8. a) 1.22. b) 12.22. c) 0.82. d) 0.08.

9. a)

x	-6	-5	-4	-3	-2	-1	0	1	2	3	4	5	6
y	3	2.5	2	1.5	1	0.5	0	-0.5	-1	-1.5	-2	-2.5	-3

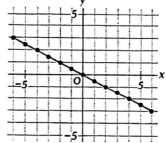

b)

x	-3	-2	-1	0	1	2	3
y	10	5	2	1	2	5	10

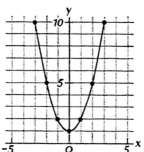

c)
x	−1	0	1	2	3
y	−7	−3	1	5	9

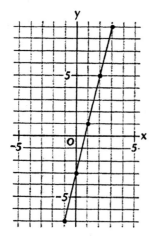

d)

x	−6	−5	−4	−3	−2	−1
y	1	1.2	1.5	2	3	6

	0	1	2	3	4	5	6
−		−6	−3	−2	−1.5	−1.2	−1

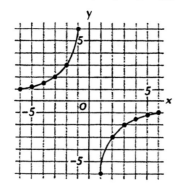

10. a) $y = -0.5x$ and $y = 4x - 3$. b) $y = -0.5x$.
c) $y = x^2 + 1$ and $y = \frac{-6}{x}$. d) $y = \frac{-6}{x}$.

Chapter 5, Lesson 1

Set I

1. a) 2. b) 1.6. c) 1.62. d) 1.618. e) 1.6180.
2. a) 4(−8). b) (−8)⁴. c) $-x + -x + -x$.
d) $(-x)(-x)(-x)$.   **3.** a) 365 days in an ordinary year
and 366 days in a leap year. b) 1,461.
c) $365(x − y) + 366y$ days or $365x + y$ days.

Set III

8. a) If a certain number is multiplied by 3, the
result is 27. b) If 4 is subtracted from a certain
number, the result is 10. c) If 18 is divided by a
certain number, the result is 6. d) If a certain
number is squared, the result is 25. e) If 4 is raised
to a certain power, the result is 16.   **9.** a) True.
b) False. c) Neither. d) True. e) False.
f) Neither.   **10.** a) 20. b) 6. c) 55. d) 12.
e) 300. f) 18. g) 6 and −6. h) 0. i) Not true for
any number, because no number is equal to 1 less
than itself. j) −1, 0, and 1. k) 21. l) −2. m) −9.
n) 0. o) 0. p) 1. q) 0. r) −4. s) −1. t) Not true
for any number, because no number squared is
negative. u) −2. v) 2. w) 4. x) True for all
positive integers.   **11.** a) 2 and 4. b) 0, 3, and −7.
c) −4 and −5. d) 2, 6, 11, −8, and −9.

Set IV

1. Guess 4: $5(4) - \frac{4}{2} = 18$, which is half of 36. So

guess is half of solution; solution is 8. $5(8) - \frac{8}{2} = 36$.

The method works.   **2.** Guess 8: $\frac{8}{4} + 4 = 6$, which

is one-fourth of 24. So guess is one-fourth of

solution; solution is 32. $\frac{32}{4} + 4 = 12$. The method

does not work.

Chapter 5, Lesson 2

Set I

1. a) The number is 0. b) The number is positive.
c) The number is not 0. d) The number is negative.
2. a)
x	−3	−1	0	2
y	5	3	2	0
b) Linear. c) $y = 2 − x$. d) 12.   **3.** a) $x + 5$.
b) $x − 15$.

Set III

11. a) Multiply by 3. b) Add 3. c) Subtract 7.
d) Divide by 7. e) Subtract −5. f) Divide by −5.
g) Add −2. h) Multiply by −2. i) Add x.
j) Multiply by x.   **12.** a) 2. b) $x − 6$. c) 7.
d) $x + 3$. e) 10. f) $\frac{x}{7}$. g) 55. h) 5x.

13. a) $4x - 9$. b) $4(x - 9)$. c) $\frac{x}{7} + 3$. d) $\frac{x + 3}{7}$.

e) $(x - 1) + y$. f) $(x + y) - 1$. g) $\frac{5x}{y}$. h) $5\left(\frac{x}{y}\right)$.

14. a) Add 9 and then divide the result by 4.
b) Divide by 4 and then add 9 to the result.
c) Subtract 3 and then multiply the result by 7.
d) Multiply by 7 and then subtract 3 from the result.
e) Subtract y and then add 1 to the result. f) Add 1 and then subtract y from the result. g) Multiply by y and then divide the result by 5. h) Divide by 5 and then multiply the result by y.

15. a)

-2	-1	0	1
1	2	3	4
-7	-6	-5	-4

b)

-2	-1	0	1
-10	-9	-8	-7
-7	-6	-5	-4

c)

-1	0	1	2
-5	0	5	10
-3	2	7	12

d)

-1	0	1	2
1	2	3	4
5	10	15	20

e)

-3	0	3	6
-13	-10	-7	-4
$-\frac{13}{3}$	$-\frac{10}{3}$	$-\frac{7}{3}$	$-\frac{4}{3}$

f)

-3	0	3	6
-1	0	1	2
-11	-10	-9	-8

g)

-15	-10	-5	0
-3	-2	-1	0
-12	-8	-4	0

h)

-15	-10	-5	0
-60	-40	-20	0
-12	-8	-4	0

i) Tables a and b; tables g and h.

16. a) False. b) False. c) True.

Set IV

1. $3 + 7 = 10$, $10 + 10 = 20$,

$20(1000 - 8) = 20(992) = 19,840$, $\frac{19,840}{992} = 20$,

$20 - 17 = 3$.   **2.** 3.   **3.** $4 + 7 = 11$,

$11 + 10 = 21$, $21(1000 - 8) = 21(992) = 20,832$,

$\frac{20,832}{992} = 21$, $21 - 17 = 4$.   **4.** 4.   **5.** The number with which we start.   **6.** We apply inverse operations. Dividing by "nine hundred and ninety two" is the inverse of multiplying by "one thousand diminished by eight." Subtracting "seventeen" is the inverse of adding "seven and ten."

Chapter 5, Lesson 3

Set I

1. a) -16. b) 24. c) -80. d) -0.2.   **2.** a) $-(-x)$ or x. b) -1. c) 0.   **3.** a) 0.2. b) $d = 0.2s$. c) A direct variation.

Set III

11. a) Five 1-pound weights were removed from each pan. b) Half the contents of each pan was removed. c) $2x + 5 = 13$. d) $2x = 8$. e) $x = 4$. f) Subtract 5 from each side of the equation. g) Divide each side of the equation by 2. h) 4 pounds.
12. a) $5x = 9$. b) $x = 0$. c) $x = -11$.
d) $8x = -12$. e) $2x = 2$. f) $x = -6$.   **13.** a) 2 was subtracted from both sides of the equation. b) 2 was added to both sides of the equation. c) Both sides of the equation were divided by 3. d) Both sides of the equation were multiplied by 3. e) 8 was subtracted from both sides of the equation. f) Both sides of the equation were multiplied by 5. g) 28 was subtracted from both sides of the equation. h) Both sides of the equation were divided by 4. i) 1 was added to both sides of the equation. j) Both sides of the equation were multiplied by 6. k) 7 was subtracted from both sides of the equation. l) Both sides of the equation were multiplied by -1.   **14.** a) -2.
b) 0.75. c) 21. d) -4. e) -10. f) -1. g) -25.
h) -9.   **15.** a) 6.6. b) 2.5. c) 15.4. d) 48.4.
e) -4.5. f) 4. g) -8.25. h) -2.   **16.** a) -4.
b) -2. c) 27. d) -56. e) 8. f) -2.   **17.** a) 5.
b) -5. c) -2. d) 28. e) 3. f) -40.

Set IV

Five glasses.

Chapter 5, Lesson 4

Set I

1. a) 21. b) –121. c) –22. d) –900.
2. a) $y = -3x$. b) $y = -5$. c) $y = x^2 - 1$.
3. a) $1,000,000x + 1 = 2$. b) There is such a number: 0.000001.

Set III

10. a) Yes. b) Yes. c) No. d) Yes. e) No. f) Yes.
g) Yes. h) No. i) Yes. j) No. k) Yes. l) Yes.
11. a) $9 + 3x$. b) $11 + 2x$. c) $x + (4 + 6)$ or
$x + 10$. d) $x^2 - 5x$. e) $(x - 5)x$. f) $x + 1$.
g) $(2 \cdot 10)x$ or $20x$. h) $1 + (x + x)$ or $1 + 2x$.
12. a) Commutative property of addition.
b) Associative property of multiplication.
c) Distributive property of multiplication over
addition. d) Commutative property of
multiplication. e) Distributive property of division
over subtraction. f) Associative property of
addition.   13. a) $13x$. b) $x + 13$. c) $15x$.
d) $5x^3$. e) $10x$. f) 0. g) $25x^2$. h) $8x$. i) $6x$. j) x.
k) $-x$.   14. a) $9x$. b) $15x^3$. c) $5 + 4x$. d) $15x^2$.
e) $x - 5$. f) $x - 9$. g) $2x - 4$. h) $10x$.
15. a) Think of a number: x
 Multiply by 3: $3x$
 Subtract 6: $3x - 6$
 Divide by 3: $x - 2$
 Add 7: $x + 5$
 Subtract the number
  first thought of: 5
b) 5.

Set IV

1.

Table 1	
1	= 1
1 + 2	= 3
1 + 2 + 3	= 6
1 + 2 + 3 + 4	= 10
1 + 2 + 3 + 4 + 5	= 15

Table 2	
1^3	= 1
$1^3 + 2^3$	= 9
$1^3 + 2^3 + 3^3$	= 36
$1^3 + 2^3 + 3^3 + 4^3$	= 100
$1^3 + 2^3 + 3^3 + 4^3 + 5^3$	= 225

2. $1^3 + 2^3 + 3^3 + \cdots + x^3 = (1 + 2 + 3 + \cdots + x)^2$
3. $1 + 2 + 3 + 4 + 5 + 6 \qquad = 21$
$1 + 2 + 3 + 4 + 5 + 6 + 7 = 28$

$1^3 + 2^3 + 3^3 + 4^3 + 5^3 + 6^3 \qquad = 441$
$1^3 + 2^3 + 3^3 + 4^3 + 5^3 + 6^3 + 7^3 = 784$
Yes.

Chapter 5, Lesson 5

Set I

1. a) $5 - x$. b) $6 - 2x$. c) $2 - 3x$. d) $-6x$.
2. a) The number negative five. b) Subtraction: x
minus five. c) The opposite of x.
3. a)

x	–3	–2	–1	0	1	2	3
y	3	2	1	0	1	2	3

b)

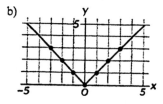

c) Yes. d) Yes. e) No.

Set III

9. a) Four boxes were removed from each pan.
b) One circle was added to each pan. c) Half the
contents of each pan was removed.
d) $4x + 3 = 6x - 1$. e) $3 = 2x - 1$. f) $4 = 2x$.
g) $2 = x$. h) $4x$ could be subtracted from both sides
of the equation. i) 1 could be added to both sides of
the equation. j) Both sides of the equation could be
divided by 2. k) $x = 2$.
10.

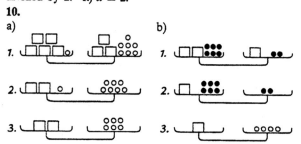

11. a) 2. b) 2.4. c) –0.4. d) –3. e) 7. f) 3.5.
g) 1. h) –10. **12.** a) 9. b) –3. c) 4. d) 0.5.
e) 1.4. f) –3.5. g) 0.1. h) 7. **13.** a) 1.5.
b) –3.2. c) 2.7. d) –4.6. e) 112.

Set IV

1. $x + 12$. **2.** $x - 72$.
3. $x + 12 = 3(x - 72)$. **4.** 114 years old.
5. $x + 12 = 126$, $x - 72 = 42$, $126 = 3(42)$.

Chapter 5, Lesson 6

Set I

1. a) Associative property of multiplication.
b) Distributive property of multiplication over
addition. c) Commutative property of addition.
d) Distributive property of division over
subtraction. **2.** a) 32.4. b) –12.5. c) 80.
d) –70. **3.** a) $n = 15t$. b) 360. c) $180x$.

Set III

10. a) $x + 11$. b) $2x + 10$. c) $4x + 8$.
d) $5x - 10$. **11.** a) $6x$. b) x^2. c) $7x + 21$.
12. $x + (x + 3) + 9 = 20$, $x = 4$; the sides are 4, 7,
and 9. **13.** a) 3, 4, 3, and 4. b) 7, 19, and 14.
c) 8, 6, 4, and 3. d) 7.5, 6.5, 4, 4.5, and 3.5.
14. a) 7 and 6. b) 6 and 8. c) 4 and 7.
15. a) AE = 9; CD = 20. b) AE = 22; EB = 17.
c) AB = 7; CE = 6; and ED = 1. d) AB = 41.4;
CE = 6.6; and ED = 34.8.

Set IV

1. No; the area is 4 times as large. **2.** The area of
one rectangle would be 9 times as large as that of the
other. **3.** The area of one rectangle would be x^2
times as large of that of the other.

Chapter 5, Lesson 7

Set I

1. a) $5x - 1$. b) $5(x - 1)$. c) $\frac{x}{3} + 8$. d) $\frac{x + 8}{3}$.
2. a) $6x^2$. b) $5x$. c) 0. d) $12x$. e) $-5x$.
f) $2x - 6$. **3.** a) $15x$. b) $\frac{y}{15}$.

Set III

10. a)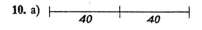

b)

c)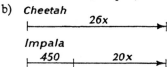

11. a)

t	0	1	2	3	4
d	0	90	180	270	360

b) Directly. c) $d = 90t$.

12. a)

r	400	500	700	800
t	3.5	2.8	2	1.75

b) $t = \dfrac{1{,}400}{r}$. c) Inversely. **13.** a) 20.

b) 45. c) $1\frac{1}{3}$. d) x. e) $\frac{y}{3}$. f) $\frac{50}{z}$.
14. a) $365x = 584{,}000{,}000$. b) 1,600,000.
15. a) 500 meters, 13.5 meters per second;
1,000 meters, 13.2 meters per second;
1,500 meters, 13.0 meters per second;
3,000 meters, 12.0 meters per second. b) 810.8
meters per minute. c) 48,648.6 meters per
hour. d) 48.6 kilometers per hour.

Set IV

1. No time. (It takes him $\dfrac{1}{25} = 0.04$ hour to cover
the first mile, but he wants to cover the two miles in
$\dfrac{2}{50} = 0.04$ hour.) **2.** No, he can't do it.

Chapter 5, Lesson 8

Set I

1. a) $>$. b) $<$. c) $>$. d) $>$. **2.** a) Yes.
b) No. c) Yes. **3.** a) $2x$ dollars. b) $8x$ dollars.
c) No money.

Set III

7. a) Cheetah, $26x$ meters; Impala, $20x$ meters.
b) *Cheetah*

c) $26x = 450 + 20x$. d) 75 seconds. e) 1,950 meters. **8.** a) $160x$ miles and $200x$ miles.

b)

1st plane — Airport — 2nd plane

160x   200x

← 900 miles →

c) $160x + 200x = 900$. d) 2.5 hours. e) 400 miles and 500 miles. **9.** a) $4 - x$ hours.

b)

5x → Downstream

3(4−x) ⊢ Upstream

c) $5x = 3(4 - x)$. d) 1.5. e) 7.5 miles.

Set IV

1. 7. **2.** 784 feet.

Chapter 5, Review

Set I

1. a) False. b) True. c) Neither. d) True.
2. a) −16. b) 13. c) 8. d) No solution because no number equals itself minus 2. e) 8 or −8. f) 6.
3. a) 3 and −1. b) 0 and −7. c) 2, 3, 4, 5, and 6.
4. a) Add 7 and then divide by 4. b) Multiply by 3 and then subtract 8. c) Add y and then multiply by 6. d) Divide by 2 and then subtract y. **5.** a) −13.
b) 13. c) −1.6. d) 2. e) 63. f) 3. **6.** a) −2.
b) 30. c) 3. d) −21. e) 18. **7.** a) $2(x - 5)$.
b) $x^2 + 8x$. c) $(3 + 7) + x$ or $10 + x$. d) $(3 \cdot 7)x$ or $21x$. e) $4x - 1$. **8.** a) $11x$. b) $18x$. c) $11 + x$.
d) $3x$. e) $6x^3$. f) $6 + 3x$. **9.** a) −3. b) −2.2.
c) 8. d) −1. e) 5. f) 0. **10.** a) 3. b) −1.5. c) 2.4.
d) −215. **11.** a) Perimeter, $2x + 16$; area, $8x$.
b) Perimeter, $12x$; area, $9x^2$. c) Perimeter, $2x + 14$; area, $9x - 18$. **12.** a) 12, 14, and 24.
b) 5.5, 4.5, 14.5, and 8.5. c) 6 and 8. d) 10 and 13. **13.** a) AE = 20; EB = 26. b) AB = 24; CE = 10. c) AE = 69; EB = 18; CD = 87.
14. a) 20. b) $20x$ miles.
15. a)

t	0	1	2	3	4
d	0	9	18	27	36

b) Directly. c) $d = 9t$. **16.** a) $2x$ feet per minute.

b)

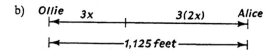

Ollie   3x   3(2x)   Alice

⊢ 1,125 feet ⊣

c) $3x + 3(2x) = 1,125$.
d) 125. e) Ollie swam 375 feet and Alice swam 750 feet.

Chapter 6, Lesson 1

Set I

1. A (3, −2); B (−2, 3); C (−4, 0); D (0, −4).
2. a) No. b) Yes; −1. c) 0.
3. a)

x	1	2	3	4
y	12	6	4	3

b) Inversely. c) $y = \dfrac{12}{x}$.

Set III

9. a) 16. b) 17. c) 16. d) 28. e) 20. f) −28.
g) −4. h) 20. i) 4. j) −74. k) −3. l) 10. m) 41.
n) 40. o) 0. p) 40. **10.** Yes. b) No. c) Yes.
d) No. e) Yes. f) No. g) No. h) Yes. i) No.
j) No. k) No. l) Yes. m) No. n) Yes. o) No.
p) Yes. **11.** a) $8x + 2y$. b) $8x + 2y = 36$.
c) Yes. d) No. e) Yes. f) $4xy$. g) $4xy = 32$.
h) Yes. i) Yes. j) No.
12. a)

x	0	3	6	9	12
y	4	2	0	−2	−4

b) y is decreased by 2.

c)

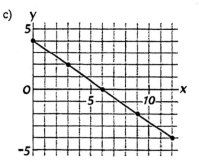

d) They lie on a straight line. **13.** a) (1, 6), (2, 5), (3, 4), (4, 3), (5, 2), and (6, 1). b) (2, 2) and (5, 1).
c) An unlimited number of solutions. d) (1, 9), (3, 3), and (9, 1). e) (1, 8) and (2, 4). f) No such solution.
g) An unlimited number of solutions. h) (1, 3) and (3, 1).

Set IV

1. Ollie bought 11 records.
$35x + 45y = 425 \rightarrow 7x + 9y = 85$. This equation has only one solution that is a pair of positive

integers: $(7, 4)$.   **2.** It isn't possible to figure out from this information how many records Alice bought. $25x + 50y = 375 \rightarrow x + 2y = 15$. This equation has seven solutions that are pairs of positive integers: $(1, 7)$, $(3, 6)$, $(5, 5)$, $(7, 4)$, $(9, 3)$, $(11, 2)$, and $(13, 1)$.

Chapter 6, Lesson 2

Set I

1. a) 81. b) 41. c) 16. d) 40.   **2.** a) 6. b) 0.5. c) 18.

3. a)

33.5x

27(x+26)

b) $33.5x = 27(x + 26)$. c) 108. d) 3,618 miles.

Set III

12. a) $y = x + 1$. b) $y = x - 6$. c) $y = 3x$.

d) $y = \dfrac{x}{4}$. e) $y = 8 - x$. f) $y = 2x + 5$.

13. a) $x = y - 1$. b) $x = y + 6$. c) $x = \dfrac{y}{3}$.

d) $x = 4y$. e) $x = 8 - y$. f) $x = \dfrac{y - 5}{2}$.

14. a) $x = 5 - y$. b) $y = 5 - x$. c) $x = \dfrac{2}{y}$.

d) $y = \dfrac{2}{x}$. e) $x = \dfrac{4y}{3}$. f) $y = \dfrac{3x}{4}$.

g) $x = 6y + 12$. h) $y = \dfrac{x - 12}{6}$.   **15.** a) $r = \dfrac{d}{t}$.

b) $t = \dfrac{d}{r}$.   **16.** a) $p = 2b + 2h$. b) $b = \dfrac{p - 2h}{2}$.

c) $h = \dfrac{p - 2b}{2}$.

17.   a)

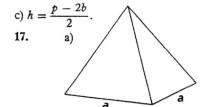

b) 7,500 cubic meters. c) $h = \dfrac{3v}{a^2}$. d) 12 meters.

18. a) 35 cents. b) $n = \dfrac{c - 2}{11}$.

c) $3 = \dfrac{35 - 2}{11} = \dfrac{33}{11} = 3$.   **19.** a) 80. b) $w = \dfrac{c\ell}{100}$.

c) $12 = \dfrac{(80)(15)}{100} = 12$. d) $\ell = \dfrac{100w}{c}$.

Set IV

Approximately 234 feet tall.

Chapter 6, Lesson 3

Set I

1. a) $2x^3$. b) $2 + 3x$. c) $6x$. d) $8x^3$.

2. a) $y = -x$. b) $y = \dfrac{60}{x}$. c) $y = 2x^2$.

3. a) $y = 1,000x$. b) $x = \dfrac{y}{1,000}$. c) Directly.

Set III

11. a) $1x + 7y = 3$; $a = 1$, $b = 7$, $c = 3$.
b) $2x + 9y = -1$; $a = 2$, $b = 9$, $c = -1$.
c) $8x + 4y = 5$; $a = 8$, $b = 4$, $c = 5$.
d) $6x + -11y = 0$; $a = 6$, $b = -11$, $c = 0$.
e) $0x + 5y = 9$; $a = 0$, $b = 5$, $c = 9$.
f) $3x + 1y = 12$; $a = 3$, $b = 1$, $c = 12$.
g) $1x + 2y = 13$; $a = 1$, $b = 2$, $c = 13$.
h) $-1.5x + 8y = 2$; $a = -1.5$, $b = 8$, $c = 2$.

12. a) $y = 2 - 5x$. b) $y = 4x + 8$. c) $y = \dfrac{x + 9}{3}$.

d) $y = \dfrac{10 - 2x}{7}$. e) $y = \dfrac{1 - 6x}{12}$.

f) $y = 9x - 5$.   **13.** a) $y = \dfrac{16 - 3x}{2}$.

b) Example table:

x	0	2	4	6
y	8	5	2	-1

c)

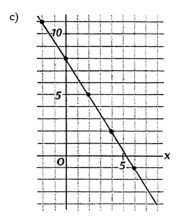

d) Yes.   e) Yes.   f) No.   g) No.
14. a)

x	–2	0	2	4
y	5	5	5	5

b)

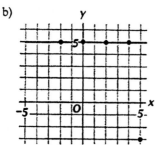

c) No.   d) No.   e) Yes.   f) Yes.
15.

a)

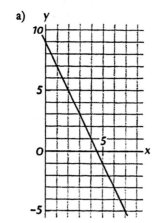

b)

c)

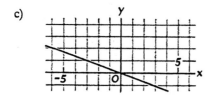

d)

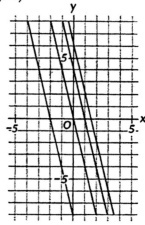

16. a through d)

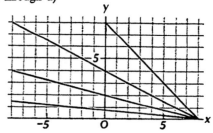

e) The lines are parallel.
17. a through d)

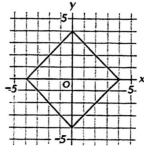

e) All the lines pass through (8, 0).

Set IV

The graph is a square.

Chapter 6, Lesson 4

Set I

1. a) True. b) Neither. c) False. d) True.
e) Neither.   **2.** a) 7, 8, and 9. b) 3 and 12. c) 8
and 13.   **3.** a) No; _12. b) [(5 − 3)7 + 8] ÷ 2.

Set III

11. a) x-intercept, 5; y-intercept, 1. b) x-
intercept, −4; y-intercept, 3. c) x-intercept, −5; y-
intercept, −2. d) x-intercept, 0; y-intercept, 0. e) No
x-intercept; y-intercept, 3.   **12.** a) x-intercept, 8;
y-intercept, 6. b) x-intercept, 7; y-intercept, 1.
c) x-intercept, 5; y-intercept, −10. d) x-intercept, 0;
y-intercept, 0. e) x-intercept, −1.5; y-intercept, 4.
f) x-intercept, 1; y-intercept, −4. g) No x-intercept;
y-intercept, 7.5. h) x-intercept, 1.8; No y-intercept.
13. a) x-intercept, −4; y-intercept, 8.
b) x-intercept, −4; y-intercept, 8. c) x-intercept, −4.
y-intercept, 8. d) x-intercept, −4; y-intercept, 8.
e) Each graph is the same line (because each line
passes through (−4, 0) and (0, 8). f) The equations
are equivalent   **14.** a) x-intercept, 4; y-intercept, 12.

b)

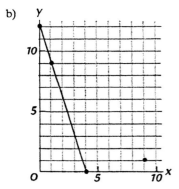

c) Yes. d) Yes. e) No. f) No.
15. a)

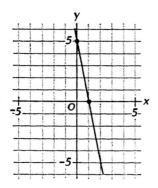

b)

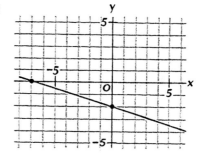

c)

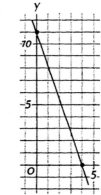

d)

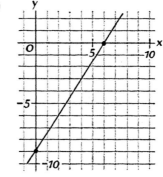

e)

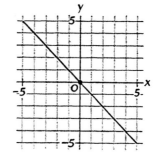

f)

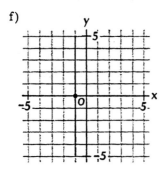

16. a, b, and c)

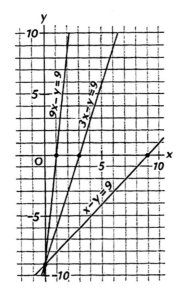

d) All have the same y-intercept, -9.

17. a, b, and c)

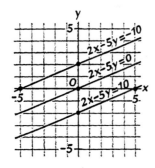

d) The lines are parallel.

Set IV

a) Graph the function.
   ANS: Graph below.

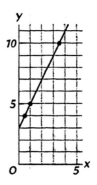

b) Write an equation for the function.
   ANS: $y = 2x + 3$.

c) Fill in the blanks in the table.
   ANS:

x	4	2
y	11	7

Chapter 6, Lesson 5

Set I

1. a) $\frac{-7}{1}$. b) $\frac{3}{2}$. c) $\frac{-4}{9}$. d) $\frac{61}{10}$.   **2.** a) $(-1, -2)$

and $(-2, -1)$. b) Infinitely many such pairs.
c) $(-1, -8)$, $(-2, -4)$, $(-4, -2)$, and $(-8, -1)$.
d) None.   **3.** a) $752 - x$. b) $5x$ dollars.
c) $2(752 - x)$ or $1504 - 2x$ dollars. d) $3x + 1504$
dollars.

Set III

10. a) 1. b) $\frac{3}{4}$. c) $-\frac{1}{2}$. d) The slope is not

defined.

11. a)

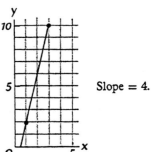

Slope $= 4$.

b)

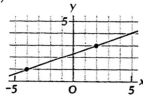

Slope $= \frac{1}{3}$.

c)

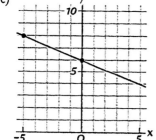

Slope $= -0.4$.

d)

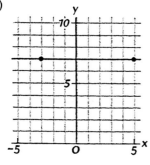

Slope $= 0$.

12. a)

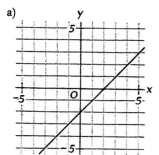

Slope $= 1$.

b)

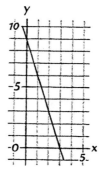

Slope $= -3$.

c)

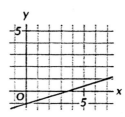

Slope $= \frac{2}{7}$.

d)

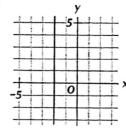

Slope is not defined.

13. a and b)

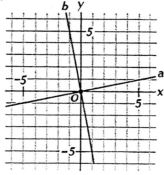

c and d)

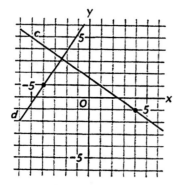

e) They are perpendicular.

14. a)

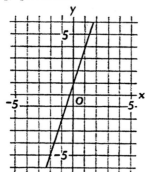

b)

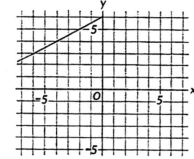

c)

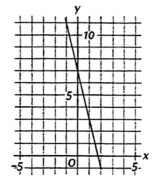

d)

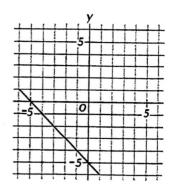

e)

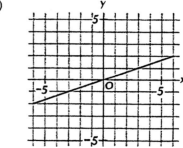

15. a) Slope, 3; y-intercept, 1. b) Slope, $\frac{1}{2}$; y-intercept, 6. c) Slope, -4; y-intercept, 7.
d) Slope, -1; y-intercept, -5. e) Slope, $\frac{1}{3}$; y-intercept, 0. f) Slope, 5; y-intercept, 3.
g) Slope, -2; y-intercept, -9.

Set IV

1. Approximately -1.8.   **2.** 180 meters.

Chapter 6, Lesson 6

Set I

1. a) $3x + 2$. b) $x + 2$. c) $4x + 2x^2$.
d) $4x - 2x^2$.   **2.** a) $x = y + 10$. b) $y = x - 10$.
c) $x = \dfrac{y + 2}{3}$. d) $y = 3x - 2$.   **3.** a) 50 and 0 are
even; -5 is odd. b) Odd. c) Even. d) Odd.
e) Even. f) Odd. g) Odd.

11. a) Direct variations. b) Straight line passing through the origin. c) The constant of variation or slope of line. d) At the origin.   **12.** a) Slope, 5; y-intercept, 2. b) Slope, $\frac{1}{3}$; y-intercept, 9.

c) $y = 1x + -4$; slope, 1; y-intercept, -4.
d) $y = 2x + -14$; slope, 2; y-intercept, -14.
e) $y = 3x + 0$; slope, 3; y-intercept, 0.
f) $y = -6x + 0$; slope, -6; y-intercept, 0.
g) $y = -1x + 8$; slope, -1; y-intercept, 8.
h) $y = 0x + 11$; slope, 0; y-intercept, 11.

13. a) $y = 4x + 7$. b) $y = \frac{3}{5}x - 2$.

c) $y = -9x + 1$. d) $y = -5$. e) $y = 1.6x$.

f) $x = 0$.   **14.** a) $y = x + 5$. b) $y = \frac{1}{2}x + 1$.

c) $y = -3x - 2$. d) $y = -6$.
15.

a)

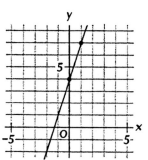

b)

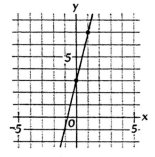

c)

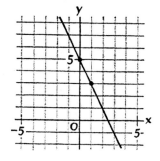

d)

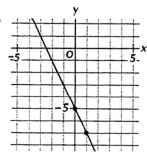

e)

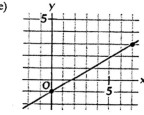

f)

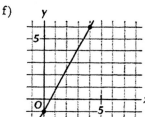

16. a) $y = x - 2$.

b)

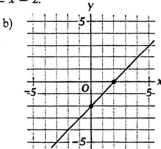

c) $(2, 0)$. d) $(0, -2)$. e) 1.
17. a) $y = 3x$.

b)

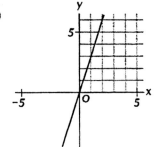

c) $(0, 0)$. d) $(0, 0)$. e) 3.

Set IV

1.

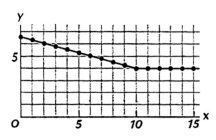

2. The slope is –0.25 from 1 to 10 and 0 from 10 on. **3.** No x-intercept; y-intercept, 6.5.

Chapter 6, Review

Set I

1. a) 12. b) –23. c) 12. d) 49. e) 17. f) 17.
2. a) Yes. b) No. c) Yes. d) No. e) Yes.
f) Yes.   **3.** a) $(1, 8)$, $(2, 4)$, $(4, 2)$, and $(8, 1)$. b) An unlimited number of solutions. c) $(1, 10)$, $(2, 6)$, and $(3, 2)$. d) None.   **4.** a) $x = \dfrac{y + 7}{3}$.

b) $y = 3x - 7$. c) $x = \dfrac{1 - y}{6}$. d) $y = 1 - 6x$.

5. a) $h = \dfrac{2a}{b}$. b) $b = \dfrac{2a}{h}$. c) 12. d) $4 = \dfrac{2 \cdot 12}{6} = 4$.

e) $6 = \dfrac{2 \cdot 12}{4} = 6$.   **6.** a) 100 dollars. b) 1.5

dollars. c) $d = \dfrac{c - 1.5m}{100}$. d) 4.

7. a) $8x + -1y = 10$; $a = 8$, $b = -1$, $c = 10$.
b) $3x + 6y = 4$; $a = 3$, $b = 6$, $c = 4$.
c) $x + -7y = 9$; $a = 1$, $b = -7$, $c = 9$.
d) $5x + 0y = -1$; $a = 5$, $b = 0$, $c = -1$.
8. a) $y = 12 - 6x$. b) $y = \dfrac{x - 8}{5}$. c) $y = \dfrac{1 - 2x}{7}$.

d) $y = 4x - 9$.   **9.** a) x-intercept, 15; y-intercept, 9. b) x-intercept, 3.5; y-intercept, –7.
c) x-intercept, 0; y-intercept, 0. d) x-intercept, 4; y-intercept, 9.

10. a)

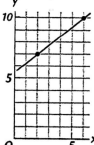

Slope $= \dfrac{3}{4}$.

b)

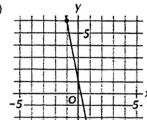

Slope $= -5$.

c)

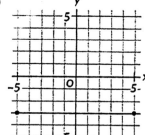

Slope $= 0$.

11. a)

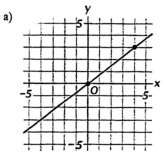

b)

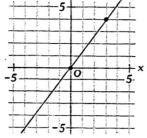

c)

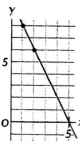

d)

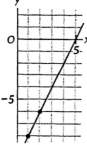

c)

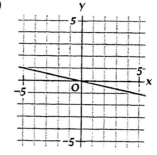

d)

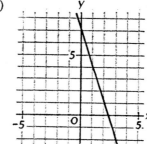

12. a) Slope, 6; y-intercept, -2.   b) Slope, $\frac{1}{4}$;

y-intercept, 0.   c) Slope, -1; y-intercept, 7.

d) Slope, 0; y-intercept, 3.   **13.** a) $y = 4x + 1$.

b) $y = -\frac{2}{3}x + 7$.   c) $y = -4$.   d) $x = 0$.

14.

a)

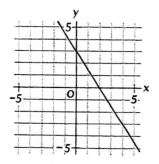

b)

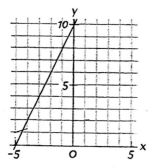

e)

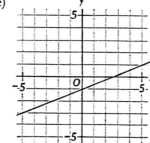

f)

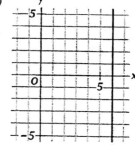

Chapter 7, Lesson 1

Set I

1. a) $6x$. b) 0. c) $8x$. d) $6x$. e) x. f) $-x$.
2. a) $>$. b) None. c) $=$. d) $<$. e) None.
f) $=$.   **3.** a) $50 + w$ meters per minute.
b) $50 - w$ meters per minute. c) $50x$ meters.

d) $\dfrac{y}{50}$ minutes.

Set III

15. a) Yes. b) No. c) No. d) No. e) Yes. f) No.
g) Yes. h) Yes. i) Yes.   **16.** a) $6x = 30$.
b) $x = 5$. c) $y = 3$. d) $(5, 3)$.
e) $3(5) - 3 = 12$, $15 - 3 = 12$, $12 = 12$.
17. a) $4x + y = 31$. b) $4x - y = 25$. c) $(7, 3)$.
d) $4(7) + 3 = 31$, $4(7) - 3 = 25$.   **18.** $(9, 5)$.
19. $(7, 8.5)$.   **20.** $(25, 12)$.   **21.** $(3, -2)$.
22. $(8, -3)$.   **23.** $(5, 18)$.   **24.** $(0, -2)$.
25. $(-0.5, 3)$.

Set IV

$x + y + z = 50$, $x + y - z = 40$, $x - y - z = 12$;
$(31, 14, 5)$.

Chapter 7, Lesson 2

Set I

1. a) 2. b) a, 3; b, -1. c) $y = 2x + 3$.
d) $y = 2x - 1$.   **2.** a) $x + -y$. b) $y + -x$.
c) $x + -y + -z$. d) $z + -y + -x$.
3. a) $p = 10h - 0.5m$. b) 385 dollars.

c) $h = \dfrac{p + 0.5m}{10}$. d) 49 hours.

Set III

17. a) 2. b) 9. c) $x + y = 11$, $x + 3y = 15$.
d) $-2y = -4$, $y = 2$, $x + 2 = 11$, $x = 9$.
18. $5x + 2y = 62$, $2x + 2y = 38$; $(8, 11)$.
19. $4x + y = 23$, $4x - 6y = 2$; $(5, 3)$.
20. a) $2x = 24$, $x = 12$;
$12 + 4y = 20$, $4y = 8$, $y = 2$; $(12, 2)$.
b) $8y = 16$, $y = 2$;
$x + 4(2) = 20$, $x + 8 = 20$, $x = 12$; $(12, 2)$.
c) $-8y = -16$, $y = 2$;

$x - 4(2) = 4$, $x - 8 = 4$, $x = 12$; $(12, 2)$.
21. $(3, 7)$.   **22.** $(2, -4)$.   **23.** $(-5, 1)$.
24. $(-4, -12)$.   **25.** $(0, -3.5)$.   **26.** $(8, -15)$.
27. $(-6, 2)$.   **28.** $(1.5, -9)$.   **29.** $(-1, -3)$.

Set IV

Tweedledum weighs 120 pounds and Tweedledee
weighs 121 pounds.

Chapter 7, Lesson 3

Set I

1.   a)

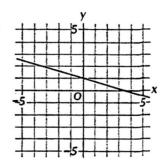

b)

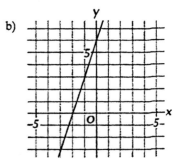

c)

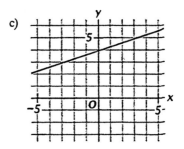

d)

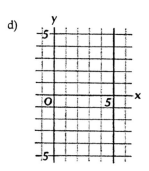

2. a) $x - y$. b) $3z$.

Set III

14. $2x + y = 6$, $6x + 3y = 18$; multiply each side of the first equation by 3.

15. $x - 4y = 11$, $2x - 8y = 22$; multiply each side of the first equation by 2.

16. $5x + 5y = 45$, $x + y = 9$; divide each side of the first equation by 5.   **17.** a) $6x + 2y = 16$.
b) $30x - 24y = 6$. c) $2x + y = 9$.
d) $-x + 9y = 10$. e) $8x - 28y = 0$.
f) $2x - 3y = 8$.   **18.** a) $20x - 4y = 28$.
b) $21x = 63$. c) $x = 3$. d) $y = 8$.
e) $3 + 4(8) = 35$, $3 + 32 = 35$, $35 = 35$;
$5(3) - 8 = 7$, $15 - 8 = 7$, $7 = 7$.
f) $5x + 20y = 175$. g) $21y = 168$. h) $y = 8$.
i) $x = 3$.   **19.** $(-2, 7)$.   **20.** $(9, 1)$.
21. $(5, -3)$.   **22.** $(11, 9)$.   **23.** $(0, 6)$.
24. $(4, -10)$.

Set IV

Add the equations to get $100x + 100y = 1{,}000$. Subtract the second equation from the first to get $10x - 10y = 40$. Simplifying these equations by dividing by 100 and 10 respectively, we get $x + y = 10$, $x - y = 4$. Solving, we get $(7, 3)$.

Chapter 7, Lesson 4

Set I

1. a) $x = \dfrac{y - b}{a}$. b) $x = \dfrac{c - by}{a}$. c) $y = \dfrac{c - ax}{b}$.

2. a) 7. b) 3.6.   **3.** a) 160. b) $8xy$. c) 25. d) $\dfrac{z}{8x}$.

Set III

13. $(-3, 5)$.   **14.** $(2, -6)$.   **15.** $(-4, -1.5)$.
16. a, c, and e)

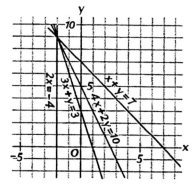

b) $4x + 2y = 10$. d) $2x = -4$. f) All four lines pass through $(-2, 9)$.

17. a, c, and e)

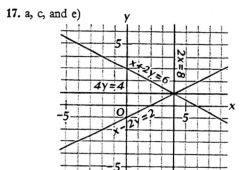

b) $2x = 8$. d) $4y = 4$. f) All four lines pass through $(4, 1)$.

18.

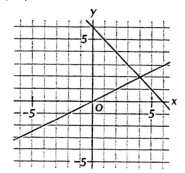

Solution: $(4, 2)$.

19.

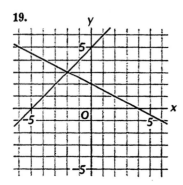

Solution: (-2, 3).

20.

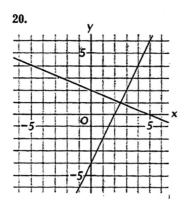

Solution: (2.5, 1).

21.

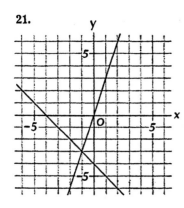

Solution: (-1, -3).

Set IV

1. The greyhound overtakes the cat after 5 seconds.   **2.** The cat is not running and the greyhound overtakes the cat after 4 seconds.
3. The cat is running as fast as the greyhound and they always stay 40 meters apart.   **4.** The cat is running faster than the greyhound and the distance between them is increasing.

Chapter 7, Lesson 5

Set I

1. a through d)

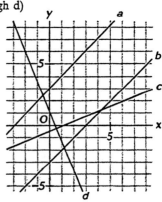

2. a) (-2, 7).   b) (5, -1).   **3.** a) 125 cubic centimeters.   b) 30°C.

Set III

14. a) There is one solution: (1, -4).   b) The equations have an unlimited number of solutions.
c) The equations have no solutions.
15. a) $x + y = 11$, $3x + 3y = 33$.   b) $3x + 3y = 33$, $3x + 3y = 33$, subtracting; $0 = 0$.   c) The equations are equivalent.   **16.** a) $4x + 2y = 15$, $2x + y = 7$.
b) $4x + 2y = 15$, $4x + 2y = 14$; subtracting, $0 = 1$.
c) The equations are inconsistent.
17. a) $6x - 6y = 25$, $4x - 4y = 16$.
b) $12x - 12y = 50$, $12x - 12y = 48$; subtracting, $0 = 2$.   c) The equations are inconsistent.
18. a) $x - 3y = 9$, $2x - 6y = 18$.   b) $2x - 6y = 18$, $2x - 6y = 18$; subtracting, $0 = 0$.   c) The equations are equivalent.

19.   a)

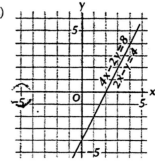

b) The graphs are the same line.   c) Infinitely many.
d) Equivalent.

20. a)

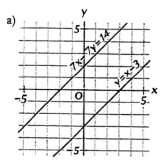

b) The graphs are parallel lines.   c) None.
d) Inconsistent.

21.

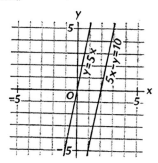

The equations are inconsistent.

22.

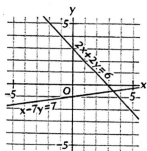

The equations have one solution: (3.5, –0.5).

23.

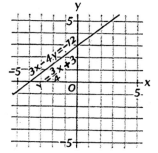

The equations are equivalent.

1. The equations in Graph A are inconsistent and have no solutions. The equations in Graph B have four solutions: (2, 2), (–2, 2), (2, –2), and (–2, –2).
2. A possible answer: Dividing the first equation in Graph A by 4 and the second by 9, we get $x^2 + y^2 = 9$ and $x^2 + y^2 = 4$. Subtracting the second of these equations from the first gives $0 = 5$.

Chapter 7, Lesson 6

Set I

1. a) 11.   b) –4.   c) 6 and 10.   d) –9 and 5.
2. a) $y = 0$.   b) 0.   c) $x = 0$.   d) It is undefined.
3. a) $d = 376g$.   b) $g = \dfrac{d}{376}$.   c) 0.04 gallon.

Set III

17. $3x + y = 30$, $y = 2x$; (6, 12).
18. $x + 5y = 55$, $x = y + 7$; (15, 8).
19. $3x + 2 = y$, $x + 2y = 25$; (3, 11).
20. $4x = y + 26$, $4x = 18 + 3y$; (7.5, 4).
21. (7, 14).   **22.** (3, –4).   **23.** (0, –1).
24. (14, 5).   **25.** a) $2(2 + 4y) - 8y = 4$,
$4 + 8y - 8y = 4$, $4 = 4$.   b) The equations are equivalent.   c) A single line.
26. a) $6x - 3(2x) = 10$, $6x - 6x = 10$, $0 = 10$.
b) The equations are inconsistent.   c) Two parallel lines.   **27.** a) $x + y = 18$, $2x = 10y$.   b) (15, 3).
c) $15 + 3 = 18$, $18 = 18$; $2(15) = 10(3)$, $30 = 30$.
28. a) $x + y = 21$, $8x = 6y$.   b) (9, 12).
c) $9 + 12 = 21$, $21 = 21$; $8(9) = 6(12)$, $72 = 72$.
29. a) $x + y = 27$, $7x = 2y$.   b) (6, 21).
c) $6 + 21 = 27$, $27 = 27$; $7(6) = 2(21)$, $42 = 42$.

Set IV

The duck weighs 20 pounds and the rabbit weighs 12 pounds. ($d + 3 + 5 + r = 40$, $9d + 7(3) = 9(5) + 13r$.)

Chapter 7, Lesson 7

Set I

1. a) $2x + y = 8$, $4x + 2y = 12$.   b) No solutions.
c) Inconsistent.   d) Two parallel lines.   **2.** a) A possible answer: $x + 2y = 6$.   b) A possible answer: $2x + 4y = 10$.   c) A possible answer: $x + y = 3$.
3. a) Inverse.   b) The wavelength decreases.

c) $f = \dfrac{300,000}{w}$. d) 150 kilocycles per second.

Set III

8. a) $x + y = 42$. b) $2x$. c) $3y$.
d) $2x + 3y = 100$. e) $(26, 16)$. f) There are 26
questions worth 2 points and 16 questions worth 3
points. **9.** a) $x + 40 = y$. b) $84x$. c) $79y$.
d) $84x + 71(40) = 79y$. e) $(64, 104)$. f) 64.
10. a) $x + y = 1{,}255$, $3x + 2y = 3{,}680$.
b) $(1{,}170, 85)$. c) 1,170 adult tickets and 85
children's tickets. **11.** a) $x + y = 10$,
$19.3x + 8.9y = 172.2$. b) $(8, 2)$. c) 154.4 grams of
gold and 17.8 grams of copper.

Set IV

The price of a citron is 8 and the price of a wood
apple is 5.

Chapter 7, Review

Set I

1. a) No. b) Yes. c) Yes. d) No. e) Yes.
2. a) $4x - 3y = 16$. b) $5y = -4$. c) $8x + 4y = 24$.
d) $x - 2y = 5$. **3.** $(51, -16)$. **4.** $(-2, 3)$.
5. $(18.5, 0.9)$. **6.** $(-1, -10)$. **7.** $(7, 1)$.
8. $(0, -8)$.
9.

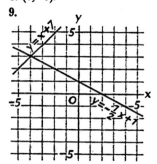

Solution: $(-4, 3)$.

10.

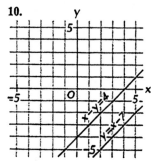

The equations have
no solutions.

11.

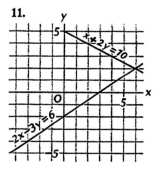

Solution: $(6, 2)$.

12.

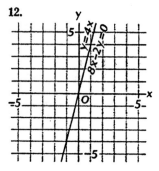

The equations have
infinitely many solutions.

13. $(3.5, 10.5)$. **14.** $(45, 9)$. **15.** $(-19, -2)$.
16. $(5, 12)$. **17.** $x + y = 21$, $4x = 10y$; $(15, 6)$.
18. $x + y = 20$, $2x = 3y$; $(12, 8)$. **19.** a) A possible
answer: $x + 4y = 7$. b) A possible answer:
$2x + 8y = 16$. c) A possible answer: $x + y = 2$.
20. a) $x + y = 85$, $10x + 13y = 1000$. b) $(35, 50)$.
c) 35 ten-cent stamps and 50 thirteen-cent stamps.

Chapter 8, Lesson 1

Set I

1. a) $s = \dfrac{p}{4}$. b) $p = \dfrac{i}{rt}$. c) $h = \dfrac{3v}{a^2}$.
2. a) Associative property of multiplication.
b) Distributive property of multiplication over
addition. c) Commutative property of addition.
3. a) $h = 1.5t$. b) A direct variation. c) 1.5. d) 36
inches or 3 feet.

Set III

16. a) One billion. b) Ten billion. c) One hundred
billion. d) One trillion. e) Ten trillion.
17. a) 10^4. b) 10^{15}. c) 10^3. d) 10^1. **18.** a) 5,100.
b) 92.3. c) 4. d) 0.0077. e) 10^{13}. f) 6×10^6 or
60×10^5. g) 2×10^8 or 0.2×10^9. **19.** a) 640.

b) 0.19. c) 0.005. d) 10^9. e) 2.7×10^3 or 27×10^2. **20.** a) 800. b) 10,000,000. c) 52,000.
d) 4,040,000,000. e) 6. f) 31,700. **21.** a) 5,000.
b) 5×10^3. **22.** a) 5×10^6. b) 6×10^2.
c) 4.8×10^4. d) 1.09×10^3. e) 7.7×10^1.
f) 3.002×10^7. **23.** a) 1×10^9. b) 8×10^5.
c) 2.5×10^{12}. d) 4×10^1. e) 6.1×10^3.
24. a) 70,000,000. b) 7×10^7. **25.** a) Five
billion. b) 5×10^9. **26.** a) 970,000. b) Nine
hundred seventy thousand. **27.** a) 141,000,000.
b) One hundred forty-one million.

Set IV

1. 2×10^6. **2.** 1×10^{12}. **3.** $1 \times 10^{6,000,000}$.

Chapter 8, Lesson 2

Set I

1. a) $y = -x^2$. b) $y = 12 - x$. c) $y = 3x - 5$.
2. a) $(17, 8)$. b) $(6.5, 1.5)$. **3.** a) 16.2. b) 0.5 hour or
30 minutes.

Set III

11. a) x^4. b) y^9. c) $x^6 y^3$. d) xy^5. e) $x^2 y^4$.
12. a) x^6. b) y^8. c) x^{10}. d) y^7. e) $x^{12} y^{12}$.
f) $x^{18} y^3$. g) $21x^{10}$. h) $36x^{13}$. **13.** a) 8. b) 16.
c) 6. d) 9. e) 12. f) 8. g) 20. h) 18.
14. a) 16,384. b) 262,144. c) 67,108,864.
d) 268,435,456. e) 1,073,741,824. f) 16,777,216.
g) False. h) True. i) False. j) True.
15. a) 6×10^7. b) 8×10^9. c) 3×10^{10}.
d) 3.6×10^{11}. e) 7×10^{10}. f) 6×10^{12}.
16. a) 3×10^{16}. b) 30,000,000,000,000,000.
17. a) 6×10^8 and 3.6×10^3. b) 2.16×10^{12}.

Set IV

6.31×10^8 $[(15)(60)(24)(365)(80)]$.

Chapter 8, Lesson 3

Set I

1. a) 1.86×10^5. b) 1.2×10^1. c) 3×10^{13}.
d) 5×10^8. **2.** a) $y = \frac{1}{2}x + 4$. b) $y = -\frac{3}{2}x$.
c) $(-2, 3)$. **3.** a) $x + y = 98$, $x + 5y = 150$. b) 85
pennies and 13 nickels.

Set III

11. a) $2^4 \cdot 2^3 = (2 \cdot 2 \cdot 2 \cdot 2)(2 \cdot 2 \cdot 2) =$
$2 \cdot 2 \cdot 2 \cdot 2 \cdot 2 \cdot 2 \cdot 2 = 2^7$. b) $\dfrac{2^4}{2^3} = \dfrac{2 \cdot 2 \cdot 2 \cdot 2}{2 \cdot 2 \cdot 2} =$
$2 = 2^1$. c) $(2^4)^3 = (2^4)(2^4)(2^4) = 2^{4+4+4} = 2^{12}$.
12. a) x^5. b) x^1 or x. c) x^3. d) x^{12}. e) x^{12}. f) x^{64}.
13. a) 8. b) 4. c) 1. d) 16. e) 3. f) 18. g) 8.
h) 15. i) 10. j) 24. k) 6. l) 4. m) 8. n) 4.
14. a) 243. b) 1,594,323. c) 243. d) 6,561.
e) 531,441. f) 6,561. g) 14,348,907. h) 14,348,907.
i) False. j) True. k) False. l) True.
15. a) 4×10^6. b) 2×10^4. c) 5×10^3.
d) 5×10^4. e) 2×10^6. f) 1.4×10^{11}.
16. a) 250,000,000,000 and 2,000. b) 2.5×10^{11} and
2×10^3. c) 1.25×10^8. **17.** 1.2×10^{25}.

Set IV

1. Yes; 10^{100} is a googol. **2.** $10^{100} = 1 \times 10^{100}$,
$100^{10} = 1 \times 10^{20}$. **3.** 10^{100} is 10^{80} times as large
as 100^{10}.

Chapter 8, Lesson 4

Set I

1. a) 1×10^{11}. b) 1.44×10^{24}. c) 6.3×10^{17}.
2. a) $(9, 0)$. b) No solutions (these equations are
inconsistent). **3.** a) $x + 1$, $x + 2$, $x + 3$.
b) $x + (x + 1) + (x + 2) + (x + 3) = 2$.
c) $-1, 0, 1, 2$.

Set III

9. a) 1. b) 8. c) 0. d) 1. e) -1. f) $\frac{1}{9}$. g) 64.
h) $\frac{1}{36}$. i) 125. j) $\frac{1}{125}$. k) -125. l) $-\frac{1}{125}$.
m) $-\frac{1}{4}$. n) 1. **10.** a) 0.5. b) 0.015625. c) 32.
d) 0.03125. e) 2. f) 256. g) 1. h) 64. i) 256.
j) 1.
11.

x	4	3	2	1	0	-1	-2	-3
y	10,000	1,000	100	10	1	0.1	0.01	0.001

12. a) $>$. b) $=$. c) $<$. d) $>$. e) $<$. f) $>$.
g) $>$. h) $>$. **13.** a) x^4. b) x^{-12}. c) x^7. d) x^0.
e) x^0. f) x^{-9}. g) x^{-9}. h) x^{20}. i) x^{-9}. j) x^0.
k) x^{-1}. l) x^{-4}. m) x^{11}. n) x^{-14}. o) x^{-6}. p) x^0.
q) x^{10}. r) x^8.

Set IV

1. 3.   **2.** The square root of x.

Chapter 8, Lesson 5

Set I

1. a) 10^3. b) 10^7. c) $10^5 - 10^2$ or 99,900.

2.

a)

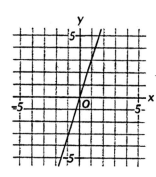

b)

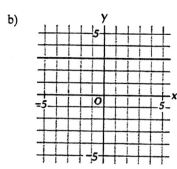

c)

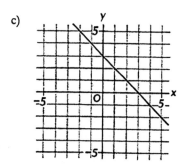

3. a) 1.4×10^5 and 3×10^8. b) 4.2×10^{13} miles.

Set III

15. a) 10^0. b) 10^{-1}. c) 10^{-2}. d) 10^{-3}. e) 10^{-4}.
16. a) 0.02. b) 11.1. c) 10^6. d) 10^{-4}. e) 40×10^{-8} or 4×10^{-7}. f) 0.3×10^{-11} or 3×10^{-12}.
17. a) 0.09. b) 1.234. c) 10^6. d) 10^{-8}.
e) 0.8×10^{-10} or 8×10^{-11}. **18.** a) 9×10^2.
b) 9×10^{-3}. c) 3.1×10^1. d) 3.1×10^{-1}.
e) 2.22×10^{-4}. f) 2.22×10^{-10}. g) 5×10^0.
h) 1×10^{-5}. **19.** a) 40. b) 0.4. c) 0.007.
d) 0.0075. e) 2,080,000. f) 0.00000208.
20. a) 0.0004. b) 4×10^{-4}. **21.** a) 5×10^3.
b) 5×10^{-1}. c) 7×10^8. d) 7×10^{-10}.
22. a) $<$. b) $>$. c) $>$. d) $<$. e) $>$. f) $=$.
23. a) 8×10^{-6}. b) 3×10^1. c) 2.7×10^{-5}.
d) 4×10^{10}. e) 2.5×10^{-11}. f) 6×10^{-1}.
24. a) 0.0000003. b) Three-ten millionths.
25. a) 5×10^4 and 1×10^{-8}. b) 5×10^{12}. c) Five trillion.

Set IV

The sun and a hydrogen atom. Sun: $\dfrac{2 \times 10^{30}}{6 \times 10^1} \approx$
3×10^{28}. Hydrogen atom: $\dfrac{6 \times 10^1}{2 \times 10^{-27}} = 3 \times 10^{28}$.

Chapter 8, Lesson 6

Set I

1. a) 1.25. b) 12.5. c) 0.8. d) 0.08.   **2.** a) -1.
b) 1. c) -7. d) $\dfrac{1}{7}$.   **3.** a) $4x - 7 = 5(1 + x)$.
b) -12.

Set III

9. a) x^5y^5. b) 1. c) $49x^2$. d) $81y^4$. e) x^{27}.
f) $x^{15}y^3$. g) $36x^{-2}$. h) $-32y^{10}$. i) $\dfrac{x^6}{y^6}$. j) $\dfrac{x^9}{64}$.
k) $\dfrac{1}{y^{32}}$. l) $\dfrac{81x^2}{y^{10}}$.   **10.** a) $<$. b) $=$. c) $<$.
d) $=$. e) $>$. f) $>$. g) $=$. h) $=$. i) $>$.
j) $>$.   **11.** a) 400. b) 0.05. c) 64,000,000.
d) 3,200,000. e) 8,000. f) 160,000. g) 0.000125.
h) 0.0025.   **12.** a) 9×10^{10}. b) 6.4×10^{13}.
c) 6.25×10^{-2}. d) 2.5×10^{-6}. e) 1×10^{-18}.
f) 2.5×10^{13}.   **13.** a) 28. b) 7. c) 12. d) 7.
e) 10. f) 5. g) 6. h) 3. i) 6. j) 4. k) 16. l) 10.

1. $y = \dfrac{6^{10}}{x}$.   **2.** Inverse variation.   **3.** Yes;

$xy = 6^{10}$.

Chapter 8, Lesson 7

Set I

1. a)

x	-4	-2	0	2	4
y	9	7	5	3	1

b)

x	0	1	2	3	4
y	not defined	36	9	4	2.25

c)

x	1	2	3	4	5
y	0	0	2	6	12

2. a) True. b) False; example:
$|-2 + 2| \neq |-2| + 2$. c) True. d) False; example:
$|2 - (-2)| \neq 2 - |-2|$. e) True.
3. a) 220,000,000 and 11,000,000. b) Two hundred
twenty million and eleven million. c) $\dfrac{1}{20}$ or 0.05.

Set III

10. a) $6 \cdot 10^4$. b) $7 \cdot 2^5$. c) $4 \cdot 5^3$. d) $9 \cdot 10^{-2}$.
e) $8 \cdot 3^{-6}$.   **11.** a) 80. b) 1,100,000. c) 7. d) 3.
e) 4. f) 0.3125.
12. a)

x	0	-1	-2	-3	-4
y	1	0.2	0.04	0.008	0.0016

b) y is multiplied by 5^{-1} or 0.2. c) y is multiplied by
5^{-2} or 0.04. d) y is multiplied by 5^{-3} or 0.008.

13. a)

x	-3	-2	-1	0	1	2	3
y	-6	-4	-2	0	2	4	6

d)

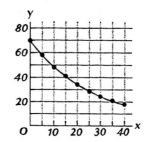

b)

x	-3	-2	-1	0	1	2	3
y	9	4	1	0	1	4	9

e)

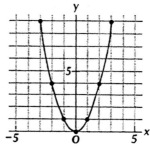

c)

x	-3	-2	-1	0	1	2	3
y	0.125	0.25	0.5	1	2	4	8

f)

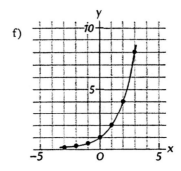

14. a) $1,060. b) $1,123.60. c) $1,191.02.
d) $1,790.85.
15. a)

x	0	1	2	3	4	5	6	7
y	80	40	20	10	5	2.5	1.25	0.625

b) One day. c) 0.625 gram. d) 0.00488 gram.

Set IV

1.

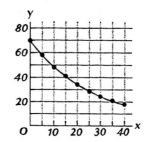

2. No; if it were, the graph would be a straight
line.   **3.** 8.5 degrees.

Chapter 8, Review

Set I

1. a) 10^5. b) 13.5. c) 10^{-7}. d) 0.0007.
e) 2×10^5. **2.** a) 2.2×10^8. b) 2.2×10^1.
c) 10,000,000. d) The exponent becomes 7 less.
3. 30,000. b) 0.0003. c) 0.62. d) 85,000,000.
4. a) 7×10^{11}. b) 4.12×10^7. c) 2×10^{-3}.
d) 1.0002×10^1. **5.** a) $\frac{1}{8}$. b) $\frac{1}{625}$. c) 1. d) 36.
e) $\frac{1}{36}$. f) 64. **6.** a) x^{-6}. b) x^{-16}. c) x^{-5}. d) x^7.
e) x^{-11}. f) x^2. **7.** a) $>$. b) $<$. c) $>$. d) $=$.
8. a) $64x^3$. b) $\frac{1}{7y}$. c) $25x^{10}$. d) $\frac{x^4}{y^8}$.
9. a) 5.4×10^{16}. b) 1.4×10^{-2}. c) 8×10^{-6}.
d) 5.5×10^{-8}. e) 8.1×10^9. f) 2×10^6.
10. a) $<$. b) $>$. c) $=$. d) $>$. e) $>$. f) $=$.
g) $<$. **11.** a) 10. b) 24. c) 6. d) 4. e) 9.
f) 4. **12.** a) 1.86×10^5 and 2×10^{-1}.
b) 9.3×10^5. c) Nine hundred thirty thousand.

13. a)

x	0	1	2	3	4
y	8	4	2	1	0.5

b) y is halved.

c)

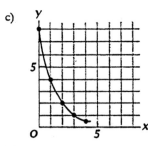

14. a) \$2,140. b) \$2,289.80. c) No; the increase the first year is \$140 and the increase the second year is \$149.80.

Midterm Review

Set I

1. $8x - 24$. **2.** True. **3.** $3y$. **4.** True.
5. -18. **6.** $\frac{80}{x}$. **7.** False. **8.** 4^{-1}.

9. $-x + 7$. **10.** True. **11.** -49. **12.** $\frac{100}{x}$.
13. -42. **14.** True. **15.** $x + -y$.
16. False. **17.** 5^{-2}. **18.** 16. **19.** x^{-10}.
20. -23. **21.** $p = \frac{i}{rt}$. **22.** False. **23.** $<$.
24. True. **25.** An even number. **26.** $\frac{x^2}{y}$.
27. $3x$. **28.** False. **29.** Add 7 to each side.
30. $\frac{1}{36}$. **31.** At -9. **32.** $=$. **33.** -1.
34. $8x^6$. **35.** 10^{-3}. **36.** $5 + x$. **37.** y^5.
38. False. **39.** $y = 4x - 3$. **40.** x^9.

Set III

1. 1.2×10^4. **2.** $x^2 < 3x$.
3.
```
oo  oo  oo  oo  oo
oo  oo  oo  oo  oo
```

4. $x + 2y = -4$. **5.** -6. **6.** -2.25. **7.** -2.
8. $-12x^2$. **9.** 120.

10.

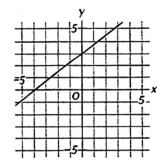

11. -2.2. **12.** 7.

13.

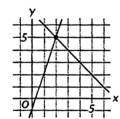

Solution: $(2, 5)$.

14. 2^6.

15.

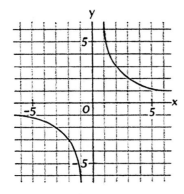

16. -2.   **17.** Add x to each side.   **18.** (1, 13),
(2, 10), and (3, 5).   **19.** $y = \dfrac{60}{x}$.

20.

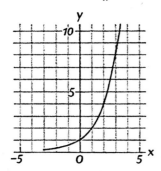

21. -2.   **22.** (2, -6).   **23.** 0.00095.   **24.** 38.
25. 2.5×10^5.   **26.** $>$.   **27.** -7.   **28.** -22.
29. 0.075.   **30.** Even.   **31.** 12^6.   **32.** 68.
33. $x = \dfrac{y + 9}{2}$.   **34.** (5, 3).   **35.** 10.
36. $y = 2x + 7$.   **37.** $y = 2 - 4x$.
38. $5(x + 40)$.   **39.** $5x + 5(x + 40) = 6,200$.
40. 600 meters per minute.

Chapter 9, Lesson 1

Set I

1.

Think of a number:	x
Add one:	$x + 1$
Multiply by three:	$3x + 3$
Subtract nine:	$3x - 6$
Divide by three:	$x - 2$
Subtract the number first thought of:	-2

The result is negative two.

2. a) (-2, -5).   b) (6, -1).   **3.** 2×10^{-3} second.

11. a) Coefficient, 4; degree, 3.
b) Coefficient, 1; degree, 7.
c) Coefficient, -6; degree, 1.
d) Coefficient, 1; degree, 0.   **12.** a) Yes. b) No.
c) Yes. d) No. e) Yes. f) No.   **13.** a) $8x$.
b) Not possible. c) Not possible. d) $3x^7$. e) $8x$.
f) Not possible. g) $12x$. h) $3x^4$. i) $9x^3$.
j) Not possible. k) Not possible. l) $4x^6$. m) $-8x$.
n) Not possible. o) Not possible.   **14.** a) $15x^2$.
b) $5x^4$. c) x^8. d) $2x^{14}$. e) $-9x^2$. f) $-81x$. g) $64x^3$.
h) x^{12}. i) $24x^9$. j) $27x^9$. k) $40x^{13}$. l) $-6x^7$.
m) $-3x^4$. n) $49x^8$. o) $-64x^{21}$.   **15.** a) $11x^3$.
b) $28x^6$. c) Not possible. d) $28x^5$. e) $9x^6$.
f) $-10x^{12}$. g) Not possible. h) $-10x^7$. i) $27x^9$
j) $729x^{27}$. k) $20x^8$. l) $625x^{32}$.   **16.** a) acx^{b+d}.
b) Not possible. c) acx^{2b}. d) $(a + c)x^b$.
17. a) Perimeter, $28x$; area, $49x^2$.
b) Perimeter, $10x^2$; area, $4x^4$.
c) Perimeter, $2x^2 + 2x^5$; area, x^7.
d) Perimeter, $6x + 2x^3$; area, $3x^4$.

Set IV

1. a) If $x = 0$, $10x^2 = 0$ and $2x^{10} = 0$. b) If $x = 1$,
$10x^2 = 10$ and $2x^{10} = 2$. c) If $x = 2$, $10x^2 = 40$ and
$2x^{10} = 2^{11}$. d) If $x = -1$, $10x^2 = 10$ and $2x^{10} = 2$.
e) If $x = -2$, $10x^2 = 40$ and $2x^{10} = 2^{11}$.   **2.** If x is
an integer larger than 2 or less than -2, $2x^{10} > 10x^2$.

Chapter 9, Lesson 2

Set I

1. a) 4×10^{-6}. b) 9×10^0. c) 1.25×10^{-5}.
d) 3.2×10^{-6}.

2.   a)

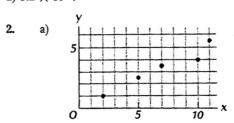

b) 4.0. c) 5.0. d) $y = \dfrac{1}{2}x$.

3. a) $x + y = 35$, $18x = 24y$; (20, 15).
b) $x + y = 70$, $18x = 24y$; (40, 30). c) The values of
x and y in part b are twice those in part a.

11. a) Yes. b) No. c) Yes. d) Yes. e) Yes.
f) No. 12. a) $3x^2 + -5$. b) $4 + -x + x^2$.
c) $-2x + -8$. 13. a) $x^4 + 4x$; degree, 4.
b) $3y^5 - 9y^2 + 2$; degree, 5. c) $x^3 + 6x^2 - 5x - 7$;
degree, 3. d) $-3y^2 + 10$; degree, 2. e) $x + 4^7$;
degree, 1. f) $-y^6 + 3y^5 - 30y + 24$; degree, 6.
14. a) $8x + 20$. b) $-3y^2 + 12$. c) $x^4 - 8x^3 + 3x^2$.
d) $2x^3 - 2x^2 + 12x$. e) $x^4 - x^3 + 6x^2$.
f) $-5y^8 + 15y^3$. 15. a) $-x^3 - 4x - 6$.
b) $-y^4 + 8$. c) $5x^2 - 10x$.
d) $-4 + 3y - 2y^2 + y^3$. 16. a) 6. b) 3. c) 2.
d) 11. e) 18. f) 20. g) 18. h) 18. i) -30.
j) -20. k) -18. l) -18. m) 30. n) -1. o) 0.
p) -4. q) 51. r) -85. s) 909. t) -1,111.
17. a) $4x^3 + 3x$. b) $2x^4 - 2x^3 + x$. c) $x^2 - 2x$.
d) $2x^2 - 7x$. e) $x^2 + 5x - 5$. f) $5x^5 - 4x^4$.

Set IV

1. 0 feet per second when $x = 0$, 7.5 feet per second when $x = 1$, 8 feet per second when $x = 2$, 7.5 feet per second when $x = 3$, and 0 feet per second when $x = 4$.
2. Yes. An elevator's speed rapidly increases at the beginning of a trip and rapidly decreases at the end.

Chapter 9, Lesson 3

Set I

1. a) -22. b) 4. c) -4. d) -4. e) 22. f) -22.
2. a) Slope, 2; y-intercept, -1. b) Slope, 1; y-intercept, 0. c) Slope, -1; y-intercept, 1.
3. a) Inverse. b) $y = \dfrac{60}{x}$. c) 0.75.

Set III

10. a) $5x + 5$. b) $8x - 2y$. c) $6x^2 - 10x$.
d) $4x^2 - 3$. e) $x^4 + 10x^2 - 2x - 2$.
11. a) $7x - 12$. b) $5y$. c) $x^2 - 4$.
d) $-x^2 + 5x - 15$. e) $2x^3 - x^2 + 2$.
12. a) Polynomial C: $5x^2 + 2x - 1$. b) Polynomial
D: $3x^2 - 4x + 7$.
c)

x	3	5
polynomial A	36	98
polynomial B	14	36
polynomial C	50	134
polynomial D	22	62

d) The values of polynomials C and D are the sums and differences of the values of polynomials A and B respectively. 13. a) $2x^2 + 2$. b) $14x - 10y$.
c) $x^3 + x^2 + x - 1$. d) $3x^3 + 5x^2 + 4$.
14. a) $5x - 5y^2$. b) 26. c) $3x^2 - 4x + 1$.
d) $x + 2y + 7$. 15. a) $2x + 10y$. b) $7x^2 + 10x$.
c) $= x^5 - x^4 - x^3 + x^2 + x - 1$.

Set IV

1. Example: 29, $9(9 - 2) = 63$, $29 + 63 = 92$.
2. The digits in the result are the digits in the original number interchanged. 3. Let $10x + y$ represent the original number. $9(y - x) = 9y - 9x$, $(10x + y) + (9y - 9x) = 10y + x$.

Chapter 9, Lesson 4

Set I

1. a) Not possible. b) $10x^3$. c) $10 + 2x^3$.
d) $250x^3$. 2. a) None; $1^3 = 1^5$, but $2^3 < 2^5$ and
$(-2)^3 > (-2)^5$. b) =. c) <. d) None; $\dfrac{0}{2} = 0$,
but $\dfrac{1}{2} < 1$ and $\dfrac{-1}{2} > -1$. 3. a) Exponential.
b) 243. c) 59,049. d) No.

Set III

9. a) The digits are not lined up correctly.
b)

$$\begin{array}{r} 22 \\ \times\ 14 \\ \hline 88 \\ 220 \\ \hline 308 \end{array}$$

	20	2
10	200	20
4	80	8

10. a)

$$\begin{array}{r} 63 \\ \times\ 24 \\ \hline 252 \\ 126 \\ \hline 1512 \end{array}$$

	60	3
20	1200	60
4	240	12

b)

$$\begin{array}{r} 5 \\ \times 197 \\ \hline 35 \\ 45 \\ 5 \\ \hline 985 \end{array}$$

	5
100	500
90	450
7	35

c)
$$\begin{array}{r} 409 \\ \times\ \ 38 \\ \hline 3272 \\ 1227 \\ \hline 15542 \end{array}$$

	400	9
30	12000	270
8	3200	72

11. a) $(3x + 2)(x + 8) = 3x^2 + 26x + 16$.
b) $5y(y^2 - y + 6) = 5y^3 - 5y^2 + 30y$.
c) $(a^3 - 4a + 9)(a^2 + 4) = a^5 + 9a^2 - 16a + 36$.
d) $(2b + 7)(2b - 7) = 4b^2 - 49$.

12. a)

	x	5
x	x^2	$5x$
6	$6x$	30

$x^2 + 11x + 30$.

b)

	$7y$	-10
$2y$	$14y^2$	$-20y$
3	$21y$	-30

$14y^2 + y - 30$.

c)

	$2a^3$	$-a$
$3a^2$	$6a^5$	$-3a^3$

$6a^5 - 3a^3$.

d)

	b^2	$5b$	2
b	b^3	$5b^2$	$2b$
-5	$-5b^2$	$-25b$	-10

$b^3 - 23b - 10$.

13. a) $x^2 + 14x + 40$. b) $x^2 + x - 42$.
c) $8x^2 - 2x - 15$. d) $64x^2 + 16x + 1$.
e) $64x^2 - 1$. f) $10x^2 - 29x + 10$. g) $-x^2 + 49$.
h) $x^4 - 5x^2 - 36$. i) $x^3 + 7x^2 + x - 5$. j) $x^4 - 1$.
k) $x^4 + 5x^3 + 6x^2 - 6x - 18$.
l) $x^4 - 16x^2 + 40x - 25$.

1.

	$3x$	2
$3x$	$9x^2$	$6x$
2	$6x$	4

Set IV

$(3x + 2)(3x + 2)$.

2.

	x	-4
x	x^2	$-4x$
4	$4x$	-16

$(x + 4)(x - 4)$.

3.

	$2x$	7
$5x$	$10x^2$	$35x$
1	$2x$	7

$(5x + 1)(2x + 7)$.

Chapter 9, Lesson 5

Set I

1. a) 12^3. b) 12^1. c) 12^0. d) 12^{-2}.
2. a) $5x - y - 2$. b) $x^2 + 2x + 5$. c) $-6x + 8y$.
d) $x^4 - 3x^2 - 4x + 1$. **3.** a) $x = 2l + 2w + 4h$.
b) $x = 14h$.

Set III

8. a) $x^2 + 8x + 15$. b) $x^2 + 2x - 15$.
c) $x^2 - 13x + 36$. d) $x^2 + x - 42$.
e) $x^2 - 16x + 64$. f) $x^2 - 64$. g) $8x^2 + 16x + 6$.
h) $8x^2 - 16x + 6$. i) $15x^2 - 11x - 12$.
j) $8x^2 - 70x - 18$. k) $x^2 + 2xy + y^2$.
l) $x^2 - y^2$. **9.** a) Polynomial C: $10x^2 + 13x - 3$.

b)

x	1	2	10
polynomial A	5	7	23
polynomial B	4	9	49
polynomial C	20	63	1,127

c) The values of polynomial C are the products of the values of polynomials A and B.
10. a) $x^4 + 4x^2 - 12$. b) $x^5 + 4x^4 - x^2 - 4x$.
c) $x^3 - 27$. d) $x^4 + x^3 - 3x^2 - 5x - 2$.
e) $3x^3 - 15x^2 - 42x$. f) $x^3 + 9x^2 + 23x + 15$.
11. a) $x^2 - 1$. b) $x^3 + 1$. c) $x^4 - 1$. d) $x^5 + 1$.
e) $x^{11} + 1$.

Set IV

1. Example: 5, 6, 7; $6^2 = 36$, $5 \times 7 = 35$. Example:
6, 7, 8; $7^2 = 49$, $6 \times 8 = 48$. **2.** The square of the
second number is 1 more than the product of the first
and last numbers. **3.** $x, x + 1, x + 2$;
$(x + 1)^2 - x(x + 2) = (x^2 + 2x + 1) - (x^2 + 2x) = 1$.

Chapter 9, Lesson 6

Set I

1. a) False. b) True. c) False. d) True. **2.** a) 8.
b) $\frac{22}{7}$. c) -8. d) -100. **3.** a) Graph C.
b) Graph A. c) Graph B.

11. a) $(a + 8)^2 = a^2 + 16a + 64$.
b) $(2b - 3)(2b + 3) = 4b^2 - 9$.
c) $(1 - x)^2 = 1 - 2x + x^2$.
d) $(9y + 4)(9y - 4) = 81y^2 - 16$.

12. a)

	x	11
x	x^2	$11x$
11	$11x$	121

$x^2 + 22x + 121$.

b)

	$6x$	y
$6x$	$36x^2$	$6xy$
y	$6xy$	y^2

$36x^2 + 12xy + y^2$.

c)

	$10a$	-2
$10a$	$100a^2$	$-20a$
-2	$-20a$	4

$100a^2 - 40a + 4$.

d)

	a	$5b$
a	a^2	$5ab$
$-5b$	$-5ab$	$-25b^2$

$a^2 - 25b^2$.

13. a) $(\square + \triangle)^2 = \square^2 + 2(\square)(\triangle) + \triangle^2$.
b) $(\square - \triangle)^2 = \square^2 - 2(\square)(\triangle) + \triangle^2$.
14. a) $x^2 + 24x + 144$. b) $x^2 - 144$.
c) $25a^2 + 30a + 9$.
d) $4b^2 - 44b + 121$. e) $x^2 - 81y^2$.
f) $16x^2 - 40xy + 25y^2$. **15.** a) 1,521. b) 819.
c) 1,521. d) 1,599. **16.** a) $4 + 4x^3 + x^6$. b) $4x^6$.
c) $25y^8 - 10y^4 + 1$. d) $25y^8 - 1$. e) $a^{10}b^{12}$.
f) $a^{10} + 2a^5b^6 + b^{12}$.
17. a) $x^2 + \boxed{12x} + 36 = (x + 6)^2$.
b) $x^2 + 20x + \boxed{100} = (x + 10)^2$.
c) $x^2 - \boxed{14x} + 49 = (x - 7)^2$.
d) $x^2 - 2x + \boxed{1} = (x - 1)^2$.

1.

Gene for eye color from father

		Blue 5	Brown 3	Green 2
Gene for eye color from mother	Blue 5	25 blue	15 brown	10 blue
	Brown 3	15 brown	9 brown	6 brown
	Green 2	10 blue	6 brown	4 green

2. Brown: 51 percent; blue: 45 percent; and green: 4 percent.

Chapter 9, Lesson 7

Set I

1. a) 3×10^8. b) 1.75×10^{-6}. c) 2.4×10^{-4}.
d) 1.25×10^2. **2.** a) The solution is $(3, -2)$.
b) The equations are inconsistent and have no solutions. c) The equations are equivalent and have infinitely many solutions. d) The equations have one solution in which x is negative and y is positive.

Set III

10. a)

	$7x$	1
$2x$	$14x^2$	$2x$
9	$63x$	9

b)

	$2x^2$	$-6x$	4
x	$2x^3$	$-6x^2$	$4x$
-5	$-10x^2$	$30x$	-20

c)

	$6x$	2
$4x^3$	$24x^4$	$8x^3$
$-x^2$	$-6x^3$	$-2x^2$
$3x$	$18x^2$	$6x$

11. a) $2x + 9$. b) $x - 5$. c) $4x^3 - x^2 + 3x$.

12. a)

	$3x$	7
$4x$	$12x^2$	$28x$
3	$9x$	21

$4x + 3$.

b)

	$2x$	5
$3x^2$	$6x^3$	$15x^2$
$-x$	$-2x^2$	$-5x$
4	$8x$	20

$3x^2 - x + 4$.

13. a) $4x + 3$. b) $3x^2 - x + 4$.   **14.** a) $5x + 2$.
b) $x^2 - 2x + 3$. c) $x - 5$. d) $2x^2 - x + 3$.
e) $4x - 5$. f) $3x^2 + 4$.   **15.** a) $5x^2 + 0x + 2$.
b) $3x^3 + 0x^2 + 4x + 0$.
c) $2x^4 - x^3 + 0x^2 + 8x + 5$.
d) $-x^6 + 0x^5 + 0x^4 + 0x^3 + 0x^2 + 0x + 1$.
16. a) $2x^2 + x - 4$. b) $3x^2 + x + 6$. c) $3x - 1$.

Set IV

```
        953
   57)54321
       513
       ───
       302
       285
       ───
       171
       171
       ───
         0
```

Chapter 9, Review

Set I

1. $(a + b)^2 = a^2 + 2ab + b^2$. It is the formula for
squaring a binomial.   **2.** a) False. b) False.
c) False. d) True. e) True.   **3.** a) 240. b) 240.
c) 28. d) 81. e) 0. f) 9,801.   **4.** a) $2x^4$. b) Not
possible. c) $-6x^4$. d) $9x^8$. e) $6 - 2x^4$. f) $162x^4$.
g) $-3x^5$. h) Not possible.

5. a)

	$3x$	5
$3x$	$9x^2$	$15x$
5	$15x$	25

$9x^2 + 30x + 25$.

b)

	$7x$	y
$7x$	$49x^2$	$7xy$
$-y$	$-7xy$	$-y^2$

$49x^2 - y^2$.

c)

	x	-4
$8x$	$8x^2$	$-32x$
1	x	-4

$8x^2 - 31x - 4$.

d)

	y	-11
x	xy	$-11x$
-9	$-9y$	99

$xy - 11x - 9y + 99$.

6. a) $x^4 + 5x^3 + 9x$. b) $3x^6 - 6x^2 + 6x$.
c) $-x^5 + 5$. d) $-3x^3 + 7x + 5$.   **7.** a) $x^2 + 4x$.
b) $x^2 + 7x + 12$. c) $8x^2 - 16x$. d) $8x^2 - 17x + 2$.
e) $x^2 + x - 30$. f) $30x^2 + x - 1$.
g) $x^2 + 14x + 49$. h) $49x^2$. i) $4x^2 - 12x + 9$.
j) $x^2 - 100$. k) $x^6 + 6x^3 + 9$. l) $x^6 - 9$.
8. a) $-4y$. b) $x^4 - x^3 + 1$. c) $x^3 + 7x^2 + 8x - 4$.
d) $15x + 15y$. e) $35x^2 + 36x - 20$.
f) $2x^5 - 2x^4 + 7x^3 + x^2 - 4x$. g) $x^2 + 3x - 1$.
h) $x^3 + 4x^2 + 14x$.   **9.** a) Perimeter, $20x^4$; area,
$25x^8$. b) Perimeter, $6x + 16$; area, $27x - 9$.
c) Perimeter, $10x - 12$; area, $4x^2 - 30x - 16$.
d) Perimeter, $24x$; area, $36x^2 - 25$.   **10.** a) $24x$.
b) 225. c) $6xy$. d) 25.   **11.** a) $9x + 4$.
b) $20x^2 + 23x - 21$. c) $x^3 + x^2 - 11x - 3$.
d) $x^2 + 4x + 2$. e) $x^4 + x$. f) $x^4 - x - 2$.
g) $x^5 + x^4 - x - 1$. h) $x^3 - x^2 + x - 1$.

Chapter 10, Lesson 1

Set I

1. a) 0. b) 1. c) -128. d) $\dfrac{1}{49}$.

2. a) $a = p - b - c$. b) $f = \dfrac{w}{d}$. c) $x = \dfrac{y - b}{a}$.

3. 91 atoms weighing 20 units and 9 atoms weighing
22 units.

9.

41	prime
42	$2 \cdot 3 \cdot 7$
43	prime
44	$2^2 \cdot 11$
45	$3^2 \cdot 5$
46	$2 \cdot 23$
47	prime
48	$2^4 \cdot 3$
49	7^2
50	$2 \cdot 5^2$

10. a) $3 \cdot 5^2$ b) $2^3 \cdot 3 \cdot 5$. c) $2 \cdot 7 \cdot 13$.
d) $2^2 \cdot 5 \cdot 7 \cdot 13$. e) $2 \cdot 3^4 \cdot 11$. f) $2^2 \cdot 3^4 \cdot 11$.
g) $3^2 \cdot 7^2$. h) $3^5 \cdot 7^5$. i) $2^4 \cdot 5^8$. j) 2^{30}.
11. a) No. b) Yes. c) Yes. d) No. e) Yes.
f) Yes. g) No. h) Yes. i) No. j) Yes. k) Yes.
l) No. **12.** a) 1, 2, 13, 26. b) 1, 3, 9, 27, 81.
c) 1, 59. d) 1, 2, 3, 4, 5, 6, 10, 12, 15, 20, 30, 60.
e) $1, 7, 7^2$. f) $1, 2, 2^2, 2^3, 2^4, 2^5, 2^6, 2^7$.
13. a) 12. b) 3. c) 1. d) 14. e) 25. f) 2. g) 1.
h) 4^3. i) 3^4. j) 1.

Set IV

1. Those numbers for which the number of digits is divisible by 3. **2.** Those numbers for which the number of digits is even. **3.** 1,111,111 because its smallest factor is 239.

Chapter 10, Lesson 2

Set I

1. a) $y = 10x - 7$. b) $y = x(x + 1)$ or $y = x^2 + x$.
c) $y = 3^x$. **2.** a) $25x^2 + 10x + 1$.
b) $25x^2 - 10x + 1$. c) $25x^2 - 1$. **3.** a) Perimeter, feet; area, 24,780 square feet. c) Perimeter, $678 + 4x$; area, $24,440 + 339x + x^2$.

Set III

10. a) $54x^2$. b) $54x^2$. c) $8x^6$. d) $-x^{14}$. e) $25x^6$.
f) $32x^{15}$. g) acx^{b+d}. h) a^3x^{3b}.
11. a) $1 \cdot 20, 2 \cdot 10, 4 \cdot 5$.
b) $1 \cdot 26x, 2 \cdot 13x, 13 \cdot 2x, 26 \cdot x$.
c) $1 \cdot 9x^2, 3 \cdot 3x^2, 9 \cdot x^2, x \cdot 9x, 3x \cdot 3x$.
d) $1 \cdot 5x^4, 5 \cdot x^4, x \cdot 5x^3, 5x \cdot x^3, x^2 \cdot 5x^2, x^3 \cdot 5x$.
12. a) 6^2. b) $(x^{18})^2$. c) $(5x^4)^2$. d) $(3xy)^2$.

13. a) 1, 2, 3, 4, 6, 9, 12, 18, 36.
b) 1, 2, 5, 10, x, $2x$, $5x$, $10x$, c) 1, 3, 9, 27, 81, x, $3x$, $9x$, $27x$, $81x$, x^2, $3x^2$, $9x^2$, $27x^2$, $81x^2$.
d) 1, x, x^2, x^3, x^4, x^5, x^6, x^7, x^8.
e) 1, 2, x, $2x$, x^2, $2x^2$, x^3, $2x^3$, x^4, $2x^4$.
14. a) $25x^5$. b) $3x^6$. c) x^7. d) $-6y$. e) x^4y^6.
f) $3x^4$. **15.** a) 4. b) $7x$. c) 1. d) x^9. e) 1.
f) 10. g) $18xy$. h) $5x^2y$. i) 1. j) 2. k) 1. l) $3x$.

Set IV

1. Example: $474{,}747$; $\dfrac{474{,}747}{3} = 158{,}249$;

$\dfrac{474{,}747}{7} = 67{,}821$; $\dfrac{474{,}747}{13} = 36{,}519$;

$\dfrac{474{,}747}{37} = 12{,}831$. **2.** Writing a two-digit number three times to form a six-digit number is equivalent to multiplying it by 10,101. The numbers 3, 7, 13, and 37 are all factors of 10,101.

Chapter 10, Lesson 3

Set I

1. a) $2^3 \cdot 7$. b) $13 \cdot 17$. c) $3 \cdot 5 \cdot 11^2$. **2.** a) $4x^2$.
b) $4x^2 - 4x - 2$. c) $8x^3 - 4x - 1$. **3.** a) 94.
b) $6x^2 - 2$. c) 3, 4, and 5. d) 94.

Set III

12. a) $27x + 9y$. b) $8x^2 - 28x$. c) $x^6 + 4x^3$.
d) $3x^2y + 3xy^2 - 3xy$. **13.** a) $3x$. b) x^4. c) 2.
d) 1.
14.

a)
$3(x + 2)$.

	x	2
3	$3x$	6

b)
$x(x + 4)$.

	x	4
x	x^2	$4x$

c)
$x^2(2x + 1)$.

	$2x$	1
x^2	$2x^3$	x^2

d)
$5x(x^2 + 3)$.

	x^2	3
$5x$	$5x^3$	$15x$

15. a) $5(x + 3)$. b) $2(x + 1)$. c) $7(x + y)$.
d) $3(x - 6)$. e) $1(3x - 16)$. f) $1(x + 2y)$.
g) $4(x - 6y)$. h) $x(8 + x)$. i) $5(x^2 + 2)$.
j) $1(2x^5 + 5)$. k) $x(x^2 - 3)$. l) $1(x^2 + y^2)$.
16. a) $8x$. b) $3x$. c) $6x^3$. d) $-5x^2$. e) $12xy$.
f) $-9x^2y$. **17.** a) 540. b) 450. c) 300.
d) 10,100. **18.** a) 5. b) $x^4 - 3x + 2$.
c) $5x^2 + 4x - 1$. d) $7 - x - y$.
19. a) $6x(5x^2 + 3x - 2)$. b) $xy(2x + 1 + 2y)$.
c) $1(5x^8 + 8)$. d) $3x^3(1 + 2x^3 + 3x^6)$.

Set IV

1. Multiply: $(6)(6 + 1) = 42$, square: $5^2 = 25$;
$65^2 = 4225$.
2. $(10x + 5)^2 = 100x^2 + 100x + 25 =$
$100(x^2 + x) + 25 = 100[x(x + 1)] + 25$.

Chapter 10, Lesson 4

Set I

1. a) 13. b) $14x$. c) x. **2.** a) $x^2 - x + 5$.
b) $x^2 + 2x + 4$. **3.** a) $d = \frac{s}{13}$. b) $s = 13d$.
c) Directly.

Set III

11. a) $x^2 + 15x + 14$. b) $x^2 - 15x + 14$.
c) $x^2 + 9x + 14$. d) $x^2 - 9x + 14$.
e) $x^2 + 13x - 14$. f) $x^2 - 13x - 14$.
g) $x^2 + 5x - 14$. h) $x^2 - 5x - 14$.
12. a) $x^2 + 7x + 12 = (x + 3)(x + 4)$.
b) $x^2 - 10x + 16 = (x - 8)(x - 2)$.
c) $x^2 + 2x - 63 = (x - 7)(x + 9)$.
d) $x^2 - x - 30 = (x - 6)(x + 5)$. **13.** a) $(2)(3)$.
b) $(-2)(-3)$. c) $(-5)(7)$. d) $(5)(-7)$. e) $(4)(11)$.
f) $(2)(22)$. g) $(-9)(9)$. h) $(-3)(27)$. **14.** a) $x + 2$.
b) $x - 11$. c) $x + 9$. d) $x - 7$. e) $x - 12$.
f) $x + 10$. **15.** a) $(x + 1)(x + 3)$.
b) $(x - 1)(x - 3)$. c) $(x + 5)(x + 11)$.
d) $(x + 6)(x + 10)$. e) $(x - 3)(x + 7)$.
f) $(x + 3)(x - 7)$. g) $(x - 2)(x - 2)$.
h) $(x + 2)(x - 2)$. i) $(x - 6)(x - 8)$.
j) $(x - 4)(x - 12)$. k) $(x - 3)(x + 16)$.
l) $(x + 2)(x - 24)$. **16.** a) $(x + y)(x + 5y)$.
b) $(x - 3y)(x + 6y)$. c) $(x + 2)(y + 10)$.
d) $(x^2 - 9)(x^2 - 4)$. **17.** a) $(x + 2y)(x + 11y)$.
b) $(x + 8y)(x - 3y)$. c) $(x + 10)(y + 5)$.
d) $(x^2 - 7)(x^2 + 4)$.

Set IV

1. $(1)^2 + (1) + 41 = 43$, $(2)^2 + (2) + 41 = 47$,
$(3)^2 + (3) + 41 = 53$, $(4)^2 + (4) + 41 = 61$,
$(5)^2 + (5) + 41 = 71$, $(6)^2 + (6) + 41 = 83$,
$(7)^2 + (7) + 41 = 97$, $(8)^2 + (8) + 41 = 113$,
$(9)^2 + (9) + 41 = 131$, $(10)^2 + (10) + 41 = 151$.
2. 41; $(41)^2 + (41) + 41 =$
$(41)(41 + 1 + 1) = (41)(43)$.

Chapter 10, Lesson 5

Set I

1. a) $5(8x + 1)$. b) $3(3x - 4y)$. c) $x(x - 4)$.
d) $x^2(2x + 3)$. **2.** a) 5.3. b) 0.526. c) 0.05.
d) 0.01.
3. a)

$12x = 10(x + 90)$.

b) 5,400.

Set III

10. a) $x^2 - 144$. b) $25x^2 - 1$. c) $x^2 - 9y^2$.
d) $x^8 - 36$. e) $x^4 - y^2$. f) $x^2y^2 - 4$. **11.** a) 9^2.
b) $(x^5)^2$. c) Not possible. d) $(x^8)^2$. e) $(5x^2)^2$.
f) Not possible. g) $(x^3y^3)^2$. h) $(8x^{32})^2$.
12. a) $x^2 - 100 = (x + 10)(x - 10)$.
b) $16x^2 - 9 = (4x + 3)(4x - 3)$.
c) $x^2 - 36y^2 = (x - 6y)(x + 6y)$.
d) $x^4 - y^8 = (x^2 - y^4)(x^2 + y^4)$.
13. a)

b)

14. a) $(x + 7)(x - 7)$.   b) Not possible.
c) $(3x + 2y)(3x - 2y)$.   d) $(10 + x)(10 - x)$.
e) $6(x + 1)(x - 1)$.   f) $6(x^2 - 2)$.   g) $4(x^2 + 9y^2)$.
h) $(x^2 + 5)(x^2 - 5)$.   i) $(8x^6 + 1)(8x^6 - 1)$.
j) $3(x^3 + 2)(x^3 - 2)$.
15. a) $(x - y + 2)(x - y - 2)$.
b) $(2x + 15)(2x + 1)$.   c) $x(6 - x)$.
d) $(1 + x)(9 - x)$.

Set IV

1. Because $x^2 - y^2 = (x + y)(x - y)$.   **2.** If $x + y$
is prime and $x - y = 1$, $x^2 - y^2$ is prime. For
example, $3^2 - 2^2 = 5$, $4^2 - 3^2 = 7$, and
$6^2 - 5^2 = 11$.

Chapter 10, Lesson 6

Set I

1. a) $8x - 4x^4$.   b) $2x^4 - x^7$.   c) $-8x^4$.   d) $-2x^7$.
2. a) 8,700.   b) 290.   c) 2,460.   d) 54,000.
3. a) 10^9 dollars.   b) 4.54×10^{11} dollars; four
hundred fifty-four billion dollars.
11. a) $x^2 + 16x + 64 = (x + 8)^2$.
b) $x^2 - 2x + 1 = (x - 1)^2$.
c) $25x^2 - 10xy + y^2 = (5x - y)^2$.
d) $x^{10} + 2x^5y + y^2 = (x^5 + y)^2$.
12. a) $x^2 + 10x + 25$.   b) $x^2 - 10x + 25$.
c) $64x^2 + 16x + 1$.   d) $16x^2 + 24xy + 9y^2$.
e) $9x^2 + 24xy + 16y^2$.   f) $x^6 + 2x^3y^4 + y^8$.
13. a) $(4x)^2$.   b) $(x^{50})^2$.   c) $(30)^2$.   d) Not possible.
e) $(3x^{32})^2$.   f) Not possible.   g) Not possible.
h) $(6x^3y^3)^2$.   **14.** a) $20x$.   b) 81.   c) $4x$.   d) 4.
15. a) $(x - 4)^2$.   b) $(x + y)^2$.   c) Not possible.
d) $(3x + 1)^2$.   e) $(2x + 7)^2$.   f) Not possible.
g) $(6x + y)^2$.   h) $(4x - 9y)^2$.   **16.** a) 0.   b) 1.
c) 16.   d) 49.   e) 100.   f) 0.   g) 1.   h) 16.   i) 49.
j) 100.   **17.** a) $(x^2 + 7)^2$.   b) $(x^4 - 2)^2$.
c) $(x^3 - y^3)^2$.   d) $3(x - 5)^2$.   e) $2(6x + 1)^2$.
f) $(x^5 - 5y)^2$.

Set IV

1. $x + 5$.
2. $(x + 5)^3 = (x + 5)(x + 5)^2 = (x + 5)(x^2 +$
$10x + 25) = x^3 + 10x^2 + 25x + 5x^2 + 50x +$
$125 = x^3 + 15x^2 + 75x + 125$.

Chapter 10, Lesson 7

Set I

1. a) $2(x - 32)$.   b) $(x + 8)(x - 8)$.   c) Not
possible.   d) $x(x + 6)$.   **2.** a) 7.   b) 12.   c) 5.
d) 7.   **3.** a) $10x + 25y$ cents.   b) $100x + 300y$
cents (or $x + 3y$ dollars).   c) $90x + 275y$ cents.

Set III

9. a) $3x^2 + 26x + 35 = (3x + 5)(x + 7)$.
b) $4x^2 - 6x + 2 = (x - 1)(4x - 2)$.
c) $10x^2 + 21x - 10 = (5x - 2)(2x + 5)$.
d) $9x^2 - 16 = (3x + 4)(3x - 4)$.
10. a) $2x^2 + 37x + 18$.   b) $2x^2 + 35x - 18$.
c) $2x^2 + 20x + 18$.   d) $2x^2 + 16x - 18$.
e) $2x^2 + 15x + 18$.   f) $2x^2 + 9x - 18$.
g) $2x^2 + 12x + 18$.   h) $2x^2 - 18$.
11. a) $(3x + 1)(x + 5)$.   b) $(3x - 1)(x + 5)$.
c) $(4x + 1)(4x + 1)$.   d) $(4x + 1)(4x - 1)$.
e) $(2x + 7)(x - 2)$.   f) $(2x - 1)(x + 14)$.
g) $(3x + 2)(5x + 3)$.   h) $(3x - 2)(5x - 3)$.
i) $(4x - 1)(2x + 7)$.   j) $(4x + 1)(2x - 7)$.
12. a) $(4x + y)(2x - y)$.   b) $(x - 2y)(5x - y)$.
c) $(2x - 1)(2y + 3)$.   d) $(x^2 + 2)(3x^2 + 4)$.
13. a) $(3x + y)(x + 7y)$.   b) $(5x - y)(2x - 3y)$.
c) $(6x - 5)(y + 1)$.   d) $(2x^2 + 3)(x^2 + 7)$.

Set IV

$1 - 4x + 4x^2 + 2y - 4xy + y^2 = (1 - 2x + y)^2$.

	1	$-2x$	y
1	1	$-2x$	y
$-2x$	$-2x$	$4x^2$	$-2xy$
y	y	$-2xy$	y^2

Chapter 10, Lesson 8

Set I

1. a) $<$.   b) $>$.   c) $=$.   d) $>$.   e) $=$.   f) $>$.
2. a) $(-1, -2)$.   b) $(3, 8)$.   **3.** a) 7.09×10^{15}.
b) 7,090,000,000,000,000.

Set III

9. a) $12x^2 - 9x - 3$. b) $12x^2 - 9x - 3$.
c) $12x^2 - 9x - 3$. d) $12x^2 - 9x - 3$.
10. a) $x^4 - 1$. b) $x^4 - 16$. c) $x^4 - 81$. d) All of
them have degree 4 and are the difference of two
fourth powers (or squares).   **11.** a) $x(x^2 - 12)$.
b) $5x(x + 1)^2$. c) $x^2(x + 5)(x + 8)$.
d) $2x(x + 7)(x - 7)$. e) $4x(x - 1)(x - 3)$.
f) $7x^2(x^2 + 3x - 5)$. g) $3x^3(x + 3)(x - 2)$.
h) $3x^2(2x + 5)(2x - 5)$. i) $(x^2 + 4)(x + 2)(x - 2)$.
j) $x^3(7x + 1)(x + 7)$.   **12.** a) 13. b) 83.
c) $2x + 1$. d) $x - 8$. e) $2x - 1$.
f) $x^2 - 2x + 4$.   **13.** a) $y(y^2 - 3x)$.
b) $5(x + y)(x - y)$. c) $xy(x + y)^2$.
d) $x(x - y)(x - 2y)$. e) $2(x - 7)(y + 3)$.
f) $(x^2 + y^2)(x + 2y)(x - 2y)$.

Set IV

$x^3 + y^3 = (x + y)(x^2 - xy + y^2)$.

	x^2	$-xy$	y^2
x	x^3	$-x^2y$	xy^2
y	x^2y	$-xy^2$	y^3

Chapter 10, Review

Set I

1. a) The number on the card held by the woman is
composite because it ends in 5 and hence has 5 as a
factor. The number on the card held by the man is
composite because it is even and hence has 2 as a
factor.   **2.** a) $2^4 \cdot 7$. b) $5 \cdot 37$. c) $3^2 \cdot 11 \cdot 17$.
d) $2^3 \cdot 7^3$. e) $2^4 \cdot 3^8$. f) 2^{18}.   **3.** a) 9. b) 1.
c) 3. d) 15. e) 8^3. f) 1.   **4.** a) 1, 2, 3, 4, 5, 6, 10,
12, 15, 20, 30, 60. b) 1, 2, 7, 14, x, $2x$, $7x$, $14x$.
c) 1, x, x^2, x^3, x^4. d) 1, 5, x, $5x$, x^2, $5x^2$, x^3, $5x^3$.
5. a) $2x$. b) x. c) x^4. d) 4. e) $7xy$. f) $3x$.
6. a) 8,300. b) 19,000. c) 880. d) 14,399.
7. a) 1. b) $42x$.   **8.** a) $2x^6$. b) $-4x^3$. c) 7.
d) $2 + x^3$. e) $x - 5$. f) $x + 3$. g) $5x - 2$.
h) $4x + 1$.   **9.** a) $3(2x - 7)$. b) $x(x + 10)$.
c) $(x + 7)(x - 7)$. d) $x^2(3x - 2)$. e) $(x + 1)^2$.
f) $(x + 3)(x + 13)$. g) $(x + 7)(x - 6)$. h) Cannot
be factored. i) $x(x + 2)(x - 2)$. j) $(4x - 5)^2$.

k) $(x + 7)(2x + 1)$. l) $(3x - 5)(x + 2)$.
m) $2(x + 6)^2$. n) $x(x + 10)(x - 9)$. o) $5x(x^2 + 1)$.
p) $3x^2(x + 1)(x + 2)$.   **10.** a) $x(x + 2y)$.
b) $5(3x - y^3)$. c) $(x + y)(x + 7y)$.
d) $(x + 8)(y - 1)$. e) $(2x + 3y)(2x - 3y)$.
f) $(x - y^2)^2$.

Chapter 11, Lesson 1

Set I

1. a) 6×10^5. b) 1×10^4. c) 3.2×10^{16}.   **2.** a) It
crosses the x-axis at $(4, 0)$ and the y-axis at $(0, -12)$.
b) 3. c) No.   **3.** a) Example: $\dfrac{44{,}444}{41} = 1{,}084$ and

$\dfrac{44{,}444}{271} = 164$. b) Each such number is a multiple of
11,111 and $11{,}111 = 41 \cdot 271$.

Set III

15. A, $\dfrac{1}{5}$; B, $\dfrac{7}{5}$; C, $-\dfrac{4}{5}$; D, $-\dfrac{10}{5}$.
16. a, b, and c)

d) $\dfrac{5}{3}$ or $1\dfrac{2}{3}$. e) $\dfrac{10}{3}$ or $3\dfrac{1}{3}$. f) $\dfrac{6}{3}$ or 2. g) $-\dfrac{1}{3}$.
17. a) 0.25. b) 0.25. c) 0.67. d) 0.67. e) 0.45.
f) 2.20. g) 0.24. h) 3.17. i) 0.49. j) 0.33.
18. a) 0 and 1. b) 4 and 5. c) -1 and 0. d) -2 and
-1. e) 16 and 17. f) 166 and 167.   **19.** a) $\dfrac{1}{3}$. b) $\dfrac{2}{5}$.
c) $\dfrac{3}{8}$. d) $\dfrac{8}{3}$. e) $\dfrac{1}{101}$. f) $\dfrac{1}{91}$. g) $\dfrac{3}{49}$. h) $\dfrac{3}{7}$.
20. a) $\dfrac{6}{8}$ and $\dfrac{5}{8}$. b) $\dfrac{4}{18}$ and $\dfrac{81}{18}$. c) $\dfrac{3}{30}$ and $\dfrac{2}{30}$.
d) $\dfrac{21}{24}$ and $\dfrac{28}{24}$. e) $\dfrac{21}{42}$, $\dfrac{14}{42}$, and $\dfrac{6}{42}$. f) $\dfrac{8}{12}$, $\dfrac{10}{12}$,
and $\dfrac{11}{12}$.   **21.** a) $\dfrac{7}{11}$, $\dfrac{8}{11}$, $\dfrac{9}{11}$. b) $\dfrac{11}{9}$, $\dfrac{11}{8}$, $\dfrac{11}{7}$.
c) $\dfrac{1}{2}$, $\dfrac{101}{200}$, $\dfrac{11}{20}$. d) $\dfrac{8}{15}$, $\dfrac{3}{5}$, $\dfrac{2}{3}$.   **22.** a) $-\dfrac{2}{9}$. b) $\dfrac{-7}{-6}$.
c) $-\dfrac{-5}{-8}$. d) $\dfrac{-1}{3}$.   **23.** a) 2. b) 0. c) -3. d) Not
possible.   **24.** a) 3. b) 2. c) 6. d) 11. e) -9.

f) 6. g) 6. h) $\frac{0}{0}$ is not defined. i) 6. j) 6.

25. a) 24. b) 30. c) 12. d) 9. e) 30. f) 16. g) 4. h) 3.

Set IV

7 apples.

$$\left(\text{First gate, } 3\frac{1}{2} + \frac{1}{2} = 4, \text{ 3 remaining;}\right.$$

$$\text{second gate, } 1\frac{1}{2} + \frac{1}{2} = 2, \text{ 1 remaining;}$$

$$\left.\text{third gate, } \frac{1}{2} + \frac{1}{2} = 1.\right)$$

Chapter 11, Lesson 2

Set I

1. a) x^1. b) x^0. c) x^{-1}. 2. a) $x^2 - 4$. b) $x^2 - 2x$. c) $x^3 - 3x^2 + 4$. d) $x + 1$.
3. a) $y = \frac{12}{x}$. b) y increases. c) Inversely.

Set III

11. a) 0. b) 4. c) No values of x. d) −3. e) 1 and −1. f) No values of x. 12. a) $\frac{4x}{7}$. b) $\frac{2y}{x}$. c) $\frac{3}{x}$.
d) $\frac{1}{x^3}$. e) $\frac{x}{y}$. f) $\frac{x-1}{6}$. g) $\frac{x}{x+3}$. h) $\frac{6}{x+6}$.
13. a) $\frac{x}{9}$ and $\frac{3x}{9}$. b) $\frac{4}{2x}$ and $\frac{5}{2x}$. c) $\frac{3}{x^2}$ and $\frac{7x}{x^2}$.
d) $\frac{8}{2x}$ and $\frac{x^2}{2x}$. e) $\frac{3}{6x}$ and $\frac{2}{6x}$. f) $\frac{8x+8}{(x+8)(x+1)}$ and $\frac{x+8}{(x+8)(x+1)}$. g) $\frac{5x}{15}$, $\frac{9x}{15}$, and $\frac{x}{15}$. h) $\frac{2}{x^3}$, $\frac{3x}{x^3}$, and $\frac{4x^2}{x^3}$. 14. a) 9. b) 3. c) 6. d) 1. e) Not possible. f) −1. 15. a) $8x$. b) 4. c) $3x + 6$.
d) $21x$. e) $4x - 16$. f) $x - 3$. 16. a) $\frac{1}{4}$. b) Not possible. c) $\frac{x}{3x-1}$. d) Not possible. e) $\frac{3}{2}$. f) $\frac{1}{7}$.
g) $\frac{x}{x-11}$. h) $\frac{x+3}{3}$. i) $\frac{1-x}{x^2}$. j) Not possible.
k) $\frac{1}{x+1}$. l) $\frac{6}{x+6}$ 17. a) 0.3. b) 0.5. c) 1.

d) 0.3. e) 0.5. f) 1. g) Yes, for all values of x except −2 because $\frac{x^2 + 5x + 6}{10x + 20} = \frac{(x+3)(x+2)}{10(x+2)} = \frac{x+3}{10}$.

Set IV

1. $\frac{x^2 - 1}{x - 1} = x + 1 = 2$ if $x = 1$;
$\frac{x^3 - 1}{x - 1} = x^2 + x + 1 = 3$ if $x = 1$;
$\frac{x^4 - 1}{x - 1} = x^3 + x^2 + x + 1 = 4$ if $x = 1$. 2. 10.

Chapter 11, Lesson 3

Set I

1. a) $5^2 \cdot 11$. b) 331 is prime. c) $2 \cdot 3^4 \cdot 31$.
2. a) (4, 3). b) (1, 39), (3, 13), (13, 3), and (39, 1).
3. Fifteen minutes.

Set III

7. a) 3. b) $\frac{2}{5}$. c) $\frac{x}{4}$. d) $\frac{16}{x}$. e) x. f) $\frac{7 + 2x}{7}$.
g) $2x$. h) $\frac{13}{6}$. i) $\frac{4x}{5}$. j) 8. k) $\frac{x+4}{2}$. l) $\frac{8}{x+4}$.
m) 1. n) $\frac{x-4}{x+4}$. o) $\frac{7x}{x+7}$. p) $\frac{-8}{x+7}$. q) 1.
r) x. 8. a) $\frac{1}{3}$. b) $\frac{1}{18}$. c) $\frac{9}{2x}$. d) $\frac{4x}{21}$.
e) $\frac{5x^2 + 2x^5}{10}$. f) $\frac{2x^3 + 5}{x^5}$. g) $\frac{64 - x^2}{8x}$. h) $\frac{4x - 1}{6}$.
i) $\frac{1}{3x + 12}$. j) $\frac{x+4}{30}$. k) $\frac{3x}{(x-1)(x+2)}$.
l) $\frac{2x^2 + 50}{x^2 - 25}$. 9. a) $\frac{x^2 + y^2}{xy}$. b) 2. c) $\frac{2x^2}{x^2 - y^2}$.
d) −1. e) $\frac{x^2 - 2xy + y^2}{xy}$ or $\frac{(x - y)^2}{xy}$.

Set IV

1. $\frac{17}{12}$. 2. It is greater than either fraction. 3. $\frac{5}{7}$ is greater than $\frac{2}{3}$ and is less than $\frac{3}{4}$. 4. Example:

$\frac{1}{2} + \frac{1}{2}$. Ollie's method: $\frac{1}{2} + \frac{1}{2} = \frac{2}{4} = \frac{1}{2}$. Correct answer: $\frac{1}{2} + \frac{1}{2} = \frac{2}{2} = 1$. **5.** The correct answer is greater than either fraction. Ollie's answer is equal to each fraction.

Chapter 11, Lesson 4

Set I

1. a) $3(13x + 15y)$. b) $2(x - 3)(x - 4)$.
c) $xy(x + y)(x - y)$. **2.** a) $x = \frac{-b}{a}$.

b) $x = \frac{1 - ab}{a}$. c) $x = \frac{c}{a + b}$. **3.** a) $\frac{5}{8}$. b) $\frac{3}{8}$.

c) $\frac{x}{x + y}$. d) $\frac{y}{x + y}$.

Set III

10. a) $\frac{22}{7}$. b) $\frac{22}{3}$. c) $\frac{22}{9}$. d) $\frac{38}{9}$. e) $\frac{63}{8}$. f) $\frac{7}{12}$.

11. a) $\frac{6x + 1}{6}$. b) $\frac{2x - 8}{x}$. c) $\frac{16x}{5}$. d) $\frac{4x^2 - x}{4}$.

e) $\frac{10x + 1}{x}$. f) $\frac{5x + 38}{x + 8}$. g) $\frac{x^2 + x}{x - 6}$. h) $\frac{3x^2 - 3}{4}$.

i) $\frac{x^2 - 24}{x + 5}$. j) $\frac{x^3}{x - 1}$. **12.** a) 2. b) 9. c) 0.

d) 2. e) 9. f) 0. g) Yes, all values of x except -4

because $x - 3 + \frac{12 - x}{x + 4} = \frac{x - 3}{1} + \frac{12 - x}{x + 4} =$

$\frac{(x - 3)(x + 4)}{x + 4} + \frac{12 - x}{x + 4} = \frac{x^2 + x - 12 + 12 - x}{x + 4} =$

$\frac{x^2}{x + 4}$. **13.** a) $\frac{9}{x}$. b) $\frac{1}{3}$. c) $\frac{1}{5}$. d) $\frac{2x}{x - 3}$.

e) $\frac{2}{x^2}$. f) $\frac{3x^2}{x^2 + 8}$. g) $\frac{9}{x - 9}$. h) 1.

14. a) $\frac{x}{y} + \frac{4}{y}$. b) $\frac{x}{2} - \frac{1}{3}$. c) $\frac{x}{x - y} + \frac{y}{x - y}$.

d) $5 - \frac{9}{x}$. e) $1 + \frac{1}{x + y}$. f) $\frac{3x^2}{x + 4} - 1$.

15. a) $2x + 3$. b) 1. c) $x + 5$ d) x^2.

Set IV

49 fish. (7 cats would eat 7 fish in a day and a half; so 7 cats would eat $7 \cdot 7 = 49$ fish in a week and a half.)

Chapter 11, Lesson 5

Set I

1. 9, $(x + 3)(x + 6)$; 11, $(x + 2)(x + 9)$; 19, $(x + 1)(x + 18)$. **2.** a) Not possible. b) $\frac{x - 4}{x}$.
c) Not possible. d) $2(x + 2)$.

3. a)

x	3	4	5	6	7	8
y	7	15	31	63	127	255

b) 7, 31, and 127 are prime. c) From this table it seems that $2^x - 1$ is prime if x is prime (or odd).

Set III

8. a) $\frac{10}{63}$. b) $\frac{1}{32}$. c) $\frac{9}{100}$. d) 1. e) 6. f) $\frac{4}{27}$.

9. a) $\frac{2}{3}$. b) $\frac{5}{81}$. c) $\frac{14}{x}$. d) $\frac{48}{x^2}$. e) $\frac{x}{2}$. f) $\frac{x^2}{25}$.

g) 1. h) $\frac{x^2 + 49}{7x}$. i) $\frac{4}{x - 2}$. j) $\frac{4}{(x - 2)^2}$.

k) $\frac{5 + 4x}{x^5}$. l) $\frac{20}{x^9}$. m) $\frac{9}{x}$. n) x. o) $\frac{27}{x^3}$. p) $\frac{x^3}{27}$.

10. a) $\frac{2}{x}$. b) $\frac{x^4}{4}$. c) $\frac{x^2 - y^2}{x^2}$. d) $\frac{x^6}{3}$. e) 1.

f) $\frac{x^2 - 8x + 16}{x^2 + 8x + 16}$. g) $\frac{x - 3}{x}$. h) $\frac{-1}{x}$.

11. a) $\frac{x^2 + y^2}{(x + y)(x - y)}$. b) $\frac{1}{x}$. c) 1. d) $\frac{x(x - 3)}{x^2 - 3}$.

Set IV

7.5×10^{22}.

Chapter 11, Lesson 6

Set I

1. a) $\frac{x^2 - 16}{4x}$. b) 1. c) $\frac{2x}{x + 2}$. **2.** a) $\frac{2}{3}$. b) $\frac{25}{4}$.
3. a) $60 - x$. b) $y = 200 + 10x - 2.5(60 - x)$.
c) $y = 12.5x + 50$.

Set III

9. a) $\frac{10}{7}$. b) -3. c) $\frac{45}{2}$. d) $\frac{21}{2}$. e) $3x$. f) -11.

g) $\frac{1}{2}$. h) $\frac{2x}{3}$. **10.** a) $2x$. b) $\frac{2x + 25}{5}$. c) $\frac{4}{x}$.

d) $\frac{x^3 - 4}{x^2}$. e) $\frac{3x}{x + 8}$. f) $\frac{2x + 24}{x + 8}$. g) $\frac{x}{6}$.

h) $\frac{1 + 6x^3}{6x}$. i) $\frac{x^2 + 2x}{2}$. j) $\frac{3x + 4}{2}$. k) 5.

l) $\frac{x^2 - 20x + 105}{x - 10}$. **11.** a) $\frac{x^2 + 3x}{3}$. b) $\frac{6x}{x - 6}$.

c) $\frac{4x^5}{x - 4}$. d) $10x^2 + 2$. e) 1. f) $\frac{7x^2 - x - 6}{7}$.

12. a) $x^2 + x + \frac{2}{9}$. b) $\frac{x^2}{64} - \frac{1}{4}$. c) $x^2 - 10 + \frac{24}{x^2}$.

d) $5x^2 + \frac{23x}{10} - \frac{1}{10}$. **13.** a) $\frac{9x^2 + 9x + 2}{9}$.

b) $\frac{x^2 - 16}{64}$. c) $\frac{x^4 - 10x^2 + 24}{x^2}$.

d) $\frac{50x^2 + 23x - 1}{10}$.

Set IV

1. $0 \left[\text{because } \left(1 - \frac{1}{1}\right) = 0 \right].$

2. $\frac{1}{99} \left[\left(1 - \frac{1}{2}\right)\left(1 - \frac{1}{3}\right)\left(1 - \frac{1}{4}\right)\left(1 - \frac{1}{5}\right) \cdots \right.$

$\left.\left(1 - \frac{1}{99}\right) = \frac{1}{2} \cdot \frac{2}{3} \cdot \frac{3}{4} \cdot \frac{4}{5} \cdots \frac{98}{99} = \frac{1}{99}. \right]$

Chapter 11, Lesson 7

Set I

1. a) $\frac{x}{4}, \frac{x}{3}, \frac{x}{2}$. b) $\frac{2}{x}, \frac{3}{x}, \frac{4}{x}$. c) $\frac{1}{x + 1}, \frac{1}{x}, \frac{1}{x - 1}$.

d) $\frac{x - 1}{5}, \frac{x}{5}, \frac{x + 1}{5}$.

2. a) $(40 + 5)^2 = 1,600 + 400 + 25 = 2,025$.

b) $(50 + 4)^2 = 2,500 + 400 + 16 = 2,916$.

c) $\left(6 + \frac{1}{3}\right)^2 = 36 + 4 + \frac{1}{9} = 40\frac{1}{9}$.

d) $\left(3 + \frac{1}{6}\right)^2 = 9 + 1 + \frac{1}{36} = 10\frac{1}{36}$.

3. a) 1.664×10^{10}. b) 2,050,000,000 dollars.

Set III

9. a) $\frac{1}{6}$. b) 8. c) $\frac{2}{3}$. d) $\frac{1}{x}$. e) $x - 1$. f) $\frac{2x}{2x + 5}$.

10. a) $\frac{9}{10}$. b) 25. c) 16. d) $\frac{7}{8}$. e) $\frac{16}{3}$. f) 6.

11. a) $\frac{x^4}{18}$. b) $\frac{1}{2x^2}$. c) $\frac{x - 2x^3}{6}$. d) 20. e) $\frac{x^2}{5}$.

f) $\frac{x}{4}$. g) $\frac{x^2 - 16}{64}$. h) $\frac{x + 4}{x - 4}$. i) $\frac{2x}{x + 7}$. j) $\frac{x + 7}{2x}$.

12. a) x^2. b) $\frac{1}{x}$. c) $\frac{x^6}{4}$. d) $\frac{x - 4}{2x}$. e) $2x - 4$.

f) $-\frac{y}{x}$. **13.** a) $\frac{3}{4}$. b) 10. c) 1. d) $x + y$.

e) $\frac{x + 1}{x}$. f) $\frac{x^2 + 1}{x^2}$.

Set IV

1. $\frac{5}{9}$. **2.** $\frac{1}{3} \times \frac{1}{2} \boxplus \frac{1}{3} \boxtimes 10$.

Chapter 11, Lesson 8

Set I

1. a) Perimeter, $\frac{7x}{6}$; area, $\frac{x^2}{12}$. b) Perimeter, 8; area,

$4 - \frac{1}{x^2}$. c) Perimeter, $14x + 2\frac{1}{3}$; area,

$6x^2 + 2x + \frac{1}{6}$. **2.** a) $x^2 + 2x + 1$. b) $x + 1$.

c) 1. d) $\frac{1}{x + 1}$.

3.

a)

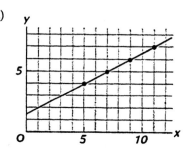

b) Linear. c) $y = \frac{1}{2}x + \frac{3}{2}$. d) 12 feet.

Set III

7. a) $\frac{2}{3}$. b) $\frac{2}{5}$. c) 24. d) $\frac{4}{25}$. e) $\frac{71}{25}$. f) $\frac{29}{21}$.

8. a) $\frac{4x - 16}{x - 1}$. b) $\frac{7x + 5}{7x - 2}$. c) $\frac{5}{x + 5}$. d) $\frac{3}{2}$. e) $\frac{x}{8}$.

f) $3x^2 + 9x$. g) $x - 1$. h) $\frac{x + 4}{x + 2}$. **9.** a) xy.

b) $\frac{1}{y^2}$. c) xy. d) $\frac{x^3}{x^2 - 1}$.

Set IV

1. $\frac{355}{113}$. **2.** 3.1415929. **3.** Pi.

Chapter 11, Review

Set I

1. a) $\frac{1}{5}$. b) $\frac{1}{3}$. c) $\frac{2}{15}$. **2.** a) False; if $x = 6$ and

$y = 1$, $\frac{6 + 4}{1 + 4} \neq \frac{6}{1}$. b) True. c) True. d) False; if

$x = 1$, $\frac{2 + 5}{2} \neq 5$. e) True. f) True. **3.** a) –5.

b) No values of x. c) 0 and 3. **4.** a) $\frac{1}{6}$. b) $\frac{x^2}{4}$.

c) Not possible. d) $\frac{3}{x}$. e) Not possible. f) $\frac{x + 7}{x - 7}$.

5. a) $\frac{4}{9}$. b) $\frac{10}{4}$. c) $\frac{7}{11}$. d) $\frac{43}{94}$. **6.** a) $\frac{4}{x}$. b) $\frac{1}{2}$.

c) $\frac{2x - 1}{x^2}$. d) $\frac{x^2 - y^2}{xy}$. e) $\frac{5x^2 + x}{5}$. f) $\frac{4x^2 + 1}{2x + 1}$.

7. a) $\frac{x}{8} + \frac{1}{2}$. b) $\frac{3x}{3x + 7} - \frac{7}{3x + 7}$. c) $1 + \frac{2}{x^2 - x}$.

d) $2x - \frac{2}{x}$. **8.** a) $\frac{x^5}{6}$. b) 1. c) $-\frac{4}{x}$. d) $\frac{x^2 - 2x}{2}$.

e) $\frac{x^2 - y^2}{x^2}$. f) $\frac{1}{4}$. **9.** a) $x^2 - x + \frac{6}{25}$.

b) $63 + \frac{2}{x} - \frac{1}{x^2}$. **10.** a) $\frac{25x^2 - 25x + 6}{25}$.

b) $\frac{63x^2 + 2x - 1}{x^2}$. **11.** a) 24. b) $\frac{2x + 28}{x + 7}$.

c) $\frac{x + y}{3}$. d) $\frac{1}{2x^5}$. e) $3x^4 + x$. f) 1.

12. a) $\frac{x + 1}{y^2}$. b) $x^2 + x$. c) $\frac{4}{7}$. d) $\frac{2x^2 + 5x}{2x + 4}$.

Chapter 12, Lesson 1

Set I

1. a) 4.2. b) –21. c) 8. **2.** a) $x^4 + 6x - 5$.
b) $x^3y^2 + x^2 + xy + y^3$. c) $y^3 + x^3y^2 + xy + x^2$.
3. a) 540. b) $y = 60x$. c) Directly.

12. a) 4 and –4. b) 3 and –3. c) 13 and –13.
d) None. e) 1 and –1. f) None. **13.** a) 1,444.
b) 6,889. c) 6.16. d) Not possible. e) 74. f) 30.
g) 3.61. h) 1.8. **14.** a) 4 and 5. b) 5 and 6.
c) –3 and –2. d) –10 and –9. e) 17 and 18. f) 54
and 55. **15.** a) 121. b) 12,100. c) 1,210,000.
d) 121,000,000. e) 2. f) 6. g) 20. h) 63.
16. a) $(13.1)^2 = 171.61$, $(13.11)^2 = 171.8721$,
$(13.115)^2 = 172.003225$. b) 13.1 and
13.11 are smaller than $\sqrt{172}$, 13.115 is larger
than $\sqrt{172}$. **17.** a) 8. b) –11. c) 9. d) 9.
e) 15. f) 15. g) 15. **18.** a) =. b) =. c) =.
d) =. e) <. f) >. g) =. h) >. i) <. j) >.
k) <. l) =. **19.** a) 5. b) 7. c) 15. d) 9.
e) 12. f) 12.

Set IV

He was born in 1806. (He was 43 years old in 1849.)

Chapter 12, Lesson 2

Set I

1. a) $\frac{5y}{6x}$. b) Not possible. c) $\frac{x + 6}{x + 2}$.
2. a) $24x^{10}$. b) $2x^3 + 3x^4 + 4x^5$.

c) $3x^4 + 10x^5 + 8x^6$. **3.** a) AB, $-\frac{3}{4}$; BC, $\frac{4}{3}$; CD,

$-\frac{3}{4}$; DA, $\frac{4}{3}$. b) The slopes of the opposite sides are

equal. The slopes of the adjacent sides are the
opposites of the reciprocals of each other.

Set III

11. a) $\sqrt{3}$. b) $\sqrt{14x}$. c) $14\sqrt{x}$. d) $\sqrt{x^5}$.
e) $7\sqrt{xy}$. f) $7\sqrt{xy}$. g) $7x\sqrt{y}$. h) $6^{\frac{1}{2}}$. i) $11x^{\frac{1}{2}}$.
j) $(11x)^{\frac{1}{2}}$. k) $(x^3)^{\frac{1}{2}}$. l) $4xy^{\frac{1}{2}}$. m) $x(3y)^{\frac{1}{2}}$. n) $2^{\frac{1}{2}}xy$.
12. a) 2. b) 9. c) 21. d) 64. e) 3. f) 11. g) 40.
h) 46. **13.** a) $3\sqrt{2}$. b) $2\sqrt{5}$. c) $4\sqrt{3}$.
d) $3\sqrt{11}$. e) $10\sqrt{6}$. f) $2\sqrt{71}$. g) $7\sqrt{7}$.
h) $5\sqrt{33}$. **14.** a) $7\sqrt{3}$, 12.124. b) $2\sqrt{37}$,
12.166. c) $10\sqrt{3}$, 17.32. d) $10\sqrt{30}$, 54.77.
15. a) 6. b) x. c) $5x^3$. d) x^3. e) $x^5\sqrt{x}$. f) $3x^8$.
16. a) $9x$. b) x^8. c) Not possible. d) $16x$.
17. a) $10\sqrt{x}$. b) x^{50}. c) $x^4\sqrt{x}$. d) $2x\sqrt{6}$.
e) $9x\sqrt{x}$. f) $2x^4\sqrt{2}$.

Approximately 3,100 meters per second.

$$v = \sqrt{\frac{4 \times 10^{14}}{4.2 \times 10^{7}}} \approx \sqrt{0.95 \times 10^{7}} = \sqrt{9.5 \times 10^{6}}$$
$$= \sqrt{9.5} \times \sqrt{10^{6}} \approx 3.1 \times 10^{3}.$$

Chapter 12, Lesson 3

Set I

1. a) $\frac{10}{x}$. b) $\frac{16}{x^{2}}$. c) $\frac{5x}{8}$. d) $\frac{x^{2}}{16}$.

2. a) $(2.3, -0.1)$. b) $(4, 3)$. **3.** a) Inversely.

b) $y = \frac{600}{x}$. c) 120 calories.

Set III

11. a) $\frac{3}{9}$. b) $\frac{28}{49}$. c) $\frac{10}{16}$. d) $\frac{2x}{4}$. e) $\frac{9x}{x^{2}}$. f) $\frac{6x}{x^{4}}$.

12. a) 7. b) $\frac{3}{11}$. c) 11. d) $\sqrt{2}$. **13.** a) $\frac{\sqrt{7}}{4}$.

b) $\frac{\sqrt{30}}{3}$. c) $\frac{\sqrt{10}}{5}$. d) $\frac{\sqrt{2}}{8}$. e) $\frac{\sqrt{33}}{6}$. f) $\frac{\sqrt{10}}{12}$.

14. a) $\frac{\sqrt{2}}{3}$, 0.471. b) $\frac{\sqrt{70}}{10}$, 0.837. c) $\frac{\sqrt{30}}{5}$,

1.095. d) $\frac{\sqrt{2}}{6}$, 0.236. **15.** a) 12. b) $\sqrt{2}$.

c) $\sqrt{15x}$. d) $\frac{x^{18}}{6}$. e) x^{2}. f) 8. **16.** a) $\frac{\sqrt{x}}{5}$.

b) $\frac{\sqrt{5}}{x}$. c) $\frac{x^{4}}{3}$. d) $\frac{\sqrt{2x}}{x}$. e) $\frac{\sqrt{2x + 6}}{2}$. f) $\frac{\sqrt{3x}}{3x^{3}}$.

17. a) 12 and 12. b) 40 and 40. c) 0 and 0. d) 8 and –8. e) 20 and –20. f) 1 and 1. g) 2 and 2. h) 1 and –1. i) Yes. j) No. k) Yes. l) No.

Set IV

Approximately 430 miles. $d = \sqrt{\frac{3}{2}(123,800)} =$

$\sqrt{185,700} = 10\sqrt{1857} \approx 10(43) = 430.$

Chapter 12, Lesson 4

Set I

1. (Answers may vary.) a) $\frac{x}{2} + 3$. b) $6 \cdot \frac{x}{2}$.

c) $\frac{x}{y} - \frac{y}{x}$. d) $\frac{x + y}{x} \cdot \frac{x - y}{y}$. **2.** a) 17. b) 60.

c) 13. d) 17. **3.** a) $p = a + b + 2c$.

b) $A = \frac{1}{2}h(a + b)$.

Set III

10. a) $6\sqrt{2}$. b) $2\sqrt{6}$. c) $5\sqrt{x}$. d) $7\sqrt{5x}$.
e) $\sqrt{7} + \sqrt{7} + \sqrt{7}$. f) $\sqrt{x} + \sqrt{x} + \sqrt{x} + \sqrt{x}$.
11. a) $9\sqrt{3}$. b) $\sqrt{3}$. c) $6\sqrt{3}$. d) $4\sqrt{3}$. e) Not possible. f) Not possible. g) $10\sqrt{2}$. h) $3\sqrt{6}$.
i) $7\sqrt{5}$. j) $3\sqrt{13}$. k) $\sqrt{7}$. l) $-\sqrt{7}$. **12.** a) 12.
b) 8.9. c) 9.3. d) 7. e) 0.4. f) 1.7. g) 4.2.
h) 3.2. i) 3. j) 1.7. **13.** a) $5 + 6\sqrt{6}$.
b) $15 + \sqrt{5} + \sqrt{10}$. c) $15 + 3\sqrt{3}$. d) 16.
e) $1 + \sqrt{7}$. f) $60 + 2\sqrt{5}$. **14.** a) $34\sqrt{x}$. b) $8x$.
c) Not possible. d) $9\sqrt{x}$. e) $72\sqrt{xy}$. f) Not possible. g) $3\sqrt{11x}$. h) $4\sqrt{3x}$. **15.** a) $\sqrt{3}$.
b) $5 - \sqrt{10}$. c) $11\sqrt{7}$. d) -2. e) $4\sqrt{5}$. f) 7.

Set IV

1. It is true if $x > 0$ and $y = 0$, if $x = 0$ and $y > 0$, or if $x = 0$ and $y = 0$. **2.** The right side, $x + y$. (If x and y are positive, $x^{2} + y^{2} < x^{2} + 2xy + y^{2}$, and $\sqrt{x^{2} + y^{2}} < x + y$.)

Chapter 12, Lesson 5

Set I

1. a) $x(4 - 7x)$. b) $(2x - 1)(x - 7)$. c) Not possible. d) $xy(3 + xy)(3 - xy)$. **2.** a) 3. b) 10.
c) 32. d) 100. e) 316.
3. a) $40,000x = 48,000(x - 10)$. b) 2,400,000.

Set III

9. a) 8. b) $5\sqrt{2}$. c) 14. d) 48. e) $32\sqrt{3}$. f) 405.
g) 225. h) 66. i) $36 + 12\sqrt{3}$. j) $6 + 6\sqrt{3}$.
k) $\sqrt{2} - 2$. l) 5. **10.** a) $25x$. b) x^{13}. c) $4x$.
d) $27x^{6}$. e) $6\sqrt{x} + 36$. f) $x + 6\sqrt{x}$. **11.** a) 9.
b) 14. c) 28. d) 98. e) $11 + 4\sqrt{7}$.
f) $51 + 14\sqrt{2}$. g) $9 + 2\sqrt{14}$. h) 15.
12. a) 118. b) 2. c) 8. d) $4 - x$. e) $5 - x^{2}$.
f) $x - y$. **13.** a) $22 + 14\sqrt{5}$. b) $19 - 5\sqrt{13}$.
c) 13. d) $12 + 14\sqrt{2}$.

1. >. **2.** >. **3.** >.

4. $\sqrt{1} + \sqrt{2} + \sqrt{3} + \cdots + \sqrt{n} > \sqrt{1} \cdot \sqrt{2} \cdot$
$\sqrt{3} \cdot \cdots \cdot \sqrt{n}$. **5.** No, $\sqrt{1} + \sqrt{2} + \sqrt{3} +$
$\sqrt{4} + \sqrt{5} < \sqrt{1} \cdot \sqrt{2} \cdot \sqrt{3} \cdot \sqrt{4} \cdot \sqrt{5}$.

Chapter 12, Lesson 6

Set I

1. a) $\dfrac{2x - 1}{6}$. b) $2x$. c) $\dfrac{2}{x^2 - 1}$. d) $\dfrac{x + 1}{x - 1}$.

2. a) =. b) <. c) >. d) =. e) =. f) =.

3. a) $x - 80$. b) $x + 8$. c) $x + 8 = 5(x - 80)$.
d) 102.

Set III

12. a) 4. b) 3. c) $\dfrac{1}{5}$. d) $\sqrt{13}$. e) $3\sqrt{2}$. f) $\dfrac{\sqrt{6}}{2}$.

13. a) $3\sqrt{3}$, 5.20. b) $\dfrac{\sqrt{38}}{2}$, 3.08. c) $\dfrac{\sqrt{5}}{10}$, 0.22.

d) $\dfrac{\sqrt{66}}{6}$, 1.35. **14.** a) 3. b) x^4. c) $\dfrac{x}{5}$. d) $\dfrac{x^3}{2}$.

15. a) $6 - \sqrt{5}$. b) $\sqrt{14} + 3$. c) $\sqrt{x} - y$.

d) $\sqrt{x} + \sqrt{y}$. **16.** a) $\dfrac{2(3 + \sqrt{2})}{7}$.

b) $10 - 3\sqrt{10}$. c) $2(\sqrt{7} + \sqrt{5})$. d) $\dfrac{6 - \sqrt{11}}{5}$.

17. a) $10(2 - \sqrt{3})$, 2.68. b) $\sqrt{5} + \sqrt{2}$, 3.65.

c) $\dfrac{4 + \sqrt{14}}{2}$, 3.87. d) $\dfrac{\sqrt{75} + 5}{10}$, 1.37.

18. a) $\sqrt{x}$. b) $\dfrac{1 + \sqrt{x}}{1 - x}$. c) $\sqrt{x} - y$.

d) $\dfrac{x + 2\sqrt{xy} + y}{x - y}$. **19.** a) $\sqrt{2}$. b) $\sqrt{6}$.

c) $2\sqrt{10}$. d) $2\sqrt{5}$. e) $\sqrt{3}$. f) $15\sqrt{5}$.

Set IV

1. 0.618. **2.** $\dfrac{2}{\sqrt{5} - 1} \approx 1.618$. Their difference

seems to be 1.

3.

$$\frac{2}{\sqrt{5} - 1} - \frac{\sqrt{5} - 1}{2} = 1$$

$$\frac{2(\sqrt{5} + 1)}{(\sqrt{5} - 1)(\sqrt{5} + 1)} - \frac{\sqrt{5} - 1}{2} =$$

$$\frac{2(\sqrt{5} + 1)}{5 - 1} - \frac{\sqrt{5} - 1}{2} =$$

$$\frac{\sqrt{5} + 1}{2} - \frac{\sqrt{5} - 1}{2} =$$

$$\frac{2}{2} = 1$$

Chapter 12, Lesson 7

Set I

1. a) $y = \dfrac{x}{6}$. b) $y = 5^x$. c) $y = \sqrt{x}$.

2. a) $\dfrac{4x + 2}{4x + 1}$. b) $\dfrac{x + y}{xy}$.

3. a) 2×10^{15} calories. b) 10^{13}.

Set III

9. a) 10. b) 2. c) 4. d) –14. e) Not possible.
f) 14. g) 7. h) 29. **10.** a) 4. b) 6. c) 10.
d) 13. e) $\sqrt{x^2 + 6x + 9} = x + 3$.
f) $\sqrt{x^2 + 6x + 9} = \sqrt{(x + 3)^2} = x + 3$. g) No;
$\sqrt{(-12)^2 + 6(-12) + 9} = \sqrt{81} = 9$ and
$(-12) + 3 = -9$. **11.** a) Yes. b) Yes. c) No.
d) Yes. **12.** a) 12. b) 4. c) 33. d) 9. e) 0.25.
f) 2. g) 1. h) 0.5. i) 4. j) No solution. k) 21.
l) 4. m) –91. n) No solution. **13.** a) $50 - \sqrt{5}$.
b) $10\sqrt{5}$. c) $5\sqrt{2}$. d) 8. e) $-\sqrt{2}$. f) $-3\sqrt{2}$.

Set IV

1. 4 miles per hour. **2.** $8 + 2\sqrt{7}$, or
approximately 13 miles per hour.

Chapter 12, Review

Set I

1. a) 6 and –6. b) 1 and –1. c) None. d) $\dfrac{1}{2}$ and

$-\frac{1}{2}$.   **2.** a) 42. b) 42. c) 20. d) 192. e) 8.
f) 20.   **3.** a) =. b) =. c) <. d) =. e) >.
f) =.   **4.** a) $5\sqrt{2}$. b) $6\sqrt{3}$. c) $3\sqrt{89}$.
d) $10\sqrt{11}$.   **5.** a) $2\sqrt{x}$. b) x^2. c) $x^3\sqrt{6}$.
d) $5x^2\sqrt{x}$.   **6.** a) $\frac{\sqrt{5}}{4}$. b) $\frac{\sqrt{6}}{10}$. c) $\frac{\sqrt{11x}}{11}$. d) $\frac{\sqrt{x}}{x^2}$.
7. a) 6. b) 4.3. c) 9.9. d) 9.
8. a) $10\sqrt{3}$. b) $3\sqrt{7}$. c) $7\sqrt{2}$. d) Not possible.
e) $\sqrt{5x}$. f) Not possible.   **9.** a) $3 + 6\sqrt{3}$.
b) $60 + 4\sqrt{6}$. c) $6\sqrt{5}$.   **10.** a) 50.
b) $27 + 10\sqrt{2}$. c) $9 - 4\sqrt{5}$. d) 27.   **11.** a) 6.
b) 24. c) $35 + 5\sqrt{7}$. d) x^3. e) $x + 5\sqrt{x}$.
f) $14 + 10\sqrt{3}$. g) $1 - x$. h) $x + 3\sqrt{x} - 10$.
12. a) 3. b) $10\sqrt{6}$. c) $12\sqrt{6} + 12$. d) $\frac{\sqrt{10} - 2}{3}$.
13. a) 1.58. b) 0.63. c) 12.96.   **14.** a) $4 - \sqrt{2}$.
b) $\sqrt{5}$. c) 63. d) 434. e) 196. f) 4. g) No
solution. h) No solution.

Chapter 13, Lesson 1

Set I

1. (Answers may vary.) a) $\frac{x}{x+9} + \frac{4}{x+9}$. b) $\frac{4}{9} \cdot \frac{1}{x}$.
c) $\left(\frac{x^2}{3}\right)^2$. d) $x^3 - \frac{9}{x}$.   **2.** a) $4x^6 - 1$.
b) $2x^2 - 5x + 6$. c) $x + 3$.   **3.** a) 1st porcupine, 3
feet per second; 2nd porcupine, 2.5 feet per second.
b) After 100 seconds.

Set III

8. a) $5x^2 - 6 = 0$; degree, 2. b) $2x + 7 = 0$;
degree, 1. c) $x^5 + x^4 - 8x - 2 = 0$; degree, 5.
d) $7x^3 - 5 = 0$; degree, 3.
e) $x^4 - 4x^3 + x - 16 = 0$; degree, 4.
f) $3x^2 + 3x = 0$; degree, 2.   **9.** a) Yes. b) Yes.
c) Yes. d) Yes. e) Yes. f) No. g) Yes. h) No.
i) Yes. j) Yes. k) Yes. l) Yes.   **10.** a) Yes.
b) Yes. c) No. d) Yes.   **11.** a) 3. b) –16. c) –2.
d) 1. e) 5.4. f) 0.9. g) –5. h) 6.1.

Set IV

1. 1, –2, –3, and 4.   **2.** A product is equal to zero
if any of its factors are equal to zero.

Chapter 13, Lesson 2

Set I

1. a) $2\sqrt{31}$. b) $5\sqrt{5}$. c) $3\sqrt{14}$.   **2.** a) $(x + 5)^2$.
b) $(2x - 1)^2$. c) $(x + 8y)^2$. d) $(3x - 4y)^2$.
3. a) Linear. b) 15. c) At the surface of the lake,
the temperature is 15 degrees Celsius. d) $-\frac{1}{3}$.
e) The temperature decreases $\frac{1}{3}$ degree for each
increase in depth of 1 meter. f) $y = -\frac{1}{3}x + 15$.

Set III

12. a) 1. b) Linear. c) A straight line.
d) | x | –3 | –2 | –1 | 0 | 1 | 2 | 3 |
   |-----|-----|-----|-----|-----|-----|-----|-----|
   | y | –13 | –10 | –7 | –4 | –1 | 2 | 5 |

e)

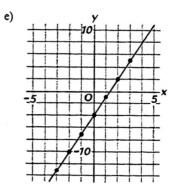

13. a) 2. b) Quadratic. c) A curved line.
d) | x | –4 | –3 | –2 | –1 | 0 | 1 | 2 | 3 | 4 |
   |-----|-----|-----|-----|-----|-----|-----|-----|-----|-----|
   | y | 8 | 3 | 0 | –1 | 0 | 3 | 8 | 15 | 24 |

e)

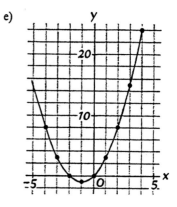

14.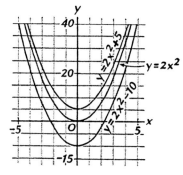

15. a)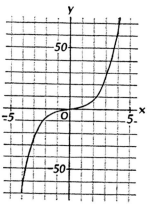

b)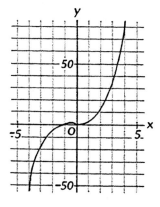

c)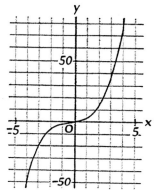

16.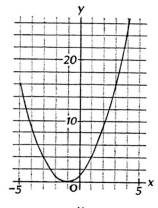

17.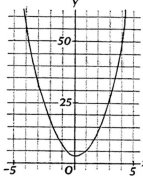

18.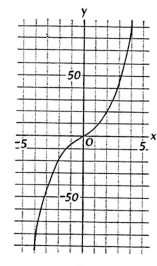

19.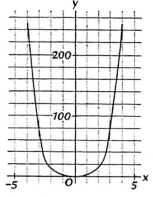

237

Set IV

1.

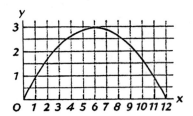

2. At 0.   **3.** At 12.   **4.** 12 units.   **5.** 3 units.

Chapter 13, Lesson 3

Set I

1. a) 25.   b) $\frac{49}{4}$.   c) 1.   d) 16.   **2.** a) 14.
b) $3\sqrt{7}$.   c) $\sqrt{7}$.   d) 2.   **3.** a) 11,400 years.
b) 17,100 years.   c) $5,700x$ years.

Set III

9. a) 2.   b) 0 and 5.   c) –5, 2, and 3.

10. a)

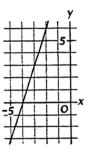

b)

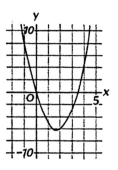

c)

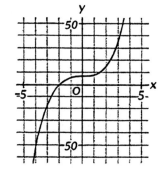

d)

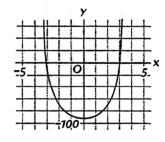

11. a) –4.   b) 0 and 3.5.   c) –2.   d) –3 and 3.

12.   a)

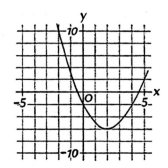

b)

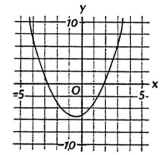

c)

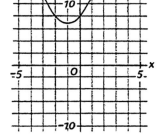

d)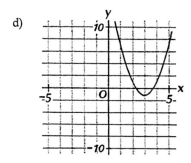

13. a) Between –1 and 0; between 4 and 5.
b) Between –3 and –2; between 1 and 2.   c) No
solutions.   d) Between 2 and 3; between 3 and 4.

Set IV

The equation $x^2 - 2x + 5 = 4$ has one solution, 1,
because the graph touches the line $y = 4$ at $x = 1$.
The equation $x^2 - 2x + 5 = 8$ has two solutions, –1
and 3, because the graph crosses the line $y = 8$ at
$x = -1$ and $x = 3$.

Chapter 13, Lesson 4

Set I

1. a) 210,000.   b) 0.018.   c) 40.   **2.** a) $\dfrac{\sqrt{3}}{6}$.

b) $\dfrac{3\sqrt{x}}{x}$.   c) $2\sqrt{5}$.   d) $4\sqrt{x}$.   **3.** a) $s = 2x + y$.

b) $y = s - 2x$.   c) $x = \dfrac{s - y}{2}$.   d) They scored an
odd number of free throws.   e) They scored an even
number of free throws.

Set III

7. a) 0 and 2.   b) –1 and –9.   c) 0.6.   d) 0 and 4.
e) –4 and 5.   f) –3 and 0.25.   g) a and b.   h) $-\dfrac{b}{a}$ and

$-\dfrac{1}{c}$.   **8.** a) –2 and 2.   b) 0 and 4.   c) 0 and –1.

d) –2 and –7.   e) 8.   f) –3 and 5.   g) 0 and –5.   h) 6
and –6.   i) –9 and 4.   j) 0.5 and –0.5.   k) 0 and
0.75.   l) 1 and –0.25.   **9.** a) 10 and –10.   b) 7 and
–3.   c) 0 and –18.   d) 2 and 5.   e) –2 and 1.   f) –5
and 3.

Set IV

1. $(x - 7)(x + 7) = 0$; $x^2 - 49 = 0$.
2. $(x - 0)(x - 2) = 0$; $x^2 - 2x = 0$.
3. $(x - 4)(x + 5) = 0$; $x^2 + x - 20 = 0$.

Chapter 13, Lesson 5

Set I

1. a) $10x$.   b) $10\sqrt{x} - 2x$.   c) $10 + 3\sqrt{x} - x$.
2. a) $y = x + 2$.   b) $y = x^2$.   c) $(-1, 1)$ and
$(2, 4)$.   **3.** a) $(x + 1)^2$ and $(x + 2)^2$.
b) $x^2 + (x + 1)^2 + (x + 2)^2 = 110$.
c) $3x^2 + 6x - 105 = 0$, $3(x + 7)(x - 5) = 0$; –7 and
5.   d) Either 5, 6, and 7 or –7, –6, and –5.

Set III

12. a) 11 and –11.   b) $\sqrt{42}$ and $-\sqrt{42}$.   c) $3\sqrt{2}$ and
$-3\sqrt{2}$.   d) $4\sqrt{5}$ and $-4\sqrt{5}$.   **13.** a) 12 and –8.

b) 5 and –13.   c) 4 and $-\dfrac{2}{3}$.   d) $6 + \sqrt{6}$ and

$6 - \sqrt{6}$.   e) $1 + 2\sqrt{10}$ and $1 - 2\sqrt{10}$.   f) $4 + \sqrt{3}$
and $4 - \sqrt{3}$.   **14.** a) $x^2 + 2x - 99 = 0$,
$(x + 11)(x - 9) = 0$; –11 and 9.   b) $x + 1 = \pm 10$,
$x = -1 \pm 10$; 9 and –11.   c) The square-root
method.   **15.** a) $x - 2 = \pm\sqrt{3}$, $x = 2 \pm\sqrt{3}$;
$2 + \sqrt{3}$ and $2 - \sqrt{3}$.   b) $x^2 - 4x + 1 = 0$, does not

factor.   **16.** a) $2x - 8 = \pm\sqrt{12}$, $x = \dfrac{8 \pm \sqrt{12}}{2}$;

$4 + \sqrt{3}$ and $4 - \sqrt{3}$.   b) 5.732 and 2.268.
c) $[2(5.732) - 8]^2 = 12$, $(11.464 - 8)^2 = 12$,
$(3.464)^2 = 12$, $11.999296 \approx 12$; $[2(2.268) - 8]^2 = 12$,
$(4.536 - 8)^2 = 12$, $(-3.464)^2 = 12$,
11.999296 $\approx$ 12.   **17.** a) $5\sqrt{2}$ and $-5\sqrt{2}$.   b) 5 and
–3.   c) –6.   d) $2 + \sqrt{15}$ and $2 - \sqrt{15}$.   e) 3 and –4.
f) 2.2 and –1.   **18.** a) 1 and –9.   b) $7 + \sqrt{10}$ and

$7 - \sqrt{10}$.   c) 4 and –5.   d) $\dfrac{2 + \sqrt{3}}{3}$ and $\dfrac{2 - \sqrt{3}}{3}$.

19. a) 7.449 and 2.551.   b) 0.123 and –8.123.
c) 3.721 and 1.613.   d) –0.143.

Set IV

$(2\sqrt{2} - 2)$ centimeters. $[(2x + 4)^2 = 2(4)^2.]$

Chapter 13, Lesson 6

Set I

1. a) B. b) A. c) A and C. **2.** a) 0.45. b) 0.24.
c) 4.24. **3.** a) (Let x = number of kilograms of
millet and y = number of kilograms of grain.)
$x + y = 500$ and $28x + 18y = 24(500)$. b) 300
kilograms of millet and 200 kilograms of grain.

Set III

11. a) $x^2 - 6x + 9 = (x - 3)^2$.
b) $x^2 + 22x + 121 = (x + 11)^2$.
c) $9x^2 + 30x + 25 = (3x + 5)^2$.
d) $4x^2 - 44x + 121 = (2x - 11)^2$.
e) $x^2 - 2ax + a^2 = (x - a)^2$.
f) $a^2x^2 + 8ax + 16 = (ax + 4)^2$. **12.** a) 19 was
subtracted from each side. b) 25 was added to each
side. c) Square roots were taken of each side. d) 5
was subtracted from each side. e) Each side of the
equation was divided by 2.
13. a) $x^2 + 12x - 45 = 0$, $(x + 15)(x - 3) = 0$; -15
and 3. b) $x^2 + 12x + 36 = 45 + 36$, $(x + 6)^2 = 81$,
$x + 6 = \pm 9$, $x = -6 \pm 9$; 3 and -15.
14. a) $1 + \sqrt{30}$ and $1 - \sqrt{30}$. b) 6.48 and -4.48.
c) $(6.48)^2 - 2(6.48) = 29$, $41.9904 - 12.96 = 29$,
$29.0304 \approx 29$; $(-4.48)^2 - 2(-4.48) = 29$,
$20.0704 + 8.96 = 29$, $29.0304 \approx 29$.
15. a)

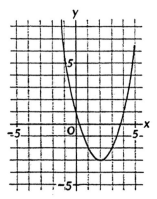

b) Between 0 and 1; between 3 and 4. c) $2 + \sqrt{3}$
and $2 - \sqrt{3}$. d) 3.73 and 0.27. **16.** a) -1 and
-9. b) 8 and -6. c) -6. d) $4 + \sqrt{13}$ and
$4 - \sqrt{13}$. e) $-9 + \sqrt{85}$ and $-9 - \sqrt{85}$.

f) $2 + 2\sqrt{5}$ and $2 - 2\sqrt{5}$. g) 4.5 and -3.5.
h) $\dfrac{-3 + \sqrt{15}}{5}$ and $\dfrac{-3 - \sqrt{15}}{5}$.
17. a) $-4 + 2\sqrt{10}$ and $-4 - 2\sqrt{10}$. b) $\sqrt{15}$ and
$-\sqrt{15}$. c) $6 + \sqrt{31}$ and $6 - \sqrt{31}$. d) $-9 + \sqrt{73}$
and $-9 - \sqrt{73}$.

Set IV

1. $x^2 + 2x + 4 = 0$, $x^2 + 2x = -4$,
$x^2 + 2x + 1 = -3$, $(x + 1)^2 = -3$, $x + 1 = \pm \sqrt{-3}$.
We obtain the square roots of -3; the equation has no
solution. **2.** The graph of $y = x^2 + 2x + 4$
(shown here) does not intersect the x-axis.

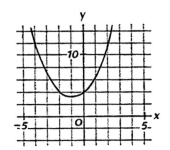

Chapter 13, Lesson 7

Set I

1. a) $4x(x^2 + 9)$. b) $8(x - 5)^2$.
c) $x(x + 10)(x - 1)$.
d) $(x + 2)(x - 2)(x + 6)(x - 6)$. **2.** a) 27.
b) 1.5. c) -12. **3.** a) $p = 28x$. b) $a = 13x^2$.

Set III

11. a) $3x^2 + x - 2 = 0$; $a = 3$, $b = 1$, $c = -2$.
b) $x^2 - 7x + 8 = 0$; $a = 1$, $b = -7$, $c = 8$.
c) $2x^2 - 4x - 1 = 0$; $a = 2$, $b = -4$, $c = -1$.
d) $4x^2 - 2x = 0$; $a = 4$, $b = -2$, $c = 0$.
12. a) $x^2 - 4x + 4 = 15 + 4$, $(x - 2)^2 = 19$,
$x - 2 = \pm\sqrt{19}$, $x = 2 \pm \sqrt{19}$; $2 + \sqrt{19}$ and
$2 - \sqrt{19}$. b) $x^2 - 4x - 15 = 0$,
$a = 1$, $b = -4$, $c = -15$.
$$x = \frac{-(-4) \pm \sqrt{(-4)^2 - 4(1)(-15)}}{2(1)} = \frac{4 \pm \sqrt{16 + 60}}{2} =$$

$\frac{4 \pm \sqrt{76}}{2} = \frac{4 \pm 2\sqrt{19}}{2} = 2 \pm \sqrt{19}$; $2 + \sqrt{19}$ and $2 - \sqrt{19}$. c) 6.36 and –2.36.

d) $(6.36)^2 - 4(6.36) = 15$, $40.4496 - 25.44 = 15$, $15.0096 \approx 15$; $(-2.36)^2 - 4(-2.36) = 15$, $5.5696 + 9.44 = 15$, $15.0096 \approx 15$.

13. a) $x^2 + 7x = 2$, $x^2 + 7x + \frac{49}{4} = 2 + \frac{49}{4}$,

$\left(x + \frac{7}{2}\right)^2 = \frac{57}{4}$, $x + \frac{7}{2} = \pm\sqrt{\frac{57}{4}} = \pm\frac{\sqrt{57}}{2}$,

$x = -\frac{7}{2} \pm \frac{\sqrt{57}}{2}$; $\frac{-7 + \sqrt{57}}{2}$ and $\frac{-7 - \sqrt{57}}{2}$.

b) $x^2 + 7x - 2 = 0$, $a = 1$, $b = 7$, $c = -2$.

$x = \frac{-7 \pm \sqrt{7^2 - 4(1)(-2)}}{2(1)} = \frac{-7 \pm \sqrt{49 + 8}}{2} = \frac{-7 \pm \sqrt{57}}{2}$. c) The quadratic formula. **14.** a) 3 and –5. b) 3.1 and –5.1. c) 3.2 and –5.2. d) –0.7 and –7.3. e) –0.6 and –8.4. f) –0.5 and –9.5.

15. a) 2 and –9. b) $-\frac{2}{3}$ and –1. c) $\frac{-3 + \sqrt{5}}{2}$ and $\frac{-3 - \sqrt{5}}{2}$. d) $\frac{\sqrt{6}}{3}$ and $-\frac{\sqrt{6}}{3}$. e) 0 and –4. f) 6.

16. a) $\frac{3 + \sqrt{57}}{2}$ and $\frac{3 - \sqrt{57}}{2}$. b) $\frac{1 + \sqrt{26}}{5}$ and $\frac{1 - \sqrt{26}}{5}$. c) 7 and 2. d) $3 + \sqrt{21}$ and $3 - \sqrt{21}$. **17.** a) $\frac{-1 \pm \sqrt{1 - 4ac}}{2a}$.

b) $\frac{d \pm \sqrt{d^2 + 4e}}{2}$. c) $\pm\sqrt{f}$. d) $\frac{b}{a}$ and 0.

Set IV

1. 1 and 6.   **2.** 6 and 1.   **3.** 0.5 and 7.5.
4. 7.46 and 0.54.

Chapter 13, Lesson 8

Set I

1. a) x^0. b) x^{-9}. c) x^6. d) $(xy)^3$. e) $\left(\frac{x}{y}\right)^3$.

2. a) $(x + 9)(x - 5) = 0$; –9 and 5.
b) $x^2 + 4x = 45$, $x^2 + 4x + 4 = 49$, $(x + 2)^2 = 49$, $x + 2 = \pm 7$, $x = -2 \pm 7$; 5 and –9.

c) $a = 1$, $b = 4$, $c = -45$;
$x = \frac{-4 \pm \sqrt{4^2 - 4(1)(-45)}}{2(1)} = \frac{-4 \pm \sqrt{16 + 180}}{2} = \frac{-4 \pm \sqrt{196}}{2} = \frac{-4 \pm 14}{2}$; 5 and –9. **3.** a) $y = 3x$.

b) $y - 5 = 4(x - 5)$.
c) $(15, 45)$. d) Fido is 15.

Set III

11. a) 16, two solutions. b) 0, one solution. c) –4, no solutions.

12.   a)

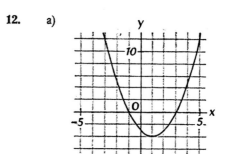

b)

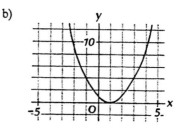

c)

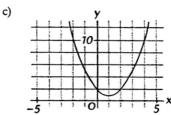

d) –1 and 3. e) 1. f) No solutions.   **13.** a) –1 and 3. b) 1. c) No solutions.   **14.** a) –1 and 3. b) 1. c) No solutions.   **15.** a) 24, two solutions. b) –24, no solutions. c) 48, two solutions. d) 0, one solution.   **16.** a) 1.5 and 6.5. b) No solutions. c) –3.8 and –0.2. d) 2.5.   **17.** a) 6.45 and 1.55. b) No solutions. c) –0.27 and –3.73. d) 2.50.

1. The folds seem to form a parabola.

2 and 4.

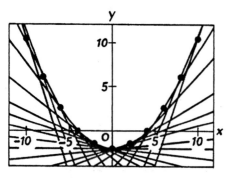

3.

x	-10	-8	-6	-4	-2	0	2	4	6	8	10
y	10.5	6	2.5	0	-1.5	-2	-1.5	0	2.5	6	10.5

Chapter 13, Lesson 9

Set I

1.

a)

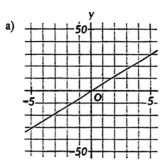

b)

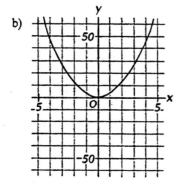

c)

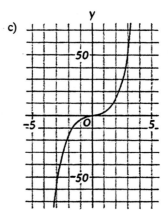

2.

	a) Example:	b)
Think of a number:	3	x
Square it:	9	x^2
Subtract four:	5	$x^2 - 4$
Divide the result by the number that is two more than your original number:	1	$x - 2$
Add nine:	10	$x + 7$
Subtract the number first thought of:	7	7
The result is seven.		

c) At the fourth step we would be dividing by 0.

Set III

9. a) $4x^2 - 11 = 0$; two solutions. b) $4x + 10 = 0$; one solution. c) $x^7 - x^4 = 0$; seven solutions. d) $x^6 - 10x^3 + 18 = 0$; six solutions. **10.** a) Yes. b) No. c) Yes. d) Yes. **11.** a) 2.7. b) -3.2, -1, 1, and 3.2.

12.

a)

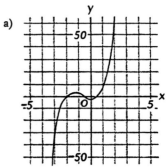

b) -2, -0.5, and 0.5. **13.** a) 0 and -1. b) 0, $-1 + \sqrt{2}$, and $-1 - \sqrt{2}$. c) 1, -1, 2, and -2. d) 0, $\sqrt{10}$, and $-\sqrt{10}$. **14.** a) 0, 5, and -0.5. b) $2 + \sqrt{2}$ and $2 - \sqrt{2}$. c) 3 and -3.

Set IV

If $x^4 + x^2 + 1 = 0$, then $x^4 + x^2 = -1$. Because the square of a number cannot be negative, x^2 and x^4 [since $x^4 = (x^2)^2$] must be either positive or zero. But the sum of two numbers that are positive or zero cannot be -1. So $x^4 + x^2 + 1 = 0$ has no solutions.

Chapter 13, Review

Set I

1. a) $x^3 + 5x - 2 = 0$, cubic, 3. b) $x^2 - x - 8 = 0$, quadratic, 2. c) $x^5 - 9 = 0$, quintic, 5.
2. a) Yes. b) Yes. **3.** a) $4x^3 + 8x^2 - 21x = 0$.
b) 3. c) -3.6, 0, and 1.6.
4.

a)

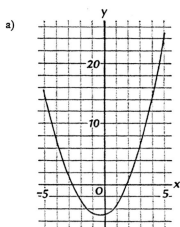

b)

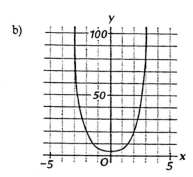

5. a) -2.8 and 1.8. b) No solutions. **6.** a) 0 and $\frac{5}{6}$.
b) 6 and -6. c) -11 and 3. d) -4 and -7.

7. a) $2\sqrt{15}$ and $-2\sqrt{15}$. b) $4 + \sqrt{7}$ and $4 - \sqrt{7}$.
c) 0 and -5. d) 1 and $-\frac{5}{3}$. **8.** a) 1 and -11.
b) $3 + \sqrt{5}$ and $3 - \sqrt{5}$. **9.** a) 0; one solution.
b) -3; no solutions. c) 24; two solutions. **10.** a) 3
and 2.5. b) $-2 + 2\sqrt{6}$ and $-2 - 2\sqrt{6}$. c) $\dfrac{2 + \sqrt{10}}{6}$
and $\dfrac{2 - \sqrt{10}}{6}$. d) $\frac{4}{3}$ and -3. **11.** a) 0, -4, and
3. b) 0, $\sqrt{5}$, and $-\sqrt{5}$. **12.** a) -1 and -7. b) 4
and -8. c) $2 + \sqrt{5}$ and $2 - \sqrt{5}$. d) -9 and 3.
e) $\dfrac{-1 + \sqrt{21}}{2}$ and $\dfrac{-1 - \sqrt{21}}{2}$. f) 0, -4, and 4.

Chapter 14, Lesson 1

Set I

1. a) Perimeter, $2\sqrt{3} + 2\sqrt{7}$; area, $\sqrt{21}$.
b) Perimeter, $4x$; area, $x^2 - 6$.
2.

a)
$$\underset{6.25x}{\rule{3cm}{0.4pt}}$$
$$\underset{5(x+4)}{\rule{3.5cm}{0.4pt}}$$

$6.25x = 5(x + 4)$, $x = 16$. b) 100.
3. a) $y = 6 - x$. b) $y = \dfrac{6}{x}$. c) $y = (x + 2)^2$.
d) $y = \sqrt{x} - 2$.

Set III

11.

x	$\dfrac{1}{x}$	$\dfrac{1}{x}$
9	0.11111	$0.\overline{1}$
10	0.10000	0.1
11	0.09090	$0.\overline{09}$
12	0.08333	$0.08\overline{3}$
13	0.07692	$0.\overline{076923}$
14	0.07142	$0.0\overline{714285}$
15	0.06666	$0.0\overline{6}$

12. a) 0.5. b) 0.25. c) 0.125. d) 0.0625. e) It "ends." f) 5. g) n. h) 0.03125. i) 0.00390625.
j) Yes. **13.** a) $\frac{1}{7} = 0.\overline{142857}$. b) $\frac{2}{7} = 0.\overline{285714}$.
c) $\frac{3}{7} = 0.\overline{428571}$. d) $\frac{4}{7} = 0.\overline{571428}$.

e) $\frac{5}{7} = 0.\overline{714285}$. f) $\frac{6}{7} = 0.\overline{857142}$. g) All six numbers have periods consisting of the same six digits. **14.** a) $\frac{17}{10}$. b) $\frac{104}{25}$. c) $\frac{5}{8}$. d) $\frac{1}{1,250}$.

15. a) $\frac{1}{3}$. b) $\frac{65}{9}$. c) $\frac{2}{11}$. d) $\frac{5}{36}$. e) $\frac{81}{55}$. f) $\frac{3}{74}$.

16. a) 23. b) It ends. c) 22. **17.** a) 2.5. b) 1.6. c) 1.3. d) 1.03. e) 1.0003. f) It decreases and gets close to 1.

Set IV

1. 16.

2. 0 5 8 8 2 3 5 2 │ 9 4 1 1 7 6 4 7

3. Each digit in the second half of the period can be obtained by subtracting the corresponding digit in the first half from 9.

Chapter 14, Lesson 2

Set I

1. a) $\frac{13}{6}$. b) $-\frac{3}{2}$. c) $\frac{11}{3}$ or $3\frac{2}{3}$. d) $\frac{1}{3}$.

2. a) True. b) False; example: $(1^2 + 1^2)^2 \neq 1^4 + 1^4$. c) False; example: $\sqrt{1^2 + 1^2} \neq 1 + 1$. **3.** a) D. b) Because it has 4 solutions, it must have degree at least 4; because it passes through the origin, the constant term must be 0.

Set III

10. a through j)

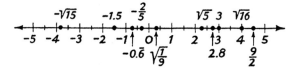

k) $\sqrt{5}$ and $-\sqrt{15}$. **11.** a) Irrational. b) Rational. c) Irrational. d) Rational. e) Rational. f) Irrational. g) Rational. h) Irrational. **12.** a) 3; rational. b) Not possible; irrational. c) 2; rational. d) Not possible; irrational. e) 4; rational. f) Not possible; irrational. g) Not possible; irrational. h) 1; rational. i) 0.05; rational. j) 0.15; rational. **13.** a) 1.75.

b) Larger. c) $\frac{49}{16}$ or 3.0625. d) Larger. e) $1.\overline{72}$.

f) Smaller. g) $\frac{361}{121}$ or 2.98. . . . h) Smaller.

14. a) 36. b) 42.25. c) 41.9904. d) 42.003361. e) 41.99947249. f) The decimal form ends. g) Smaller: 6, 6.48, and 6.4807; larger: 6.5 and 6.481. h) No, because $\sqrt{42}$ is irrational. **15.** a) 1 and $\frac{1}{9}$. b) $\frac{4 + \sqrt{7}}{9}$ and $\frac{4 - \sqrt{7}}{9}$. c) $\frac{1}{3}$.

d) $\frac{3 + \sqrt{5}}{6}$ and $\frac{3 - \sqrt{5}}{6}$. e) The discriminant must be the square of an integer.

Set IV

If $\frac{x^2}{y^2} = 2$, then $\frac{x}{y} = \sqrt{2}$. Because x and y are integers, $\frac{x}{y}$ is rational; so $\sqrt{2}$ would be rational.

Chapter 14, Lesson 3

Set I

1. a) $\frac{1}{3}$. b) $\frac{3x^2}{5}$. c) Not possible. **2.** a) $\left(0, \frac{1}{3}\right)$. b) $(-2, -4)$ and $(1, -1)$. **3.** a) 45 meters. b) After 1 second and after 5 seconds.

Set III

11.

x	x^2	x^3	x^4	x^5
1	1	1	1	1
2	4	8	16	32
3	9	27	81	243
4	16	64	256	1024
5	25	125	625	3125

12. a) Rational; 3. b) Irrational. c) Irrational. d) Rational; 3. e) Rational, 4. f) Rational; 2. g) Irrational. h) Irrational. i) Rational; $\frac{1}{3}$.

j) Rational; $\frac{4}{5}$. **13.** a) $\sqrt[n]{2}$ gets smaller. b) 1. c) $\sqrt[n]{0.2}$ gets larger. d) 1. **14.** a) The y-coordinate is the cube of the x-coordinate. b) The x-coordinate is the cube root of the y-coordinate.

c)

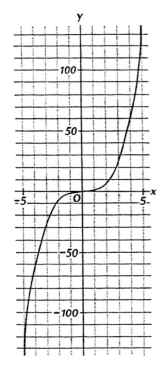

d) 3.4. e) –4.3. f) 4.8. g) One. h) One.

15.

a)

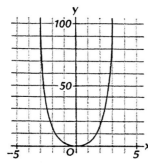

b) –1.8 and 1.8. c) –2.5 and 2.5. d) –2.7 and 2.7.
e) Two. f) None.   **16.** a) –3. b) 0. c) 10 and
–10. d) None. e) –1. f) 4.   **17.** a) $>$. b) $<$.
c) $=$. d) $<$. e) $>$. f) $<$. g) $>$. h) $>$.

Set IV

36 million miles.
$[y = \sqrt[3]{6(88)^2} = \sqrt[3]{46,464} \approx 36.]$

Chapter 14, Lesson 4

Set I

1. a) $\dfrac{3x + 1}{x}$. b) $\dfrac{3}{x}$. c) $\dfrac{2x^2 + 4}{x^4}$. d) $\dfrac{8}{x^6}$.

2. a) $\dfrac{1}{5}$. b) $\dfrac{2}{9}$. c) $\dfrac{27}{100}$. d) $\dfrac{3}{11}$.

3. a) $9x = 7x + 150$. b) 75.

Set III

11. a) $\dfrac{377}{120}$. b) $3.141\overline{6}$. c) Larger. d) $3\dfrac{17}{120}$ is
rational and π is irrational.

12.

a) Radius	1	2	3	4	5
Area	π	4π	9π	16π	25π

b) No. c) No.   **13.** a) Circumference, 0.2π cm;
area, 0.01π cm². b) Circumference, π^2 cm; area,
$\dfrac{\pi^3}{4}$ cm². c) Circumference, $2\pi\sqrt{3}$ cm; area, 3π cm².

d) Circumference, 12 cm; area, $\dfrac{36}{\pi}$ cm².

14. a) 12.5. b) $\dfrac{4}{\pi}$. c) 5. d) $2\sqrt{3}$.

15. a) $100 - 25\pi$. b) $4x^2 - \pi x^2$. c) $40 - 10\pi$.

d) $8x - 2\pi x$.   **16.** a) 32π. b) $2\pi x^2$. c) $\dfrac{1}{2}$.

17. a) 85 inches. b) 6 centimeters. c) 1,017 square
centimeters. d) 20 feet.

Set IV

The number of digits in each word is
3 1 4 1 5 9 2 6 and $\pi \approx 3.1415926$.

Chapter 14, Lesson 5

Set I

1. (Answers may vary.) a) $\dfrac{x}{x + 2} + \dfrac{6}{x + 2}$.

b) $2 - \dfrac{x}{4}$. c) $\dfrac{5}{x} \cdot \dfrac{1}{y}$.   **2.** a) $\dfrac{2}{3}$. b) 0.5 and 2.

c) $\dfrac{1 + \sqrt{11}}{5}$ and $\dfrac{1 - \sqrt{11}}{5}$.   **3.** 16 monkeys and
21 weasels.

11. a) 25, $\sqrt{25}$, and 2^5. b) –25. c) $\frac{1}{25}$, 2.5, $2.\overline{5}$, and $-\frac{2}{5}$. d) $\sqrt[3]{25}$ and 25π. **12.** a) Yes. b) Yes.

c) No; example: $\frac{1}{2}$ is not an integer.

d) Yes. e) No; example: $\sqrt{2}$ is not an integer.
13. a) No; example: The sum of 3 and 5 is not an odd integer. b) No; example: The difference of 5 and 3 is not an odd integer. c) Yes. d) No; example: $\frac{3}{5}$ is not an odd integer. e) Yes.

14. a) No; example: $\sqrt{2} + (1 - \sqrt{2}) = 1$. b) No; example: $\sqrt{2} - (\sqrt{2} - 1) = 1$. c) No; example: $\sqrt{2} \cdot \sqrt{2} = 2$. d) No; example: $\frac{\sqrt{2}}{\sqrt{2}} = 1$. e) No; example: $(\sqrt{2})^2 = 2$. f) Yes. **15.** a) 0.0005, 0.002, 0.04, 0.3. b) 0.719, $0.7\overline{19}$, $0.\overline{719}$, $0.71\overline{9}$. c) $\frac{2}{3}, \frac{4}{5}, \frac{6}{7}, \frac{8}{9}$. d) $-\sqrt{5}, -\sqrt[4]{5}, \sqrt[4]{5}, \sqrt{5}$. e) $\sqrt[3]{\pi}, \sqrt{\pi}, \pi$.

16. a) 1.32. b) 0.63. c) 0.11. d) 0.71. e) 0.18.

17. a) $-\frac{14}{3}$; rational. b) $\sqrt{5}$; irrational. c) 3; counting number. d) $\frac{\sqrt{15}}{2}$ and $-\frac{\sqrt{15}}{2}$; irrational. e) –7 and 3; integers. f) 81; counting number.

Set IV

1. $1^{2^3} = 1^{3^2}(1^8 = 1 \text{ and } 1^9 = 1)$, $2^{1^3}(2^1 = 2)$, $3^{1^2}(3^1 = 3)$, $2^{3^1}(2^3 = 8)$, $3^{2^1}(3^2 = 9)$.
2. $4^{2^3}(4^8 = 2^{16})$, $4^{3^2}(4^9)$, $3^{2^4} = 3^{4^2}(3^{16})$, $2^{4^3}(2^{64})$, $2^{3^4}(2^{81})$.

Chapter 14, Review

Set I

1. a) False. b) False. c) True. d) True. e) True. **2.** a) 3.2. b) 0.225. c) $0.\overline{72}$. d) $0.208\overline{3}$. **3.** a) $\frac{103}{10}$. b) $\frac{31}{3}$. c) $\frac{17}{20}$. d) $\frac{85}{99}$. **4.** a) Rational. b) None, because it is not an integer. **5.** a) –5. b) Not possible. c) 15. d) 10. **6.** 0.345, $0.3\overline{45}$, $0.\overline{345}$, $0.34\overline{5}$. **7.** a) 5. b) None. c) –10. d) –1 and 1. **8.** a) $>$. b) $<$.

c) $<$. d) $>$. **9.** a) $\sqrt[3]{0.9}$ gets larger. b) 1. c) $\sqrt[3]{1.1}$ gets smaller. d) 1. **10.** a) $\frac{157}{50}$. b) 3.14. c) Smaller. **11.** a) 1.6π. b) 12π. c) $\frac{15}{\pi}$. d) $\sqrt{17}$. **12.** a) 32. b) $8x^2$. c) $32 - 8\pi$. d) $8x^2 - 2\pi x^2$. **13.** a) 25,120 miles. b) 28 square meters. **14.** a) $\sqrt[3]{27}$ and 2^7. b) –27. c) 2.7, $2.\overline{7}$, and $\frac{2}{7}$. d) $\sqrt{27}$ and 27π. **15.** a) Yes. b) No; example: $(-1) - (-2) = 1$. c) No; example: $(-1)(-2) = 2$. d) No; example: $(-1)^2 = 1$. e) Yes. **16.** a) $-\frac{19}{4}$; rational. b) $5\sqrt{3}$; irrational. c) $\frac{\sqrt{7}}{6}$ and $-\frac{\sqrt{7}}{6}$; irrational. d) No solution.

Chapter 15, Lesson 1

Set I

1. 3.14, π, 3.1416, $3\frac{1}{7}$. **2.** a) $3x^2 - 4x + 1$. b) $x^2 + 4x$. **3.** a) $193,600\pi$. b) 608,000.

Set III

9. a) $\frac{3}{13}$. b) $\frac{1}{13}$. c) 2. d) 1. **10.** a) $\frac{3}{5}$. b) Not possible. c) $\frac{3x-1}{5x-1}$. d) $\frac{3}{5}$. e) Not possible. f) $\frac{1}{x^4}$. g) $\frac{6}{x^5}$. h) Not possible. **11.** a) 8. b) $\frac{14}{9}$. c) –5. d) No solution. e) 61. f) 6 and –6. g) $-\frac{3}{7}$. h) No solution. i) 7 and –7. j) 4. **12.** a) $\frac{5}{4}$. b) $\frac{6}{5}$. c) $\frac{3}{2}$. d) 352 cycles per second. e) 440 cycles per second.
13. a)

Number of clicks per minute	90	120	180	225
Speed of train in miles per hour	30	40	60	75

b) $\frac{1}{3}$. c) Directly. d) $s = \frac{1}{3}n$.

$\frac{1}{42,240} \cdot \left(\frac{3}{4} \text{ inch represents 2,640 feet; so 1 inch} \right.$
represents 3,520 feet or 42,240 inches.$\left. \vphantom{\frac{3}{4}} \right)$

Chapter 15, Lesson 2

Set I

1. a) 8×10^4. b) 1.5×10^9. c) 5.4×10^4.
d) 2×10^8. **2.** a) 0.027. b) $0.02\overline{7}$. c) $0.0\overline{27}$.
d) $0.0\overline{27}$. e) $0.02\overline{27}$. **3.** a) $x = y + 42$;
$x - 9 = 3(y - 9)$. b) Mr. Dithers is 72 and
Dagwood is 30.

Set III

8. a) $x + 21$. b) $2 - 5x$. c) $3x + 4$. d) $x^2 - 12$.
e) $5x + 19$. f) $x^2 - 8x - 12$. **9.** a) $12x - 1 = \frac{2x}{3}$.

b) $18x - \frac{3}{2} = x$. c) $36x - 3 = 2x$. d) The one in

part c. e) It contains no fractions. **10.** a) 0.25.

b) $-\frac{7}{6}$. c) 2. d) 0.4. e) 20. f) 3.5. g) No

solution. h) 5. i) 7. j) –36. **11.** a) (36, 10).
b) (–2, 11).

Set IV

9 pounds. $\left(\text{Let } x \text{ equal the weight of the watermelon;} \right.$
$x = \frac{9}{10}x + \frac{9}{10}, x = 9.$$\left. \vphantom{\frac{9}{10}} \right)$

Chapter 15, Lesson 3

Set I

1. a) 1. b) –1. c) 1. d) $\frac{1}{100}$ or 0.01. **2.** a) –0.5.

b) 1. c) $\frac{-5 + \sqrt{3}}{4}$ and $\frac{-5 - \sqrt{3}}{4}$. **3.** a) 3π

liters. b) 9.4 liters.

Set III

8. a) 6 and –6. b) $\frac{1}{3}$ and $-\frac{1}{3}$. c) –4 and 2.

d) 4. **9.** a) 5.61 and –1.61. b) 5.11 and 0.39.

c) 10.48 and –0.48. d) 2.54 and –3.54. **10.** a) Mr.
Gildersleeve, $\frac{1}{20}$; Leroy, $\frac{1}{15}$. b) Mr. Gildersleeve,

$\frac{x}{20}$; Leroy, $\frac{x}{15}$. c) $\frac{x}{20} + \frac{x}{15} = 1$. d) $8\frac{4}{7}$ minutes.
11. a) $18 + x$ kilometers per hour with the current
and $18 - x$ kilometers per hour against the current.

b) $\frac{8}{18 + x}$ hours with the current and $\frac{8}{18 - x}$ hours

against the current. c) $\frac{8}{18 + x} + \frac{8}{18 - x} = 1$. d) 6

kilometers per hour.

Set IV

$x + \frac{2}{3}x + \frac{1}{2}x + \frac{1}{7}x = 97$. The number is 42.

Chapter 15, Lesson 4

Set I

1. a) $\frac{x - 2}{2x^2}$. b) $\frac{1}{2x^3}$. c) $\frac{7x}{7 - x}$. d) $\frac{x}{7 - x}$.
2. a) 2. b) No solution. c) 0 and 3.5. d) No
solution. **3.** 8.5×10^5 kilometers.

Set III

9. a) $m = \frac{pi}{100}$. b) 19.8 years. **10.** a) $x = \frac{b}{a}$.

b) $x = \frac{6}{3} = 2$. c) 4.5. **11.** a) 15. b) 77.

c) $ab + a$. d) –2. e) –8. f) $-a$. **12.** a) $x = \frac{a^2}{b}$.

b) $\frac{-b}{a}$. c) $x = \frac{b^2 - a^2}{a}$. d) $x = a + b$.

e) $x = \frac{a - b}{a}$. f) $x = \frac{a - 1}{b}$. g) $x = \pm a$.

h) $x = \frac{a^2 - b^2}{ab}$. **13.** a) 16. b) $n = 7(C - 4)$.

c) 112. d) 4.

Set IV

The temperature was the same number on both
scales: –40. $\left[\text{Using the formula } C = \frac{5}{9}(F - 32), \text{ we} \right.$

have $x = \frac{5}{9}(x - 32), x = -40.$$\left. \vphantom{\frac{5}{9}} \right]$

Chapter 15, Lesson 5

Set I

1. a) Cubic.

b)

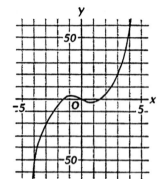

2. a) 3. b) –1.4, 0, and 1.4. c) 0, $\sqrt{2}$, and $-\sqrt{2}$.

3.

a)

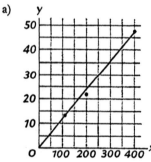

b) A direct variation. c) 200 meters. d) 110 meters, 8.4; 200 meters, 9.1; 400 meters, 8.4.

Set III

9. a) 3. b) –5. c) $a - b$. d) $\frac{1}{4}$. e) $\frac{1}{10}$. f) $\frac{1}{a+1}$.

10. a) $x = \frac{b}{a-1}$. b) $x = \frac{a^2 + b^2}{a - b}$. c) –1.

d) $x = \frac{ab}{a+b}$. e) $x = \pm\sqrt{b^2 - a^2}$.

f) $x = \frac{b}{ab-1}$. **11.** a) 2. b) $r = \frac{a+b-c}{2}$.

c) $c = a + b - 2r$. d) 29.

12. a)

A	20	35	50	65
p	122	127	134	144

b) It increases. c) No. **13.** a) 6.3 centimeters.

b) $b = \frac{af}{a-f}$. c) 5 centimeters.

1. $\frac{l-x}{x} = \frac{x}{l}$. **2.** $x = \frac{-l \pm l\sqrt{5}}{2}$. **3.** 6.18.

Chapter 15, Review

Set I

1. a) $\frac{2}{13}$. b) $\frac{4}{26} = \frac{2}{13}$. c) 26 and 2 (for proportion as written). d) 4 and 13 (for proportion as written).

e) The products are equal. **2.** a) $\frac{1}{3}$. b) $\frac{2}{3}$.

c) Not possible. d) $\frac{4}{5x}$. **3.** a) $x - 16$.

b) $4 + 3x$. c) $7x - 3$. d) $x^2 + 5x + 20$.

4. a) 2.5. b) 7.5. c) True for all values of x except 1.

d) 4. e) 30. f) –16. g) –1. h) $2\sqrt{6}$ and $-2\sqrt{6}$.

i) $\frac{5 + \sqrt{5}}{10}$ and $\frac{5 - \sqrt{5}}{10}$. j) $\frac{2}{3}$ and 1.

5. a) 180π cubic centimeters. b) $h = \frac{3v}{\pi r^2}$. c) 12

centimeters. **6.** a) $x = \pm\sqrt{a^2 - 1}$. b) $x = \frac{1 - a}{a}$.

c) $x = \frac{a}{2}$. d) $x = \frac{a}{a - b}$. e) $x = \frac{ab}{1 - ab}$.

7. a) 37 degrees. b) $F = \frac{9C + 160}{5}$. c) 320

degrees. **8.** a) 190. b) $x = \frac{1 \pm \sqrt{1 + 8n}}{2}$. c) 80.

Chapter 16, Lesson 1

Set I

1. a) $x^2(x - 4)$. b) $x(x + 2)(x - 2)$. c) $x^2(4x - 1)$.

d) $x(2x + 1)(2x - 1)$. **2.** a) $y = x(x + 1)$ or

$y = x^2 + x$. b) $y = x^3 + 1$. c) $y = \frac{1}{x^2}$.

d) $y = \frac{x - 3}{x}$. **3.** a) $40 - x$.

b)

$$Up \quad 6\left(\frac{x}{60}\right)$$

$$Down \quad 54\left(\frac{40-x}{60}\right)$$

c) $6\left(\frac{x}{60}\right) = 54\left(\frac{40 - x}{60}\right)$, $x = 36$. d) 3.6 miles.

11. a) $<$. b) $>$. c) $>$. d) $<$. e) $>$. f) $>$.
g) $>$. h) $=$. i) $<$. j) $<$. k) $<$. l) $>$. m) $=$.
n) $>$. o) $>$.   **12.** a) $1 < 6$ and $6 > 1$.
b) $-10 < 10$ and $10 > -10$. c) $4 < x$ and $x > 4$.
d) $y < 0$ and $0 > y$.   **13.** a) Figure F. b) Figure
A. c) Figure E. d) Figure B. e) Figure C.
f) Figure D.   **14.** a) $>$. b) $>$. c) $=$. d) $<$.
e) $<$. f) $>$. g) $<$. h) $>$. i) $>$. j) None,
because neither expression is defined. k) $<$. l) $<$.
m) $>$. n) $<$.   **15.** a) $>$. b) $>$. c) $>$. d) $>$.
e) $>$. f) $>$. g) $>$. h) $>$. i) $>$. j) $>$. k) $>$.
l) $>$. m) $>$. n) $>$.   **16.** a) $>$. b) None.
c) $\geq$. d) None. e) $=$. f) $<$.   **17.** a) $>$. b) $<$.
c) None. d) $>$. e) $<$. f) None.

Set IV

The life-span of a horse is more than 36 years.
$(h > 2d, d > c, d > p, c \geq p + 8, p = 10; c \geq 18,$
$d > 18, h > 36.)$

Chapter 16, Lesson 2

Set I

1. a) x^6. b) Not possible. c) x^8. d) Not possible.
e) $-x^6$. f) x^{-2}.   **2.** a) $y = 2x + 3$.
b) $y = 2x - 1$. c) $y = -x + 3$. d) Their slope.
e) Their y-intercept.   **3.** a) $850 in 7 percent
account and $350 in 5 percent account.

Set III

10.

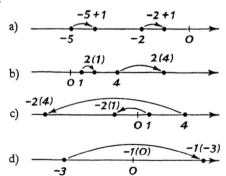

a)

b)

c)

d)

11. a) The direction remains the same. b) The
direction remains the same. c) The direction is

reversed.   **12.** a) Yes. b) No. c) Yes. d) Yes.
e) Yes. f) No. g) Yes. h) Yes. i) No. j) Yes.
k) No. l) No. m) No. n) Yes. o) Yes.   **13.** a) 3
was subtracted from each side. b) Each side was
divided by 7. c) 9 was added to each side. d) Each
side was divided by -2. e) Each side was multiplied
(or divided) by -1. f) Each side was multiplied by
-5.   **14.** a) $x > -3$. b) $x > 0.7$. c) $x < -9$.
d) $x > 4$. e) $x \geq -20$. f) $x \leq -20$. g) $0 < x < 3$.
h) $5 \leq x \leq 10$.   **15.** a) $x < -3$. b) $x \leq -3$.
c) $x \leq -6$. d) $x > 10$. e) $5 < x < 7$. f) $x > 1$.
g) $30 \geq x \geq 0$. h) No solutions.

Set IV

1. $3 \leq s \leq 18$.   **2.** 16.   **3.** 16.
4. $n \leq s \leq 6n$.   **5.** $5n + 1$. $[6n - (n - 1).]$

Chapter 16, Lesson 3

Set I

1. a) $3\sqrt{5x}$. b) $\sqrt{5x}$. c) $10x$. d) 2.
2. a) $(15, 1.5)$. b) $(-4, -5)$.   **3.** a) Quadratic.
b) 1,800 dollars. c) 4,800 dollars.

Set III

8. a) $x \geq -1$; Figure A. b) $x < 3$; Figure C.
c) $x < 3$; Figure C. d) $x > 3$; Figure D. e) $x \geq -1$;
Figure A. f) $-1 \leq x \leq 3$; Figure B.

9. a) $x > 4$.

b) $x < 12$.

c) $x \geq 5$.

d) $x \leq -10$.

e) $x < 0$.

f) $x > 1$.

g) $x > 15$.

h) $x \leq 6$.

10. a) $x < \dfrac{1-b}{a}$. b) $x < \dfrac{b}{a}$. c) $a - b \geq x \geq 0$.

d) $x \geq \dfrac{1}{a-b}$.   **11.** a) $x < 4 + 6$, $4 < x + 6$,

$6 < x + 4$; $2 < x < 10$. b) $x < (x + 3) + 11$,
$x + 3 < x + 11$, $11 < x + (x + 3)$; $x > 4$.
c) $20 < x + 3x$, $x < 20 + 3x$, $3x < 20 + x$;
$5 < x < 10$. d) $1 + x < 2 + (7 - x)$,
$2 < (1 + x) + (7 - x)$, $7 - x < (1 + x) + 2$;
$2 < x < 4$.

Set IV

The potential worth of the pot should be greater than twice the potential cost of playing the hand. $\left[\dfrac{wp}{c} > 1, \right.$

$\dfrac{w(0.5)}{c} > 1, \dfrac{w}{c} > 2, w > 2c. \left. \right]$

Chapter 16, Lesson 4

Set I

1. a) 0 and –5. b) $2\sqrt{3}$ and $-2\sqrt{3}$. c) 4 and –2.

2. a) $x + y$. b) $\dfrac{x - y}{x + y}$.   **3.** a) 37 dollars.

b) $m = 7 + 3a$. c) $a = \dfrac{m - 7}{3}$. d) 31.

Set III

11. a) –9 and 9. b) 6 and 14. c) –6 and 14. d) –7
and –3. e) $x - 7$ and $x + 7$. f) $z - y$ and
$z + y$.   **12.** a) Figure B. b) Figure D. c) Figure
A. d) Figure C.   **13.** a) Yes. b) Yes. c) No.
d) Yes. e) Yes. f) No. g) No. h) Yes. i) No.
j) Yes. k) Yes. l) No.   **14.** a) 6 and –6. b) No
solutions. c) $-10 < x < 10$. d) $-4 \leq x \leq 4$.
e) $x \geq 1$ or $x \leq -1$. f) True for all x. g) True for
all x except 0. h) $x \geq 28$ or $x \leq -28$.

15.

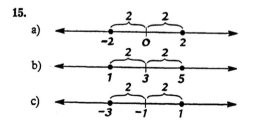

16. a) 9 and –9. b) 9 and 1. c) $-13 \leq x \leq 13$.
d) $7 \leq x \leq 13$. e) $x > 8$ or $x < -8$. f) $x > 8$ or
$x < 4$. g) No solution. h) $x \geq 6$ or $x \leq -8$.
17. a) $103 \leq x \leq 107$. b) $|x - 105| \leq 2$.

Set IV

1. Yes. $[|3 + (-1) + 4 + (-3) + 2| = 5 \leq 8.]$
2. No. $(|2 + 1 + 5 + 1 + 0| = 9 > 8.)$ **3.** Yes.
$[|-1 + 0 + 5 + (-3) + e| \leq 8, |e + 1| \leq 8,$
$-9 \leq e \leq 7.]$ **4.** No; only those parts for which
$-2 \leq e \leq -1$. $(|4 + 1 + 6 + -2 + e| \leq 8,$
$|e + 9| \leq 8, -1 \leq e \leq -17.)$

Chapter 16, Review

Set I

1. a) $>$. b) $<$. c) $<$. d) $>$. e) $=$. f) $>$.
g) $=$. h) $>$.   **2.** a) $x \geq -3$. b) $2 < x < 7$.
c) $x < 9$. d) $-4 \leq x \leq 1$.   **3.** a) $>$. b) None.
c) $\geq$. d) None.   **4.** a) $<$. b) $>$. c) None.
d) $>$.   **5.** a) No. b) Yes. c) Yes. d) Yes.
e) No. f) Yes. g) Yes. h) Yes. i) No.

6. a) $x > -1$. b) $x > \dfrac{3}{4}$. c) $x > -8$. d) $x \geq 8$.

e) $2 < x < 6$. f) $x > -2$. g) $x \leq 24$. h) $x > 3$.
i) $x < -5$. j) $x > 0$. k) $x < -3$. l) $x \leq 22$.

7. a) $x < \dfrac{b + 1}{a}$. b) $x < \dfrac{ab}{c}$. c) $x \leq \dfrac{c}{a - b}$.

d) $a < x < 2a + b$.   **8.** a) 5 and –5. b) 11 and
–7. c) 4 and –16. d) $x + 4$ and $x - 4$.

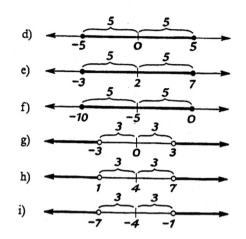

9.

a)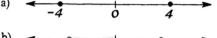

b)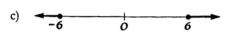

c)

d)

e)

f)

10. a) –5 and 5. b) $-15 < x < 15$. c) 3 and –3.
d) 3 and –17. e) $x \geq 8$ or $x \leq -8$. f) $x \geq 8$ or
$x \leq -4$. g) No solution. h) True for all x except –1.

Chapter 17, Lesson 1

Set I

1. a) $x > 4$. b) $-5 \leq x \leq 2$. c) $x \leq -3$.
2. a) 14. b) –3. c) $4 + 4\sqrt{2}$ and $4 - 4\sqrt{2}$.

3. a) $y = \dfrac{16x}{(0+4)^2} = \dfrac{16x}{16} = x$. b) 50 pounds.

c) 8 pounds. d) $x = \dfrac{y(d+4)^2}{16}$. e) 108 pounds.

Set III

10. a) Geometric; common ratio, 3. b) Arithmetic;
common difference, 9. c) Neither. d) Arithmetic;
common difference, ̄4. e) Geometric; common ratio,
0.1. f) Neither. g) Arithmetic; common difference,
2.5. h) Geometric; common ratio, 1.5. i) Both;
common difference, 0; common ratio, 1.
j) Geometric; common ratio, –4. k) Neither.

l) Neither. m) Arithmetic; common difference, $\dfrac{1}{6}$.

n) Geometric; common ratio, 0.5. o) Geometric;

common ratio, $\dfrac{3}{4}$. **11.** a) 112. b) 27. c) 25.

d) 2. e) 14. f) 31. g) $\dfrac{1}{4}$. h) $\dfrac{1}{7}$. i) –512.

j) $\sqrt{29}$. k) 0.042. l) 625. **12.** a) 7, 10, 13, 16, 19.

b) By adding 3. c) Arithmetic. **13.** a) 12, 36,
108, 324, 972. b) By multiplying by 3.
c) Geometric. **14.** a) –6, 30. b) 2, 54. c) 44, 86.
d) 8, 32. **15.** a) $t_n = 2n$. b) $t_n = n - 1$.
c) $t_n = n^3$. d) $t_n = n + 8$. e) $t_n = 5^n$. f) $t_n = 6$.

g) $t_n = 10n + 7$. h) $t_n = \dfrac{n}{4}$.

Set IV

1. 111, 181, 609. **2.** It consists of the numbers that
look the same when turned upsidedown.

Chapter 17, Lesson 2

Set I

1. a) $x < -6$. b) $x < -5$. c) $6 < x < 12$.
d) $x \geq 2$. **2.** a) True. b) True. c) True.
d) False. **3.** a) 22.86. b) –0.26, –0.05, –0.11, 0.01,
0.41. c) 0. d) No. (Example: 1, 2, 3, 7, 10;
average = 4.6; –3.6, –2.6, –1.6, 2.4, 5.4; sum = 0.) It
is always 0.

Set III

14. a) 1, 4, 7, 10, 13. b) 3.

c)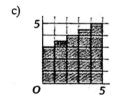

d)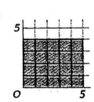

e)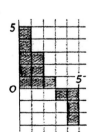

15.

a)

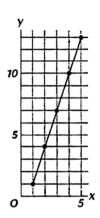

b)

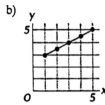

c)

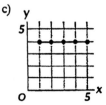

d)

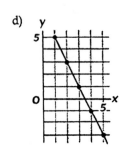

e) Linear.   **16.** a) 7, 12, 17. b) 5. c) 52.
d) 502. e) 295. f) 25,450.   **17.** a) 9, 5, 1. b) 4.
c) –27. d) –387. e) –90. f) –18,900.   **18.** a) 29.

b) 14. c) 17, 28. d) 19, 7. e) 1.5, –1.5. f) $4\frac{1}{3}$, $4\frac{2}{3}$.

g) 4, 9, 14. h) $\frac{2}{5}$.   **19.** a) $t_n = 1 + (n-1)4$; 29.

b) $t_n = 23 + (n-1)7$; 93. c) $t_n = 80 - (n-1)2$;
32. d) $t_n = 15 - (n-1)9$; –120.
e) $t_n = -100 + (n-1)13$; 563. f) $t_n = -11$; –11.
20. a) 153. b) 225. c) 360. d) 528. e) 2,275.
f) –1,683.   **21.** a) $t_1 = 9$, $t_{10} = 63$; $S_{10} = 360$.
b) $t_1 = 31$, $t_{13} = 55$; $S_{13} = 559$. c) $t_1 = 14$,
$t_{21} = -146$; $S_{21} = -1,386$. d) $t_1 = 60$, $t_{100} = 10.5$;
$S_{100} = 3,525$.   **22.** a) 21. b) 45.   **23.** a) 900
dollars. b) 945 dollars.

20,200 yards.

$$\left\{ 2(2 + 4 + 6 + \cdots + 200) = 2\left[\frac{(2 + 200)100}{2}\right] \right.$$
$$\left. = 20{,}200. \right\}$$

Chapter 17, Lesson 3

Set I

1. a) >. b) >. c) <. d) Not possible.
2. a) $t_n = -3n$. b) $t_n = n + 7$. c) $t_n = 2^n$.

d) $t_n = \dfrac{n-1}{n+1}$.   **3.** a) $5x + 5y = 100$,

$11x + y = 100$. b) (8, 12). c) $1.08.

Set III

13. a) 3, 6, 12, 24. b) 2.

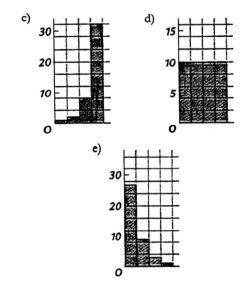

14.

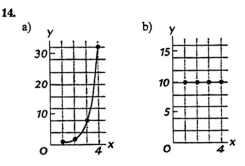

c)

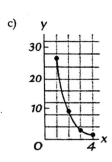

d) Exponential.   **15.** a) $\frac{1}{3}$, 2, 12.   b) 6.

c) $\frac{1}{3} \cdot 6^9$.   d) $\frac{1}{3} \cdot 6^{99}$.   **16.** a) 10, 5, 2.5.

b) $\frac{1}{2}$.   c) $10 \cdot \left(\frac{1}{2}\right)^9$.   d) $10 \cdot \left(\frac{1}{2}\right)^{99}$.   **17.** a) 0.5.

b) 40.   c) 2, –128.   d) 14, 28.   e) $\frac{1}{30}$.   f) 6^{-1}, 6^0.

18. a) $t_n = 5(3)^{n-1}$.   b) $t_n = (-7)^{n-1}$.

c) $t_n = 32\left(\frac{1}{4}\right)^{n-1}$.   d) $t_n = 4\left(\frac{9}{2}\right)^{n-1}$.

e) $t_n = 9\left(-\frac{2}{3}\right)^{n-1}$.   f) $t_n = 3\left(\sqrt{5}\right)^{n-1}$.

19. a) 1,023.   b) 279,935.   c) –728.   d) 1,953,120.
20. a) $r = 1$ and $r - 1 = 0$; we cannot divide by 0.
b) 108.   c) $S_n = 12n$.   **21.** a) 1, 4, 16, 64, 256,
1,024.   b) 1,365.   c) $t_n = 4^{n-1}$.   d) $S_n = \dfrac{4^n - 1}{3}$.

Set IV

1. 127 dollars.   **2.** 128 dollars.   **3.** Ahead 1
dollar.   **4.** $3(2^z - 1)$.   **5.** $3(2)^z$.   **6.** 3
dollars.   **7.** The person may run out of money
before winning.

Chapter 17, Lesson 4

Set I

1. a) $-1 < x < 1$.   b) 3 and –3.   c) 10 and –2.
d) $-4 \le x \le 4$.   **2.** a) $t_n = 7 + (n - 1)4$.   b) 87.

c) 987.   **3.** a) Inversely.   b) $w = \dfrac{3,000}{n}$.

c) 375 kilograms.

11. a) 0.8.   b) 1.1.   c) –0.6.   d) 1.   e) 0.5.   f) –1.
g) 1.4.   h) 0.99.   **12.** a) Sequences b and g; they
are more than 1.   b) Sequences a, c, e, and h; they
are less than 1 (and more than –1).

13. a)

1	2	3	4	5	6
6	6.6	7.3	8	8.8	9.7

b)

1	2	3	4	5	6
6	5.4	4.9	4.4	3.9	3.5

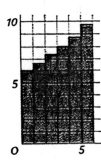

c)

1	2	3	4	5	6
3	–3.6	4.3	–5.2	6.2	–7.5

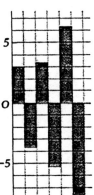

d)

1	2	3	4	5	6
3	-2.4	1.9	-1.5	1.2	-1.0

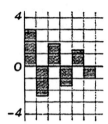

14. a) 64. b) $\frac{1}{2}$. c) 125. d) $\frac{4}{25}$.   **15.** a) $\frac{2}{3}$.

b) $\frac{8}{11}$. c) $\frac{4}{27}$. d) $\frac{1}{45}$.   **16.** a) 12, 3, $\frac{3}{4}$, $\frac{3}{16}$, $\frac{3}{64}$.

b) $\frac{1,023}{64}$. c) 16.   **17.** 210 meters.

Set IV

128 chocolates.

$$\left(64 + 32 + 16 + \cdots + 2^{-57} \approx \frac{64}{1 - \frac{1}{2}} = 128.\right)$$

Chapter 17, Review

Set I

1. a) Geometric; common ratio, 4. b) Arithmetic; common difference, 2.5. c) Neither. d) Arithmetic; common difference, $-\frac{1}{12}$. e) Geometric; common ratio, 10. f) Neither.   **2.** a) 6. b) $\frac{5}{36}$. c) 40.5.

d) 21. e) 1. f) $\frac{1}{3}$.   **3.** a) 48. b) 98. c) 5.

d) 3.   **4.** a) 2, 5, 8, 11. b) 1, 16, 81, 256. c) 4, 16, 64, 256. d) 15, 20, 25, 30. e) 10, 20, 40, 80. f) 2, $2\frac{1}{2}$, $3\frac{1}{3}$, $4\frac{1}{4}$. g) a and d. h) c and e.

5. a) $t_n = n + 4$; 104. b) $t_n = -2n$; -30. c) $t_n = 5^n$; 78,125. d) $t_n = \frac{n}{6}$; 4. e) $t_n = n^3$; 1,000.

f) $t_n = 4 \cdot 7^{n-1}$; 67,228. g) $t_n = \frac{1}{n^2}$; $\frac{1}{196}$.

h) $t_n = \sqrt{n-1}$; 7.   **6.** a) 208. b) -63. c) 2,583.
d) 1,740.   **7.** a) 65,535. b) 156,248. c) -2,184.

8. a) 0.01. b) -0.7. c) 1.5.   **9.** a) 18. b) 625.

c) $\frac{2}{9}$.   **10.** a) $\frac{8}{9}$. b) $\frac{4}{11}$. c) $\frac{1}{333}$.   **11.** a) 4.

b) 9. c) 16. d) 25. e) Squares. f) $t_n = 2n - 1$.

g) $S_n = n^2$.   **12.** a) $1,392, $1,643, $1,939, $2,288, $2,700, $3,185, $3,759, $4,435, $5,234.

b)

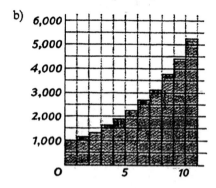

c) $t_n = 1,000(1.18)^n$.
13. After 8 seconds.

Final Review

Test I

1. a) 0.003, 0.01, 0.2.   **2.** $x + -y$.   **3.** $6x$.
4. $\geq$.   **5.** $6x^5$.   **6.** True.   **7.** $5\sqrt{2}$.

8. $a > c$.   **9.** $2x + 8$.   **10.** 13 and 3.   **11.** $\frac{21}{5}$.

12. $10x$.   **13.** $4x^2$.   **14.** $(2x + 1)(2x - 1)$.

15. $3x^2$.   **16.** False.   **17.** $(x - 3y)^2$.   **18.** $\frac{2}{3}$.

19. $4x^2 - 20x + 25$.   **20.** $\frac{2x}{5}$.   **21.** True.

22. $\frac{7}{5}$.   **23.** $5\sqrt{2}$.   **24.** $\frac{x + 2}{x - 2}$.   **25.** 121.

26.

27. $\dfrac{x^2}{8}$.　**28.** $x + \sqrt{y}$.　**29.** 73.　**30.** True.

31. 4.4.　**32.** $\dfrac{\sqrt{2x}}{4}$.　**33.** $-4x^2 + x + 8$.

34. $x < -5$.　**35.** False.

36. $xy(x + y)(x - y)$.　**37.** $x^2 + 4x + 2$.

38. 100.48.　**39.** $\dfrac{3}{11}$.　**40.** $-2 < x < 3$.　**41.** $\dfrac{1}{5}$.

42. $(x + 7)(x - 6)$.　**43.** 1.　**44.** $\sqrt{5}$.

45. $\dfrac{x}{x + y} - \dfrac{y}{x + y}$.　**46.** $0.\overline{45}$.　**47.** $y = |x|$.

48. 5.1×10^{-3}.　**49.** 0 and 3.　**50.** 1.4×10^4.

51. $x^3 + 3x^2 + 2x$.　**52.** $14x$.　**53.** $x^2 - 5x - 14$.

54. 4.8.　**55.** 14.　**56.** $1, \dfrac{1}{4}, \dfrac{1}{9}$.　**57.** $(7, 4)$.

58. Irrational.　**59.** $x = ab$.　**60.** $\dfrac{4}{3x}$.

61.
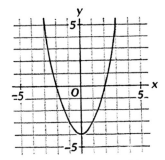

62. $\dfrac{x}{2}$.　**63.** The direction remains the same.

64. 9 and -9.　**65.** $2x^2(2x - 3)$.　**66.** 820.

67. True.　**68.** $-2 + \sqrt{5}$ and $-2 - \sqrt{5}$.

69. $x < -\dfrac{b}{a}$.　**70.** $y = \dfrac{100}{x}$.　**71.** Inversely.

72.

　73.

$$\overset{1.5x}{\longleftrightarrow}$$

$$\overset{1.2(x+20)}{\longleftrightarrow}$$

74. $1.5x = 1.2(x + 20)$.　**75.** 120 meters.

76. $l + 2h + 2w \leq 100$.　**77.** $l \leq 30$.

78. 10.　**79.** 320.　**80.** 1 hour.

SOURCES OF ILLUSTRATIONS

Transparency 1-10
Puzzle adapted from *More Posers: 80 Intriguing New Hurdles for Reasonably Agile Minds* by Philip Kaplan. Copyright © 1964 by Philip Kaplan. Reprinted by permission of Harper & Row, Publishers.

Transparency 1-14
After Sam Loyd's *Cyclopedia of Puzzles.*

Transparency 1-15
Reprinted from *The Saturday Evening Post* © 1959 The Curtis Publishing Company.

Transparency 1-18
"You owe $0.00" reprinted with the permission of the *Saturday Review.*

Transparency 1-26
Photograph from *Geology Illustrated* by J. S. Shelton. W. H. Freeman and Company. Copyright © 1966.

Transparency 1-28
From *Vision in Elementary Mathematics* by W. W. Sawyer. Penguin, 1964. Copyright © by W. W. Sawyer.

Transparency 2-1
Photograph courtesy of H. H. Ninninger, American Meteorite Museum.

Transparency 2-2
After *The Search for the Perfect Swing* by Alastair Cochran and John Stobbs. Copyright © 1968 by the Golf Society of Great Britain. Reproduced by permission of J. B. Lippincott Company.

Transparency 2-6
Photograph from Wide World Photos.

Transparency 2-9
"New York Set to Music" reprinted with the permission of Mrs. Joseph Schillinger.

Transparency 2-15
Photograph courtesy of the Museum of Vertebrate Zoology, University of California, Berkeley.

Duplication Master 4
From *Tested Demonstrations in Chemistry* (1965) from the *Journal of Chemical Education.*

Transparency 2-19
Photograph reproduced by permission of the University Museum, University of Pennsylvania.

Transparency 3-3
© 1929 by The New York Times Company. Reprinted by permission.

Transparency 3-6
Photograph by Robert Wenkam. Graph after "Beaches" by Willard Bascom. Copyright © 1960 by Scientific American, Inc. All rights reserved.

Transparency 3-9
Photograph courtesy of Hale Observatories.

Transparency 3-19
From "Can Time Go Backward?" by Martin Gardner. Copyright © 1967 by Scientific American, Inc. All rights reserved. Cartoon by R. O. Blechman.

Transparency 3-20
Cartoon by R. O. Blechman. From "Can Time Go Backward?" by Martin Gardner. Copyright © 1967 by Scientific American, Inc. All rights reserved.

Transparency 3-21
Cartoon by R. O. Blechman. From "Can Time Go Backward?" by Martin Gardner. Copyright © 1967 by Scientific American, Inc. All rights reserved.

Worksheet 4
Adapted with permission from Shirley Cox.

Transparency 4-1
Photograph of Felix the Cat on a turntable courtesy of The National Broadcasting Company, Inc. Photograph of Felix the Cat on video screen © King Features Syndicate, Inc. 1969.

Transparency 4-3
After a graph by Robert Ritter.

Transparency 4-5
Lithograph by Maurits Escher. Reproduced with the permission of the Escher Foundation, Haags Gemeentemuseum, The Hague.

Transparency 4-11
Photograph from *Mount Everest: Formation and Exploration of the Everest Region* by Toni Hagen, Günter-Oskar Dyhrenfurth, Christoph von Fürer-Harmendorf, and Erwin Schneider. © Orell Füssli Verlag, Zürich 1959.

Transparency 4-13
LIFE Science Library/Water. Published by Time-Life Books Inc.

Transparency 4-18
Photograph by Berenice Abbott reprinted with the permission of William Collins Publishers, Inc., from *The Attractive Universe* by E. G. Valens (text and diagrams) and Berenice Abbott (photographs).

Transparency 4-19
Cartoon from *Saturday Review/World*.

Transparency 5-1
From J. Maynard Smith, *Mathematical Ideas in Biology*. Cambridge University Press, 1968.

Transparency 5-2
From *Music by Computers*, edited by Heinz von Foerster and James W. Beauchamp. John Wiley & Sons, Inc., 1969.

Transparency 5-3
From *Total Poker*. Copyright © 1977 by David Spanier. Reprinted with the permission of Simon & Schuster, a division of Gulf & Western Corporation.

Transparency 5-9
From *The Hunting of the Snark* by Lewis Carroll.

Transparency 5-12
Reprinted with the permission of Sidney Harris.

Transparency 5-14
From Sam Loyd's *Cyclopedia of Puzzles*.

Transparency 5-16
From "Mathematical Games" by Martin Gardner. Copyright © 1973 by Scientific American, Inc. All rights reserved.

Transparency 5-21
Courtesy of Pratt & Whitney Aircraft.

Transparency 5-22
From the *Guinness Book of World Records*.

Transparency 5-24
Photograph by Robert Ishi.

Transparency 5-25
Courtesy of Griffith Observatory, photographed by Paul Roques.

Transparency 5-28
National Air and Space Museum, Smithsonian Institution.

Transparency 5-29
Photograph by Judy MacCready.

Transparency 5-30
From the *Guinness Book of World Records*.

Transparency 5-31
Photograph by David Harris/NEWORLD.

Transparency 6-1
Tom Hollyman, Photo Researchers, Inc. NYC.

Transparency 6-8
Photograph from the *Guinness Book of World Records*.

Transparency 6-14
Photograph courtesy of KARN-AM.

Transparency 6-23
Photograph from the *Guinness Book of World Records*.

Transparency 6-30
CBS Television Network.

Transparency 7-1
American Broadcasting Company.

Transparency 7-4
B.C. by permission of Johnny Hart and Field Enterprises, Inc.

Transparency 7-9
Illustration from *Through the Looking Glass* by Lewis Carroll.

Transparency 7-12
Photograph courtesy of Time-Life Multimedia.

Transparency 7-21 and overlay
From *Posers: 80 Delightful Hurdles for Reasonably Agile Minds* by Philip Kaplan. Copyright © 1963 by Philip Kaplan. By permission of Harper & Row, Publishers, Inc.

Transparency 7-26
From the *Los Angeles Times*.

Transparency 7-27
From the *Los Angeles Times*.

Transparency 8-1
Courtesy of Hale Observatories.

Transparency 8-4
Courtesy of NASA-Ames Research Center, Moffett Field, California.

Transparency 8-5
Courtesy of the Sierra Club.

Transparency 8-12
U. S. Bureau of the Census.

Transparency 8-14
B.C. by permission of Johnny Hart and Field Enterprises, Inc.

Transparency 9-1
Diagram from *California Driver's Handbook*.

Transparency 9-3
B.C. by permission of Johnny Hart and Field Enterprises, Inc.

Transparency 9-16
After Sam Loyd's *Cyclopedia of Puzzles*.

Transparency 9-17
From *Approved Crossword Puzzles*. February 1979 (Penny Press, Inc.).

Transparency 10-3
From "Science and the Citizen." Copyright © 1978 by Scientific American, Inc. All rights reserved.

Transparency 10-6
From *Famous Problems of Mathematics* by Heinrich Tietze. Graylock Press, 1965.

Transparency 10-9
Photograph from the *Guinness Book of World Records*.

Transparency 10-14
"Beavers" copyright L. Mezei 1967.

Transparency 10-16
From *Recreations in the Theory of Numbers*, second edition, by Albert H. Beiler. Dover, 1966.

Transparency 10-24
From *Algebra*, Part I, by G. Chrystal.

Transparency 10-28
After "Mathematical Games" by Martin Gardner. Copyright © 1970 by Scientific American, Inc. All rights reserved.

Transparency 11-2
After "Mathematical Games" by Martin Gardner. Copyright © 1965 by Scientific American, Inc. All rights reserved.

Transparency 11-3
"Relative Primes" courtesy of M. R. Schroeder, Drittes Physikalisches Institut, Universität Göttingen, Germany.

Transparency 11-15
Reproduced by Courtesy of the Trustees of the British Museum.

Transparency 11-17
Photograph courtesy of Eastman Kodak.

Transparency 11-20
Photograph courtesy of AGFA-Gevaert N.V.

Transparency 11-28
National Archives.

Transparency 12-7
Photograph courtesy of National Aeronautics and Space Administration.

Transparency 12-10
Photograph by Leonard Lee Rue III © 1971.

Transparency 12-16
Photograph courtesy of Moody Institute of Science.

Transparency 12-18
Drawing by Sidney Harris.

Transparency 12-23
Photograph from *Geology Illustrated* by John S. Shelton. W. H. Freeman and Company. Copyright © 1966.

Transparency 12-24
Photograph from the John E. Allen Inc. Collection.

Transparency 13-19
National Air and Space Museum, Smithsonian Institution. Courtesy of Mrs. Esther Goddard.

Transparency 13-25
"George Washington" from the *National Enquirer*.

Transparency 14-9
Photograph from the *Guinness Book of World Records*.

Transparency 14-13
Courtesy of R. Wm. Gosper.

Transparency 14-16
Photography courtesy of the National Aeronautics and Space Administration.

Transparency 14-20
Drawing by Sidney Harris.

Transparency 15-1
Reprinted with the permission of RKO General Pictures.

Transparency 15-2
Photograph by Marcel Delgado.

Transparency 15-6
Map courtesy of U.S. Geological Survey.

Transparency 15-7
From *Topsys & Turvys* by Peter Newell (Century Company, 1893, 1894; Dover, 1964).

Transparency 15-11
Photograph courtesy of Robert T. Orr.

Transparency 17-1
From the *Guinness Book of World Records*.

Transparency 17-14
Photograph reprinted with the permission of Ralph Morse.

Transparency 17-18
M. C. Escher's *Development II*. Escher Foundation, Haags Gemeentemuseum, The Hague.

Transparency F-1
From *Taking the SAT*, College Entrance Examination Board, New York, 1978. Reproduced by permission of Educational Testing Service, copyright owner of the sample questions.